T0320497

Examining Developments and Applications of Wearable Devices in Modern Society

Saul Emanuel Delabrida Silva
Federal University of Ouro Preto, Brazil

Ricardo Augusto Rabelo Oliveira
Federal University of Ouro Preto, Brazil

Antonio Alfredo Ferreira Loureiro
Federal University of Minas Gerais (UFMG), Brazil

A volume in the Advances in Wireless Technologies and Telecommunication (AWTT) Book Series

Published in the United States of America by
IGI Global
Information Science Reference (an imprint of IGI Global)
701 E. Chocolate Avenue
Hershey PA, USA 17033
Tel: 717-533-8845
Fax: 717-533-8661
E-mail: cust@igi-global.com
Web site: http://www.igi-global.com

Library of Congress Cataloging-in-Publication Data

Names: Delabrida Silva, Saul Emanuel, 1984- editor. | Rabelo Oliveira, Ricardo Augusto, 1978- editor. | Loureiro, Antonio Alfredo Ferreira, 1961- editor.
Title: Examining developments and applications of wearable devices in modern society / Saul Emanuel Delabrida Silva, Ricardo Augusto Rabelo Oliveira, and Antonio Alfredo Ferreira Loureiro, editors.
Description: Hershey, PA : Information Science Reference, [2018]
Identifiers: LCCN 2017015186| ISBN 9781522532903 (hardcover) | ISBN 9781522532910 (ebook)
Subjects: LCSH: Medical technology. | Medical care--Technological innovations. | Wearable technology.
Classification: LCC R855.3 .E93 2018 | DDC 610.285--dc23 LC record available at https://lccn.loc.gov/2017015186

This book is published in the IGI Global book series Advances in Wireless Technologies and Telecommunication (AWTT) (ISSN: 2327-3305; eISSN: 2327-3313)

British Cataloguing in Publication Data
A Cataloguing in Publication record for this book is available from the British Library.

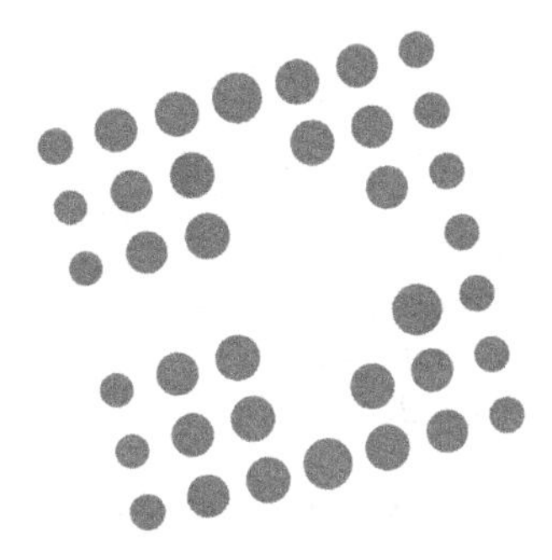

Advances in Wireless Technologies and Telecommunication (AWTT) Book Series

ISSN:2327-3305
EISSN:2327-3313

Editor-in-Chief: Xiaoge Xu, The University of Nottingham Ningbo China, China

MISSION

The wireless computing industry is constantly evolving, redesigning the ways in which individuals share information. Wireless technology and telecommunication remain one of the most important technologies in business organizations. The utilization of these technologies has enhanced business efficiency by enabling dynamic resources in all aspects of society.

The **Advances in Wireless Technologies and Telecommunication Book Series** aims to provide researchers and academic communities with quality research on the concepts and developments in the wireless technology fields. Developers, engineers, students, research strategists, and IT managers will find this series useful to gain insight into next generation wireless technologies and telecommunication.

COVERAGE

- Mobile Web Services
- Radio Communication
- Cellular Networks
- Telecommunications
- Wireless Sensor Networks
- Mobile Communications
- Global Telecommunications
- Network Management
- Mobile Technology
- Wireless Broadband

The Advances in Wireless Technologies and Telecommunication (AWTT) Book Series (ISSN 2327-3305) is published by IGI Global, 701 E. Chocolate Avenue, Hershey, PA 17033-1240, USA, www.igi-global.com. This series is composed of titles available for purchase individually; each title is edited to be contextually exclusive from any other title within the series. For pricing and ordering information please visit http://www.igi-global.com/book-series/advances-wireless-technologies-telecommunication-awtt/73684. Postmaster: Send all address changes to above address.

Titles in this Series

For a list of additional titles in this series, please visit:
http://www.igi-global.com/book-series/advances-wireless-technologies-telecommunication-awtt/73684

Routing Protocols and Architectural Solutions for Optimal Wireless Networks and ecurity
Dharm Singh (Namibia University of Science and Technology, Namibia)
Information Science Reference • ©2017 • 277pp • H/C (ISBN: 9781522523420) • US $205.00

Sliding Mode in Intellectual Control and Communication Emerging Research and Opportunities
Vardan Mkrttchian (HHH University, Australia) and Ekaterina Aleshina (Penza State University, Russia)
Information Science Reference • ©2017 • 128pp • H/C (ISBN: 9781522522928) • US $140.00

Emerging Trends and Applications of the Internet of Things
Petar Kocovic (Union – Nikola Tesla University, Serbia) Reinhold Behringer (Leeds Beckett University, UK) Muthu Ramachandran (Leeds Beckett University, UK) and Radomir Mihajlovic (New York Institute of Technology, USA)
Information Science Reference • ©2017 • 330pp • H/C (ISBN: 9781522524373) • US $195.00

Resource Allocation in Next-Generation Broadband Wireless Access Networks
Chetna Singhal (Indian Institute of Technology Kharagpur, India) and Swades De (Indian Institute of Technology Delhi, India)
Information Science Reference • ©2017 • 334pp • H/C (ISBN: 9781522520238) • US $190.00

Multimedia Services and Applications in Mission Critical Communication Systems
Khalid Al-Begain (University of South Wales, UK) and Ashraf Ali (The Hashemite University, Jordan & University of South Wales, UK)
Information Science Reference • ©2017 • 331pp • H/C (ISBN: 9781522521136) • US $200.00

Big Data Applications in the Telecommunications Industry
Ye Ouyang (Verizon Wirless, USA) and Mantian Hu (Chinese University of Hong Kong, China)
Information Science Reference • ©2017 • 216pp • H/C (ISBN: 9781522517504) • US $145.00

For an enitre list of titles in this series, please visit:
http://www.igi-global.com/book-series/advances-wireless-technologies-telecommunication-awtt/73684

701 East Chocolate Avenue, Hershey, PA 17033, USA
Tel: 717-533-8845 x100 • Fax: 717-533-8661
E-Mail: cust@igi-global.com • www.igi-global.com

Table of Contents

Detailed Table of Contents

Virtual Reality and Augmented Reality Head-Mounted Displays (HMDs) have been emerging in the last years. These technologies sound like the new hot topic for the next years. Head-Mounted Displays have been developed for many different purposes. Users have the opportunity to enjoy these technologies for entertainment, work tasks, and many other daily activities. Despite the recent release of many AR and VR HMDs, two major problems are hindering the AR HMDs from reaching the mainstream market: the extremely high costs and the user experience issues. In order to minimize these problems, we have developed an AR HMD prototype based on a smartphone and on other low-cost materials. The prototype is capable of running Eye Tracking algorithms, which can be used to improve user interaction and user experience. To assess our AR HMD prototype, we choose a state-of-the-art method for eye center location found in the literature and evaluate its real-time performance in different development boards.

With the type of ailments increasing and with the methods of diagnosis improving day by day, wearable devices are increasing in number. Many times, it is found to be beneficial to have continuous diagnosis for certain type of ailments and for certain type of individuals. One will feel uncomfortable if a number of needles are protruding out of one's body for having continuous diagnosis. From this point of view, wearable diagnosis systems are preferable. With Internet of Things (IoT), it is possible to have a number of diagnostic sensors as wearable devices. In addition, for a continuous monitoring, the information from these wearable devices must be transferring information to some central location. IoT makes this possible. IoT brings full range of pervasive connectivity to wearable devices. IoT of wearable devices can include additional intelligence of location of the person wearing the device and also some biometric information identifying the wearer.

In this chapter, wearables are presented as assistive technology to support persons with disabilities (PwD) to face the urban space in an autonomous and independently way. In the Inclusive Smart City (ISC), everyone has to be able to access visual and audible information that so far are available just for people that can perfectly see and listen. Several concepts and technologies – such as Accessibility and Universal Design, Pervasive Computing, Wearable Computing, Internet of Things, Artificial Intelligence, and Cloud Computing – are associated to achieve this aim. Also, this chapter discusses some examples of use of wearables in the context of Smart Cities, states the importance of these devices to the successful implementation of Inclusive Smart Cities, as well as presenting challenges and future research opportunities in the field of wearables in ISC.

Wearable devices have increasingly become popular in recent years. Devices attached to users body remotely monitor his daily activities/health. Although some of these devices are pretty simple, others make use of an operating system to manage memory, resources, tasks, and any user interaction. Some of them were not initially designed and developed for this purpose, having a poor performance requiring the use of more resources or better hardware. This chapter presents a characterization of wearable devices considering the operating systems area. Some constraints of this context were designed to analyze the operating system's execution when inserted into a wearable device. Data presented at the end shows that there is a lack of performance in specific areas, letting to conclude that improvements should be made.

Advances in energy harvesting hardware have created an opportunity for realizing self-powered wearables for continuous and pervasive Human Context Detection (HCD). Unfortunately, the power consumption of the continuous context sensing using accelerometer is relatively high compared to the amount of power that can be harvested practically, which limits the usefulness of energy harvesting. This chapter employs and infers HCD directly from the Kinetic Energy Harvesting (KEH) patterns generated from a wearable device that harvests kinetic energy to power itself. This proposal eliminates the need for accelerometer, making HCD practical for self-powered devices. The authors discuss in more details the use of KEH patterns as an energy efficient source of information for five main applications, human activity recognition, step detection, calorie expenditure estimation, hotword detection, and transport mode detection. This confirms the potential sensing capabilities of KEH for a wide range of wearable applications, moving us closer towards self-powered autonomous wearables.

The thermophysiological comfort is one of the aspects of the human comfort. It is related to the thermoregulatory system of the body and its reactions to the temperature of the surrounding air, activity and clothing. The aim of the chapter is to present the state of the art in the wearable technologies for helping the human thermophysiological comfort. The basic processes of body's thermoregulatory system, the role of the hypothalamus, the reactions of the body in hot and cold environment, together with the related injuries, are described. In the second part of the chapter smart and intelligent clothing, textiles and accessories are presented together with wearable devices for body's heating/cooling.

Preface

In the broadest sense, a wearable, also known as a body-borne computer, is a device comprised of electronics, sensors, software and connectivity, worn by a person or even an animal, which allow these objects to exchange data through Internet with a manufacturer, operator and/or other connected devices, without requiring human intervention. Wearables have been designed for general or special purpose applications and are especially useful in scenarios that require some sort of sensing technology, such as accelerometers or gyroscopes. Activity trackers and smartwatches are two of the most popular wearables used nowadays. Moreover, wearables can be used by a user to act as a prosthetic, becoming an extension of the user's mind or body.

There are some common features of different wearables. One of them is the persistence of activity, i.e., the interaction between the wearable and user is constant and, thus, there is no need to turn the device on or off all the times. Another feature is the ability of a person using a wearable to perform different tasks without the need to stop what one is doing to use the device. This particular feature allows the wearable's functionality to blend smoothly into all other user actions.

Wearable technology is closely related to ubiquitous computing. Wearables make technology pervasive by interweaving it into our daily life. For instance, the sensing and recording of an activity such as movement, steps and heart rate of a user, enabling to quantify the self-movement. This is an important distinction between wearable devices and portable computers (handheld and laptop computers, for instance). In wearable computing, the person and device are intertwined, ultimately leading to the so-called humanistic intelligence, i.e., intelligence that arises by having the human being in the feedback loop of the computational process.

Wearable devices are quickly advancing in terms of technology, use and size, with more real-time applications being available. It is on the rise in both personal and business use. In healthcare, wearables are now used for different sorts of problems, such as to detect health disorders (e.g., sleep apnea). In professional sports, wearable technology has applications in monitoring and real-time feedback for athletes. In business, wearables are being used to remotely manage equipment, such as machinery on an assembly line, making the workplace safer for employees.

We can expect to have wearables augmenting ourselves by integrating technology as a part of who we are, endowing us with new and enhanced senses, skills and capabilities. This will change the way people enhance their capabilities and extend how they interact with both the physical and digital worlds and will definitely impact every aspect of our lives.

This book aims at present some of the most innovative research on wearable computing and applications.

In Chapter 1, "Development of a Low-Cost Augmented Reality Head-Mounted Display Prototype", authors present a low-cost augmented reality and provide a performance evaluation of the prototype according to the eye location algorithm. That application can be used in industrial environments as suggested by the authors. Also, an in-depth description of the HDM construction techniques is presented.

In Chapter 2, "Wearable Internet of Things", the author provides an extensive review of the use of wearable as a component of IoT purposes. Examples of mobile wearable applications for health benefits are presented as well as the issues and challenges for providing advances in this research area.

Chapter 3, "When Wearable Computing Meets Smart Cities: Assistive Technology Empowering Persons With Disabilities", brings a study of the use of wearable in the inclusive smart cities for people with disabilities. A discussion of the most important aspects of the inclusive smart cities is presented. Also, the authors make a relationship of the role of wearable for this scenario.

Chapter 4, "Wearables Operating Systems: A Comparison Based on Relevant Constraints", presents a study about wearable operating systems and their evaluations under different platforms. The authors discuss current limitations and make suggestions for operating system implementations based on the results they acquired. This chapter can be a guide for researchers and practitioners who want to develop wearable applications for the main gadgets available on the market.

Chapter 5, "Human Context Detection From Kinetic Energy Harvesting Wearables", discusses the potential use of KEH as a source of information for wearable applications. Authors discuss a set of applications that can benefit from this type of sensor for identifying the human context. Furthermore, the they enhance the that this sensor does not require any power supply to operate, which might be a differential for wearable applications.

Chapter 6, "Wearable Health Care Ubiquitous System for Stroke Monitoring and Alert", proposes the use of a prototype based on neural networks for monitoring and alerting a stroke. The authors present a compatible background and a model that makes use of embedded platforms for evaluating the proposed system. The system emits alerts via SMS informing the GPS position of the user. Authors consider the use of mobile devices for improving the communication, as well as, apply machine learning techniques.

Chapter 7, "Wearable Antennas: Breast Cancer Detection" presents a broad discussion of wearable antennas providing basics concepts, advantages, and disadvantages. The authors also show an innovate prototype of an antenna for early breast cancer detection.

Chapter 8, "Wearable Technologies for Helping Human Thermophysiological Comfort", presents an extensive description of a set of wearables, which contribute to providing the thermophysiological comfort of the body. This work represents an excellent guide for researchers and practitioners to make studies and evaluation of new applications.

Chapter 1
Development of a Low-Cost Augmented Reality Head-Mounted Display Prototype

Thiago D'Angelo
Federal University of Ouro Preto, Brazil

Ricardo A. R. Oliveira
Federal University of Ouro Preto, Brazil

Saul Delabrida
Federal University of Ouro Preto, Brazil

Antonio A. F. Loureiro
Federal University of Minas Gerais, Brazil

ABSTRACT

Virtual Reality and Augmented Reality Head-Mounted Displays (HMDs) have been emerging in the last years. These technologies sound like the new hot topic for the next years. Head-Mounted Displays have been developed for many different purposes. Users have the opportunity to enjoy these technologies for entertainment, work tasks, and many other daily activities. Despite the recent release of many AR and VR HMDs, two major problems are hindering the AR HMDs from reaching the mainstream market: the extremely high costs and the user experience issues. In order to minimize these problems, we have developed an AR HMD prototype based on a smartphone and on other low-cost materials. The prototype is capable of running Eye Tracking algorithms, which can be used to improve user interaction and user experience. To assess our AR HMD prototype, we choose a state-of-the-art method for eye center location found in the literature and evaluate its real-time performance in different development boards.

DOI: 10.4018/978-1-5225-3290-3.ch001

INTRODUCTION

Head-Mounted Displays (HMDs) that provide Virtual Reality (VR) and Augmented Reality (AR) capabilities have been emerging in the last years. These technologies sound like the new hot topic for the next years. Users have had the opportunity to enjoy these technologies for entertainment, work tasks, retrieving health and body information, and other daily activities.

For years, engineers and researchers have been developing technologies to build the Head-Mounted Displays for many different purposes. Manufacturers are seeking to design hardware and software that improve the user experience and the user immersion in augmented and virtual worlds, in order to increase the user adoption and transform these devices into products of the mainstream market.

Despite the recent release of many AR and VR HMDs, two major problems are hindering the AR HMDs from reaching the mainstream market: the extremely high costs (from US$ 800 to US$ 3000 (Mirza & Sarayeddine, 2015), (Microsoft, 2016)) and the user experience issues such as the vergence-accommodation conflict (Kramida, 2016).

In order to minimize these problems, we have developed a simple AR HMD prototype based on a smartphone and on other low-cost materials, such as a beam splitter, a webcam (pointed to the user's eye, in order to perform the eye center location task) and a development board capable of running digital image processing algorithms.

The smartphone generates the 3-D (stereoscopic) virtual objects images and displays these images into the beam splitter. The beam splitter combines the virtual objects, formed by the smartphone, with the real ones, which are in the real environment. Therefore, the smartphone and the beam splitter are responsible for providing the Augmented Reality visualization to the user.

The prototype is also capable of running an eye center location algorithm, which can be used to improve the user experience. The eye center location information can be used to correct the right and left images' position of the virtual object in the stereoscopic view generated by the smartphone application. The correction of the image position using this approach guarantees the successful 3-D visualization of the virtual object independently of the user's gaze, improving the user experience.

Furthermore, the prototype was built thinking on its application in industrial environment scenario. General AR HMDs are used in several industrial applications. Thus, the prototype was developed in such a way that it can be coupled to a safety helmet, which is in accordance with the industrial environment.

We implemented the state-of-the-art algorithm proposed by Valenti and Gevers (2012), which performs the eye center location task in low resolution images. We had to adapt that algorithm to work with the HMD setup since the original version did not perform well in this scenario. A sequential version and a parallel version of the algorithm were developed. Both versions of the algorithm were evaluated in different embedded platforms. The results show that our implementation of the algorithm is in accordance with the system requirements, but it still needs some adjustments.

This work is based on an earlier work: Towards a Low-Cost Augmented Reality Head-Mounted Display with Real-Time Eye Center Location Capability (D'Angelo, Delabrida, Oliveira, & Loureiro, 2016), in Proceedings of the 6th Brazilian Symposium on Computing System Engineering. ©IEEE, 2016.

The main contributions of this work are:

- An overview of the main concepts of HMD development.
- A review of methods for HMD development which were found in the literature.
- To introduce a low-cost AR HMD prototype with eye center location capability.
- To perform an evaluation of different hardware platforms for the eye center location algorithm.

This work is organized as follows: Section "Augmented Reality Head-Mounted Display Overview" presents an overview of the HMD's classification, features, and user experience issues. Section "Overview of Eye Center Location Methods" presents an overview of the eye center location problem and the methods for finding the eye center location. Section "Related Work" presents a review of HMD development, HMD calibration and Eye and Gaze Tracking methods which were found in the literature. Section "Low-cost AR HMD Prototype with Eye Center Location Capability" introduces the AR HMD prototype with the eye center location capability. Section "Experiments" describes the experiments performed and the corresponding results. Finally, Section "Conclusion" presents the conclusions.

AUGMENTED REALITY HEAD-MOUNTED DISPLAY OVERVIEW

This section presents an overview about Augmented Reality Head-Mounted Displays (ARHMDs). Firstly, it highlights the main differences between Augmented Reality (AR) and Virtual Reality (VR). Then, it presents an HMD classification as well as its characteristics and limitations. Furthermore, it discusses user experience issues with ARHMDs systems.

Since the late 1960s, when the first Head-Mounted Display (HMD) was released, researchers and manufacturers have made many attempts to develop a variety of

HMDs aimed for Virtual Reality (VR), Augmented Reality (AR) and Wearable Computing applications. HMDs have a wide range of applications in AR, including military, industrial, medical, educational, training, navigation and entertainment applications. The development of an ideal HMD to all situations is extremely hard. Therefore, the identification of the target application requirements and restrictions is crucial to define the technologies to be employed in the development of a specific HMD for a given target application. Some of the main characteristics and aspects that need to be observed when developing an HMD for an application are: type of see-through display and ocularity demanded by the application; optical design to be employed; resolution that the application requires; field of view amplitude; occlusion capability and depth of field requested by the application; latency, parallax effect, distortions and aberrations introduced by the chosen architecture; and matters related to the user experience and acceptance.

Difference Between Virtual Reality and Augmented Reality

Both virtual reality and augmented reality are technologies that have made great strides in recent years. Both technologies seek to present virtual contents to users, but they differ fundamentally in the methodology used to expose these contents. Next, these two technologies are characterized and exemplified.

Virtual Reality

On the one hand, virtual reality provides a complete immersion of the user into a virtual world. In this virtual world, the user is surrounded by virtual objects, being able to interact with them as if they are real. In general, through a Virtual Reality HMD, the user has the experience of being transported to a completely computer generated virtual world and no longer has access to the real world where he actually is. Oculus Rift, Samsung VR Gear, Google Cardboard, HTC Live, Playstation VR are some examples of Virtual Reality HMDs which are already on the market.

Augmented Reality

On the other hand, augmented reality provides the overlap of virtual world on the real world. The user can view computer-generated virtual objects and, at the same time, can visualize the real world. Ideally, the user can also interact with these virtual objects. Besides that, these virtual objects must also be positioned and rendered correctly in real world, in order to be as consistent as the real ones. In general, through an Augmented Reality HMD, the user has the sensation of being present in the real world, which is only "augmented" by the overlapping virtual objects and information. Microsoft Hololens, Daqri Smart Helmet, Metavision Meta 2 and Magic Leap are the most prominent examples of Augmented Reality HMDs.

HMD Classification

Augmented Reality Head-Mounted Displays can be classified according to various parameters. Therefore, we choose to classify HMDs according to three main parameters, as suggested in Kiyokawa (2015) and Kress (2015): the type of see-through display, the type of ocularity and the optical design employed in the HMD development.

See-Through Displays

In general, there are two main types of see-through displays in AR: optical see-through and video see-through displays.

Optical See-Through (OST) Displays

Through an optical system, the real and virtual images are combined using an optical device that is partially transmissive and reflective. The real-world image is fully seen through this optical combiner while the virtual image overlays the real one. The advantages of the optical system see-through include: natural and instantaneous view of the real world, and its structures are usually light and simple (Kiyokawa, 2015). Most of HMDs use an optical see-through display to provide an augmented view for users. Some examples of optical see-through HMDs are Google Glass, Optinvent Ora, Epson Moverio and Microsoft HoloLens (Kiyokawa, 2015), (Kress, 2015), (Mirza & Sarayeddine, 2015), (Microsoft, 2016).

Video See-Through (VST) Display

When using a video see-through display, the real-world image is first captured by a video camera, then, the captured image and the virtual one are digitally combined. Finally, the combination of the images is displayed to the user through a video display, as an LCD or LED screen. The advantages of the video system in relation to the optical system include a pictorial consistency (precise overlay of the virtual image on the real one) and the availability of countless image processing techniques (Kiyokawa, 2015). The authors of Takagi, Yamazaki, Saito, and Taniguchi (2000) and State, Keller, and Fuchs (2005) show HMDs that use video see-through systems to display an augmented view for users. Steve Mann's EyeTap HMD (Mann, 2002) can also be considered as a video see-through HMD.

Ocularity

Another criterion used for categorizing HMDs is the ocularity, a measure of the number of eyes needed to see something. There are three types of ocularity: monocular, bi-ocular and binocular. A monocular HMD, as the name suggests, has a single viewing device and is recommended for applications in which stereoscopic view is not required, such as general purpose and daily usage HMDs. Google Glass, Optinvent Ora and EyeTap are examples of monocular HMDs (Mirza & Sarayeddine, 2015), (Kress, 2015), (Mann, 2002), (Mann, 2013). A biocular HMD provides a single image to both eyes while a binocular HMD has two separate displays with two input channels, one for each eye (Kiyokawa, 2015). A binocular HMD can function as a stereoscopic HMD only when two different image sources are properly provided. For AR, binocular video see-through HMDs are highly recommendable due to their capability of generating stereoscopic images (Kiyokawa, 2015). Epson Moverio and Microsoft HoloLens are examples of binocular HMDs (Microsoft, 2016), (Kress, 2015).

Optical Design

Regarding the optical designs, HMDs can be divided into two categories: pupil-forming and non-pupil-forming. The architecture that represents the pupil formation has frequently been used since the first HMDs allow a wide field of view, despite presenting greater size and weight. This architecture generates, at least, one intermediate image and the exit pupil is collimated by the eyepiece. In relation to the size of the device that creates the images, the existence of an intermediate image offers a flexible optical design (Kramida, 2016), (Kiyokawa, 2015).

With the emergence of high resolution displays and small imaging devices, the architectures without pupil formation have become more common (Kiyokawa, 2015). Besides, high resolution and small imaging devices allowed a moderate field of view in a light and compact structure (Kramida, 2016), (Kiyokawa, 2015). On the other hand, they have a less flexible optical design and do not generate any intermediate image (Kiyokawa, 2015). Free-form prisms, holographic optical elements and optical waveguide are some of the technologies used in architectures without pupil formation. Some recent HMDs like Google Glass, Optinvent Ora, Epson Moverio and Microsoft HoloLens use optical designs based on waveguides (Mirza & Sarayeddine, 2015), (Microsoft, 2016), (Kiyokawa, 2015), (Kress, 2015)}.

HMD Characteristics and Limitations

The main characteristics of HMDs (such as image resolution, field of view amplitude, occlusion capability, depth of field and optical design) are intrinsically related to the current limitations of the technology (such as pictorial consistency, vergence-accommodation conflict, latency, parallax effect, distortions, and aberrations). After analyzing the constructive aspects of the HMDs, it is important to define the characteristics that are demanded by the target application and try to minimize the technology limitations related to it.

Resolution

The resolution of a see-through determines the integrity of the virtual image in relation to the real image. The resolution of the whole system is limited by the optical system, by the image generator device, and possibly, by the camera resolution (in the case of video see-through HMD). Regarding the resolution of the virtual image, an ideal HMD will need to have up to 12,000 × 7,200 pixels to compete with human view, which has an angular resolution of 60 pixels per degree (PPD), considering the human total field of view of 200° (horizontal) per 120° (vertical) (Kiyokawa, 2015). As it is not possible to achieve this value of PPD with the current technologies, it is necessary to make a trade-off between the angular resolution and the amplitude of the field of view to achieve a viable solution. However, as the resolution of the screen tends to keep increasing, this trade-off between angular resolution and amplitude of the field of view must disappear in the future (Kiyokawa, 2015). It is important to note that only the augmented view suffers from limited resolution issues in optical see-through HMDs system while in video see-through HMDs both, the real and augmented views, suffer from resolution limit (Kramida, 2016).

Field of View, Depth of Field, Vergence-Accomodation Conflict and Occlusion Capability

In Augmented Reality Head-Mounted Displays, the Field of View (FOV) is an important parameter, which is typically measured in degrees and gives us an idea of how wide is the augmented view that the user sees. Generally, HMDs for AR applications, such as Microsoft HoloLens, require wide and stereoscopic FOV, while HMDs for smart glasses applications, such as Google Glass, can have narrower and monocular FOV (Kiyokawa, 2015), (Kress, 2015).

Meanwhile, the depth of field refers to the set of distances in relation to the eye (or to the camera) in which a given object remains focalized into the FOV. In real life, the accommodation of the eye is automatically adjusted to focus on an object according to the distance, and objects outside this depth of field seem to be distorted. On the other hand, the virtual image is usually observed from a fixed distance. This focal distance of virtual image represents a problem because the accommodation and the convergence of the human view system are intrinsically linked. This way, adjusting only one of these aspects and keeping the other fixed might cause ocular fatigue (Kramida, 2016), (Kiyokawa, 2015), (Kress, 2015). This problem is also known as vergence-accommodation conflict (Kramida, 2016). It can be minimized by using a new technology known as light-field display, but this technology demands high-cost hardware with high computational resources for rendering the light-field images (Ackerman, 2015), (Huang, Luebke, & Wetzstein, 2015). Therefore, it is not feasible to use this kind of display with regular embedded systems.

Another desirable characteristic for AR HMDs is the occlusion capability. An HMD with occlusion capability can introduce a virtual object between real objects providing important depth information about the augmented view (Kiyokawa, 2015), (Rolland & Fuchs, 2000). The occlusion occurs in such way that the real object in front occludes part of the virtual object, and the virtual object occludes part of the real object behind it. Occlusion capability is more easily achievable with video see-through HMDs than with optical see-through HMDs (Kiyokawa, 2015), (Rolland & Fuchs, 2000), (Billinghurst, 2014).

HMD Limitations

Some of the main limitations in Augmented Reality Head-Mounted Displays are latency, parallax effect, distortions, and aberrations. These restrictions are related to the chosen optical design, as well as to other hardware issues, and must be minimized. Some of these problems are harder to deal with in an optical see-through design than in a video see-through design, and vice-versa.

User Experience in AR HMD's Systems

Opposed to smartphones or smartwatches, an HMD can be a discomfort when the users need to wear it and take it off frequently. A future perspective for HMDs is that they will become light, small and comfortable, in a way that users will be able to use them continuously for a long period during the day for diverse purposes. Nevertheless, the HMD might be useless, or even harmful if the content it shows is irrelevant to the current context of the user. This issue is least noticeable in HMDs for AR, once it is expected to have a wide field of augmented view covering the central field of view of the user. In such situations, an AR system must be aware of the user's environment contexts, so it must change its content and presentation style correctly and dynamically according to the context (Kiyokawa, 2015), (Mann, 2002).

HMDs used inappropriately might induce undesirable symptoms like headache, shoulder stiffness, nausea, or even more severe harm to the user's health. From an ergonomic point of view, HMDs must be as light, small and comfortable as possible during usage. Besides, the look and appearance of HMDs must satisfy the application's requirements. The center of mass in the HMD must be placed as close as possible to the user's head. A heavy and well balanced HMD might seem lighter to the user than a light and unbalanced HMD (Kiyokawa, 2015), (Billinghurst, 2014).

The matters of safety have equal importance. Because of their nature, AR applications tend to distract the user's attention from what happens around him/her due to virtual images overlaying the real environment. AR applications must present minimum information to avoid catastrophic results and at the same time, help during the realization of the target task satisfactorily. When the matter of security is considered a top priority, HMDs with optical see-through systems are recommended over HMDs with video see-through systems. This happens because video see-through HMDs restrict the user's peripheral view. Moreover, in case of failure, the central view of the user would be lost (Kiyokawa, 2015), (Rolland & Fuchs, 2000), (Billinghurst, 2014).

OVERVIEW OF EYE CENTER LOCATION METHODS

This section presents an overview of the methods for eye center location and shows the importance of the eye center location for Augmented Reality Head-Mounted Displays systems. Furthermore, a classification of eye center location methods is provided. The key features, advantages and disadvantages of each type of method are briefly described.

Eye Center Location Methods for AR HMDs

The eye center location is a relevant information for several applications and researches related to computer vision. This information is usually used in applications such as face alignment, face recognition, human-computer interaction, device control for handicapped, user's gaze and attention detection (Valenti & Gevers, 2012). Moreover, with the arising of several models of Augmented Reality and Virtual Reality Head-Mounted Displays, the eye center location information plays a significant role in gaze tracking and point-of-regard detection systems (Kramida, 2016), (Hansen & Ji, 2010). These two systems, along with other systems such as Inertial Measurement

Units (IMUs), are responsible for improving the virtual image overlay into the real environment, providing the correct registration of the virtual objects according to the user's point-of-view. These systems can also be used to minimize the vergence-accommodation conflict (Kramida, 2016), which occurs in most AR and VR devices with stereoscopic displays currently existing on the market.

The use of pattern recognition and computer vision techniques in applications, which involve embedded systems and demand real-time processing using low computational resources, like AR HMD's applications, is a complex task. Therefore, the use of low-resolution images as the input to the computer vision and pattern recognition algorithms is a feasible way to enable real-time processing without increasing the computational resources needs and the hardware costs.

Classification of Eye Center Location Methods

Several methods for eye center detection using low-resolution images have been proposed in the literature and they can be grouped into basically three categories: model-based methods, feature-based methods and hybrid methods (Valenti & Gevers, 2012), (Hansen & Ji, 2010), (Valenti & Gevers, 2008).

Model-Based Methods

These methods use global information related to the eyes and face appearance. These approaches often use the classification of a set of attributes or the learning of a given model to estimate the eye location. By using the global appearance, the model-based methods present advantages such as accurate and robust detection of eye position. However, as the success of these methods requires the correct detection of different features or the convergence of a complete model, the importance of eye center location in the global information is often reduced, due to its variability. Thus, the eye center location is usually interpreted just as the central point of the eye model or as the midpoint between the two corners of the eye. Therefore, these methods are not very accurate in situations of sudden motion of the eye center (Hansen & Ji, 2010).

Feature-Based Methods

These methods use known characteristics for eyes to detect possible eye center locations through simple local features in the image (like the eye corner, borders and image gradient). Such methods do not require models usage nor any other way of learning. This way, noises or other features overlap do not disturb feature-based methods. Besides, they can achieve very precise results when locating the eye center. However, a lot of times the detected features may be wrong, so feature-based methods are less stable than model-based methods (Valenti & Gevers, 2012), (Hansen & Ji, 2010).

As a way of combining advantages from both previous methods, researchers have developed some methods named hybrid methods.

Hybrid Methods

In these methods, a classifier (trained using a given eye modeling) receives multiple candidates to locate the eye center (these candidates are obtained from the feature-based method). The classifier is responsible for determining which is the correct eye center. This way, hybrid methods can achieve a better precision and greater robustness than previous methods (Valenti & Gevers, 2012), (Hansen & Ji, 2010).

RELATED WORK

This section presents some studies related to this work. First, we discuss about the development of Augmented Reality HMDs. Then, we present studies about Augmented Reality HMDs calibration. Finally, we present some eye tracking and gaze tracking methodologies and the importance of using them in Augmented Reality HMD applications.

AR HMD Development

There are several methodologies for developing an Augmented Reality HMD. An HMD can be developed using the different existing optical elements described in the previous section, or by adapting Virtual Reality HMDs to provide Augmented Reality.

The manufacture of optical elements involves the development of optical design and the creation of a mold for the lens manufacturing process. Although this methodology allows the construction of extremely light and compact devices with the appearance of ordinary glasses, it increases the overall development cost, reflecting the high market cost of HMDs which use such technology.

Besides, existing technology does not allow the creation of optical designs that provide wide field of view and low latency (Fuchs, State, Dunn, & Keller, 2015). Then, researchers began to develop methodologies to adapt some virtual reality HMDs to provide augmented reality, maintaining the desired characteristics of wide field of view and low latency.

Fuchs et al. (2015) argue that the virtual reality HMDs market has grown in a much larger proportion than the augmented reality market. Thus, in contrast to VR HMDs, most of commercially available Optical See-Through (OST) HMDs does not present a mature technology and generally has some issues like high latency and narrow field of view, limiting their applications and the user adoption.

Thus, they propose modifying the Oculus Rift to provide augmented reality with wide field of view and low latency. Through the adaptation of Oculus Rift and its integration with an optical combiner, the authors developed a prototype of Optical See-Through Head-Mounted Display for augmented reality. It is worth mentioning that this architecture can also be used to build a smartphone-based OST-HMD by replacing the Oculus Rift with a smartphone and making the necessary changes in the physical structure of HMD, so that it remains mechanically stable.

The authors tested two different optical designs for constructing this device. First, the combination of a flat and semi-transparent mirror with a pair of Fresnel lenses was evaluated to perform the optical combiner function. This strategy allowed the virtual image magnification, but limited the field of vision of the HMD in 35 degrees, due to the diameter of the Fresnel lens which was used in this architecture. Second, the usage of curved and semi-transparent mirrors was evaluated, without the need for lenses. This strategy allows for greater field of view, although it introduces small optical distortions. Using this latter configuration it was possible to construct an augmented reality OST-HMD with wide field of view and low latency.

Itoh, Orlosky, Huber, Kiyokawa, and Klinker (2016) performed an evaluation of the temporal consistency in the positioning of virtual objects in an OST-HMD whose architecture is identical to the first proposal of Fuchs et al. (2015). Through this evaluation, it was confirmed that the low latency of Oculus Rift allows the temporarily consistent positioning of virtual objects in the augmented reality environment provided by the OST-HMD. Despite this fact, due to the architecture used in the prototype, both the accommodation distance and the field of view range were reduced.

The OST-HMD prototype developed in our work and described in the Section "Section Name" presents some advantages and disadvantages when compared to the related these works. First, because it is based on a smartphone, the prototype has the advantage of mobility, since it does not have to be connected to a personal computer in order to function. For the same reason, the resolution and field of view of HMD are dependent on the smartphone screen, which may represent an advantage or a disadvantage, according to which device is used in the project. On the other hand, the smartphone has a higher latency than Oculus Rift, so, in this case, dependence on the smartphone represents a drawback. Like the related works, our prototype can be classified as binocular HMD, which is most recommended architecture for augmented reality applications. The reduced cost of our prototype, due to low-cost materials usage, can be considered an advantage if the smartphone used in the project has a lower price than the Oculus Rift.

AR HMD Calibration

In Augmented Reality applications, the Head-Mounted Display is not always positioned in the same way on the user's head and the information displayed on the display must always be generated according to the user's perspective. Therefore, the augmented reality system must be calibrated frequently and depends, among other factors, on unique characteristics of each user. If this calibration requires user interaction, this task can become tiresome and may distract the user from his primary activity. In addition, introducing user-dependent errors in the system can reduce user acceptance of OST-HMDs (Itoh & Klinker, 2014).

As a way to overcome this problem, Itoh and Klinker (2014) proposed a method which uses the dynamic measurement of the 3D location of the user's eye through an Eye Tracker combined with static and pre-computed display calibration parameters. In this way, once the eye tracker detects the change of the 3D location of the eye, it is possible to readjust the calibration parameters according to this new location, without the need for user interaction. The results show that the proposed method is more stable than a state-of-the-art method. However, the proposed method still does not fulfill the requirement of real-time execution, essential for its use in embedded systems with low computational power. Therefore, by requiring more advanced hardware, its practical use in low cost AR HMDs is not feasible.

Plopski et al. (2015) also address the problem of HMDs automatic calibration. In this work the reflection of an image in the user's cornea is used to enable the calibration of the augmented reality HMD. The results are slightly higher than the previous work. However, since the method also requires a lot of processing power, it becomes impractical for application in embedded systems with low computational power.

Eye and Gaze Tracking Methods

From the analysis of the calibration methods proposed in the previous works, it was possible to perceive that the positioning of the eye and the direction of the user's gaze contain important information for the AR HMD calibration. Thus, it is necessary to study the methods for performing eye and gaze tracking that are feasible for the application in embedded systems with low computational power.

The authors of Gupta, Lee and Billinghurst (2016) show that, in addition to assisting in the HMD calibration, eye and gaze tracking methods can also provide meaningful information for remote collaboration applications in HMDs. Then, the study of these methods applied to HMDs becomes increasingly important and necessary.

The state-of-the-art method proposed by Valenti and Gevers (2008) and Valenti and Gevers (2012), addresses the eye center location problem in low resolution images by using the isocentric invariant patterns. This method also meets the requirements of low computational power. Therefore it can be applied in embedded systems for performing eye tracking.

Basically, this method uses the image's luminosity level curves, also known as isophotes, as the local attribute for determining the eye center location. From the isophote calculation, it is possible to estimate the eye center as the center of the level curves of the same luminosity. A voting system is used to calculate the coordinates which represent the eye center. In this voting system, the image's coordinate which accumulates more votes is considered the coordinate of the eye center. The image map that accumulates the voting results is known as the centermap, each centermap coordinate is called isocenter, and each isocenter stores a value that corresponds to the amount of votes received by its coordinate. The voting system is weighted according to the degree of curvature of the isophotes. Thus, as the degree of curvature of an isophote increases, the vote's weight of the pixels belonging to that curve in the voting system also increases. Valenti and Gevers (2008) and Valenti and Gevers (2012) proposed three variants of this algorithm: a basic method, an intermediate method and an improved method.

The basic method is based on the voting system of isophotes' center and it considers the most voted isocenter as coordinate of the eye center. This method is also known as Maximum Isocenter (MIC) method.

The intermediate method consists of a union of the basic method (MIC) with the Mean-Shift (MS) algorithm (Bradski, 1998). In this method, the algorithm searches for the region of the centermap that has the highest density of votes and, within that region, it stipulates which isocenter corresponds to the coordinates of the eye center location in the image.

The improved method uses the MIC method to calculate all the isocenters in the image's centermap. Then, it uses a variant of the Scale Invariant Feature Transform (LOWE, 2004) to identify the isocenters that are most likely to be the eye center. From this point, a previously trained k-NN (k-Nearest Neighbor) classifier is used to determine which of these isocenters corresponds to the real coordinate of the eye center location in the image.

Among all methods, only the improved method does not meet the constraints of low computational cost and, therefore, it cannot be easily applied in embedded systems with low processing power.

Although the algorithm was initially proposed to address the eye tracking problem, Valenti, Staiano, Sebe, and Gevers (2009) proposed an adaptation that allowed its application in the gaze tracking problem for low resolution images. However, the use of the gaze tracking algorithm was unstable if compared to state-of-the-art methods. Therefore, in our future works, we want to improve this algorithm to obtain more stable results for the gaze tracking problem.

An important observation lays in the fact that both versions of the algorithm, for eye and gaze tracking problems, were not proposed for use in wearable devices such as HMDs. Thus, in the present work several adaptations were made in these algorithms so that they could be used in the scenario of eye and gaze tracking in Head-Mounted Displays.

LOW-COST AR HMD PROTOTYPE WITH EYE CENTER LOCATION CAPABILITY

This section presents the assembled AR HMD prototype and its features. The system is composed of an AR display module working as user interface and an Intelligent Sensor Board (ISB) (Delabrida, D'Angelo, Oliveira, & Loureiro, 2016) module running a digital image processing algorithm for eye center location. The AR display was assembled with an Android smartphone running a simple application with stereoscopic images, a beam splitter, an eye splitter, and a camera pointed to the user's eye. Figure 1 shows the design of the AR display module. The ISB is composed of a processing unit, which is implanted to the internal structure of

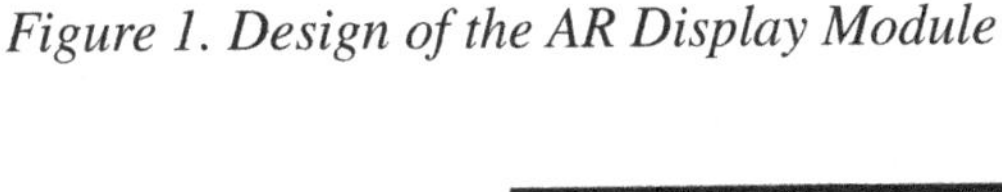

Figure 1. Design of the AR Display Module

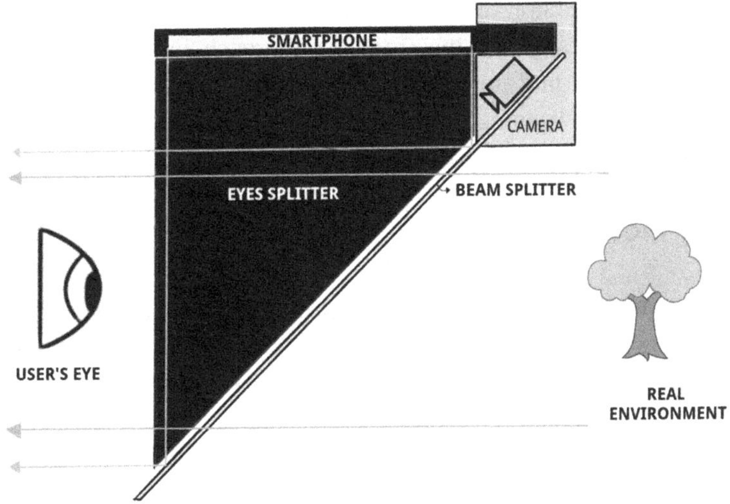

a safety helmet. This processing unit is responsible for estimating the user's eye center location through the images captured by the camera in AR display module.

The eye center location information will be used to correct the right and left images position of the virtual object in the stereoscopic view generated by the smartphone application. After obtaining this eye center information, it is possible to infer the user's gaze direction for each image of the stereoscopic view, and then correct the binocular disparity between these images. The correction of the binocular disparity using this approach guarantees the successful 3-D visualization of the virtual object independently of the user's gaze. Therefore, we can minimize some known problems of AR and VR HMDs, like the vergence-accommodation conflict (Kramida, 2016), using a feasible and low-cost solution.

The AR HMD prototype developed in this work can be classified as a binocular optical see-through HMD without pupil formation according to the HMD classification presented in Section "Augmented Reality Head-Mounted Display Overview". This architecture is better than the video see-through approaches used in (Delabrida et al., 2016) and (Delabrida, D'Angelo, Oliveira, & Loureiro, 2015), because it enables the user to see virtual and real objects at the same time without blocking the user's peripheral vision or reducing the user's mobility.

The current version was assembled using pieces of Styrofoam material mounted on top of a safety helmet. As future work, we plan to design and print a more compact and lightweight version of this prototype using a 3-D printer for usage in industrial environment.

We have implemented a state-of-the-art algorithm available in the literature, which was firstly proposed by Valenti & Gevers (2008) and later improved by Valenti & Gevers (2012), to perform the eye center location task. The eye center location algorithm was implemented in Python using the OPENCV Library. A sequential version and a parallel version of the algorithm were developed.

The prototype was designed using a modular approach. First, we created a display module capable of providing augmented reality information. Then, we designed the ISB module, capable of running digital image processing algorithms.

Next, each module of the prototype is described.

AR Display Module

The AR display module is composed of four principal components. Figure 2 shows all these components where each component is identified using the enumeration described in the following.

Figure 2. Components of the AR Display module prototype

Android Smartphone (1)

It is responsible for providing the virtual images of the augmented reality system through an application showing stereoscopic images. We use the demo application of Google Cardboard API, which originally supports the building of 3D VR applications for Android Smartphones, to provide this stereoscopic view. Figure 3 shows the application screen, which is vertically divided into two regions. The right region must show the virtual object image from the point of view of the user's right eye, whereas the left region from the point of view of the user's left eye. In this way the user can view a 3-D virtual object overlaid on the real world. In this prototype, we use the Samsung Galaxy Note 3 smartphone.

Eyes Splitter (2)

It is responsible for ensuring that the right eye sees only the virtual image of the right region and the left eye sees only the virtual image of the left area. Thus, it is

Figure 3. Stereoscopic view of a virtual object (cube) provided by Google Cardboard demo application

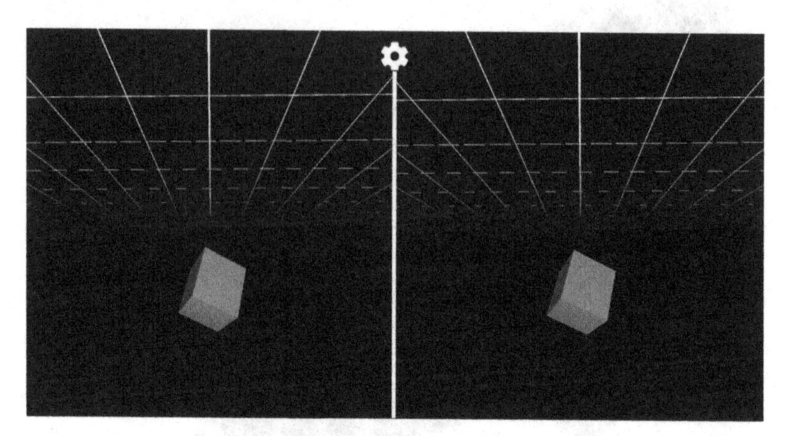

responsible for ensuring the correct visualization of stereoscopic images. The eyes splitter is placed between the user's eyes and is aligned with the application screen division.

Beam Splitter (3)

It is responsible for combining the virtual images provided by the smartphone with the real-world view. Figure 1 also shows the beam splitter functioning. The yellow arrows represent the beam of light emitted by the smartphone screen, i.e., the beam of light which forms the virtual image. When this light beam reaches the beam splitter surface, part of it is reflected directly into the user's eyes and the another part pass through the beam splitter. The green arrows represent the beam of light emitted by the real-world environment, i.e., the beam of light which forms the real-world view. When this light beam reaches the beam splitter surface, part of it is reflected directly into the user's eyes and the another part pass through the beam splitter. Therefore, by using the beam splitter, it is possible to overlay the real-world scene with virtual images and objects, enabling augmented reality features in this device. Figure 4 shows the user's point-of-view when wearing the prototype. In Figure 4, it is possible to see at the same time a real-world object (chair) and a virtual object (cube). Moreover, it is worth noting that the user's peripheral vision is not blocked by the prototype structure.

Figure 4. Augmented reality view using the beam splitter

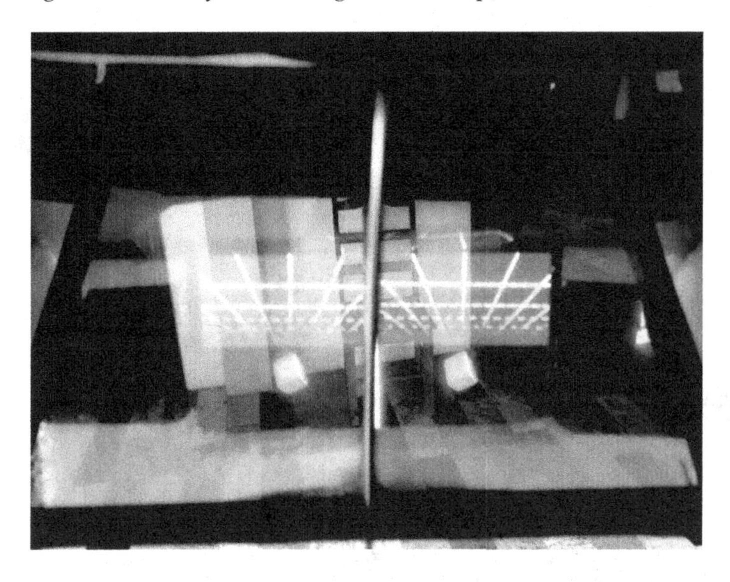

Camera (4)

It is responsible for capturing the frontal image of the user's eye and sending these images to the processing unit located inside the safety helmet. We used Logitech C270 webcam in the prototype.

ISB Module

The AR HMD prototype uses a camera pointed to user's eyes in order to estimate each eye center location using a digital image processing algorithm. This algorithm runs on the processing unit of the ISB module. Each eye center position calculated by the algorithm can be used to estimate the user's gaze direction. This information can be used as a feedback data to the smartphone application. We use Bluetooth communication to send the eye center data from the ISB module to the smartphone. Then, the smartphone application will provide the correct stereoscopic view according to user's gaze, which will guarantee the proper 3-D visualization.

As shown in Section "Development Boards Performance", different development boards were tested as the processing unit of the ISB module. Figure 5 shows the processing unit attached to the structure of a safety helmet. Together, the processing

Figure 5. The ISB module processing unit sewed into the structure of a safety helmet

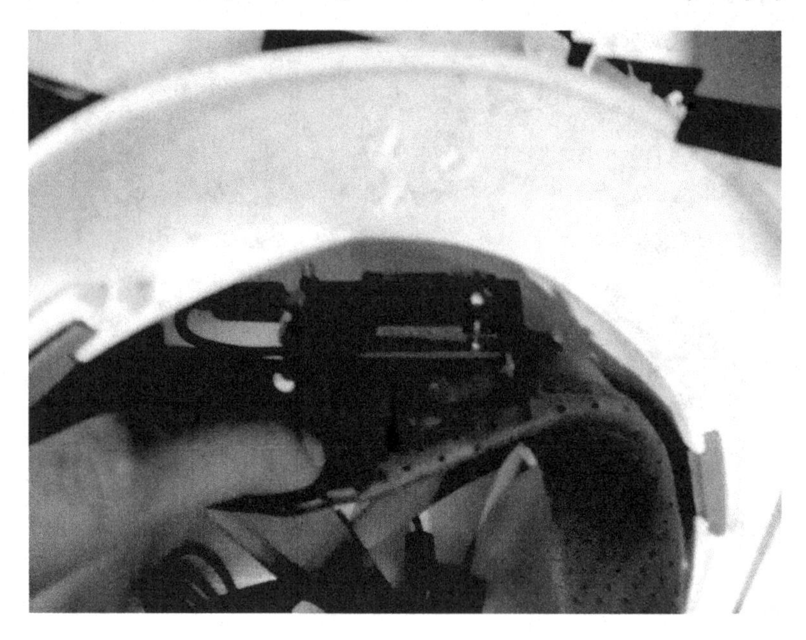

unit and the safety helmet compose the ISB module. We use a safety helmet to build this prototype because general AR HMDs are used in several industrial applications and this safety helmet is accordance with the industrial environment.

Figure 6 shows the complete AR HMD prototype and a user wearing the prototype. The total prototype cost is less than US$ 100 (Smartphone cost is not considered), and the greater part of this value is related to the processing unit of the ISB module, which costs approximately half of the total value.

Figure 6. Complete AR HMD Prototype (left) and an user wearing it (right)

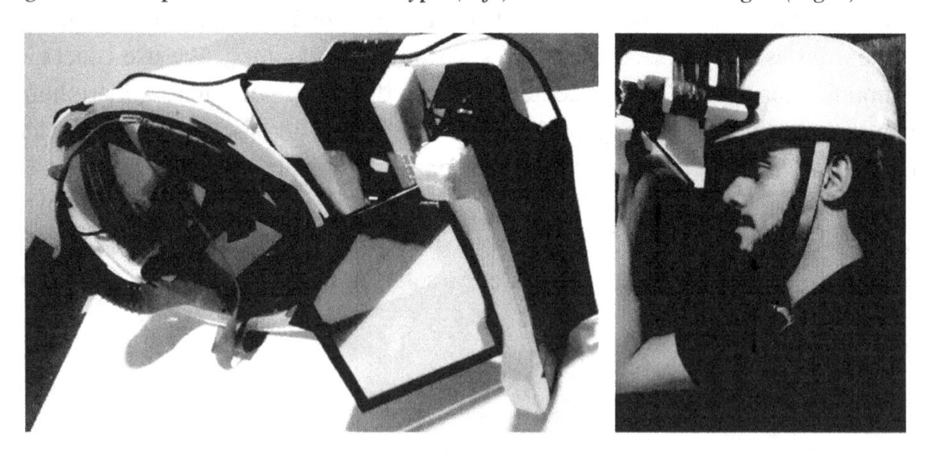

Eye Center Location Algorithm

In order to provide the eye center location information, we have implemented the eye center location algorithm proposed by Valenti & Gevers (2012) and Valenti & Gevers (2008). This algorithm provided high accuracy results and was developed using digital image processing and pattern recognition techniques. Our implementation of the algorithm does not achieve a high accuracy as the original does, but it does produce a satisfactory result.

This eye center location algorithm is based on invariant isocentric patterns formed by isophotes, which can be obtained from the gradient of an eye image in grayscale, and has three variants. The three variants of the algorithm consists of: (i) a basic feature-based method in which the eye center is estimated as the isophotes center; (ii) an intermediate feature-based method that uses the basic method along with Mean Shift (MS) algorithm (Bradski, 1998) to improve the method's stability and precision; and (iii) an enhanced hybrid method that uses the basic method with a k-Nearest Neighbor (k-NN) classifier and with the Scale Invariant Feature Transform (SIFT) algorithm (Lowe, 2004) to make the system's accuracy and robustness even better.

The results from experimentation and analysis of the method proposed by Valenti & Gevers (2012), and Valenti & Gevers (2008) indicate that the most basic variants (in other words, the basic method and the basic method with Mean Shift algorithm) can achieve good results while keeping a performance that obeys real-time requirements. Meanwhile, the enhanced version of the method, in spite of achieving even better results, follows neither real-time requirements nor low computational cost, which are both demanded by the present work. Therefore, our implementation consists of the intermediate method variant, which uses the basic method along with Mean Shift (MS) algorithm to achieve satisfactory results for the eye center location problem even with low computational cost and real-time restrictions.

Figure 7 depicts our implementation of the algorithm process. The difference between the sequential and the parallel implementations of the algorithm is related to the processing of each eye in the image. In the sequential version, each eye is processed per time. In the parallel version, both eyes are processed at the same time each one, in a processor core. Next, the six stages of the algorithm are explained.

Step 1: Frame Capture and Grayscale Conversion

In this stage, the image frame of the camera is captured and converted from a three channel RGB image to a single channel grayscale image.

Figure 7. The algorithm stages

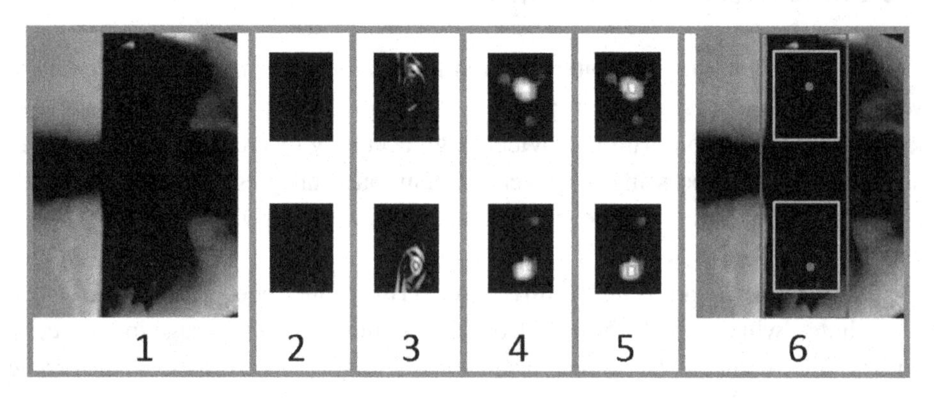

Step 2: Crop Image

In this stage, the grayscale image is cropped into two regions. As the camera and eyes have a fixed position in this head-mounted device, each eye will always appear in the same area of the image. Then, we have two regions of interest in each frame, one for each eye. We chose this strategy to reduce the computational cost of the algorithm, by limiting the eye center search area.

Step 3: Isophote Calculation

In this stage, the algorithm performs the calculation of the isophote curvatures, for each eye, using the method proposed by Valenti & Gevers, (2008).

Step 4: Center Voting and Centermap Calculation

In this stage, the algorithm performs the center voting mechanism, proposed by Valenti & Gevers (2008), to calculate the centermap. The centermap acts like 2-dimensional probability distribution function, where the most voted coordinates have a greater chance of being the real eye center location coordinate. This stage is also performed for each eye. The fourth stage of Figure 7 shows the centermap image. The centermap's brightest regions have a greater probability of being the real eye center location.

Step 5: Mean-Shift Algorithm

In this stage, the MS algorithm is applied over the centermap image of each eye. The MS's sliding window iterates over the centermap image looking for the area with high density of votes. The MS's sliding window is represented by a blue rectangle in the fifth stage of the algorithm in Figure 7.

Step 6: Eye Center Location

In this stage, each eye center location is calculated as being the most voted coordinate, in the centermap, close to the center of MS's sliding window. The sixth stage of Figure 7 shows the estimated eye center location for each eye.

EXPERIMENTS

This section presents the experiments we conducted to evaluate our implementation of the eye center location algorithm in different development boards. We evaluate the algorithm performance in different development boards. The goal of the experiments is to define the best development board for this AR HMD with eye center location capability. Also, it is essential the matching of the hardware with some of the wearable and the real-time requirements. For instance, the user interface accessibility, lightweight and compact size are some requirements needed for the wearable AR HMD. Meanwhile, to identify the user eye center in real time, the algorithm must provide the center eye information at least twice per second, in case of low eye motion speed, or at least four times per second, in case of high eye motion speed (Al-Rahayfeh, 2013). Therefore, the hardware and software need to have a minimum frame rate between 2 and 4 FPS to provide this requirement. All these requirements were considered in the selection of the hardware.

Next, we present the experiments scenarios and results.

Development Boards Performance

We chose four embedded platforms to assess the hardware and software performance. Each one is in accordance with the wearable/HMD requirements and can work with a battery as power supply (Delabrida et al., 2016). The following embedded platforms were selected: *Intel Edison* (Dual-threaded Intel Atom CPU at 500 MHz, a 32-bit Intel Quark microcontroller at 100 MHz and 1 GB LPDDR3 RAM, 1 MB cache L1), *Wandboard Quad* (Freescale i.MX6 Quad core processor at 1 GHz and 2 GB DDR3 RAM), *Raspberry Pi 3 Model B* (Broadcom Quad-Core BCM2837 64 bits, 1.2 GHz and 1 GB SDRAM), *Cubie board* (ARM Cortex A8 at 1 GHz, 1 GB DDR3).

For each development board, we used the Linux as the operating system. For Intel Edison, we used the Yocto framework to build the operating systems. Also, each one uses the minimal core building, since our intention is to optimize the hardware performance. After the operating systems installation, we build the OpenCV with Python dependencies to execute the algorithm.

The frame per second is the metric defined to evaluate this system. We use a video stream to evaluate the hardware performance. Although the system uses a camera as sensor, we use this approach in the first evaluation to avoid camera interference on the algorithm performance. The equipment used executes at maximum 30 FPS.

Figure 8 shows the results obtained after the evaluation with 99% of confidence. We execute the sequential and parallel versions of the algorithm. The parallel one divides the image into two parts (one region of interest for each eye) and each one is sent to a core of the board. All development boards had a satisfactory frame rate for this application.

All development boards increased the frame rate when compared to the single-core and multi-core versions, except Cubieboard that is a single core device. This is an expected result due to the number of context switch occurences while the algorithm is in execution.

The frame rate difference between single-core and multi-core versions is not significant. Intel Edison has the biggest difference with two frame rate more in the multi-core version than single-core version. The performance evaluation shows that all of the hardware, except Cubieboard, has computing power enough to receive new applications.

Figure 8. Hardware Performance

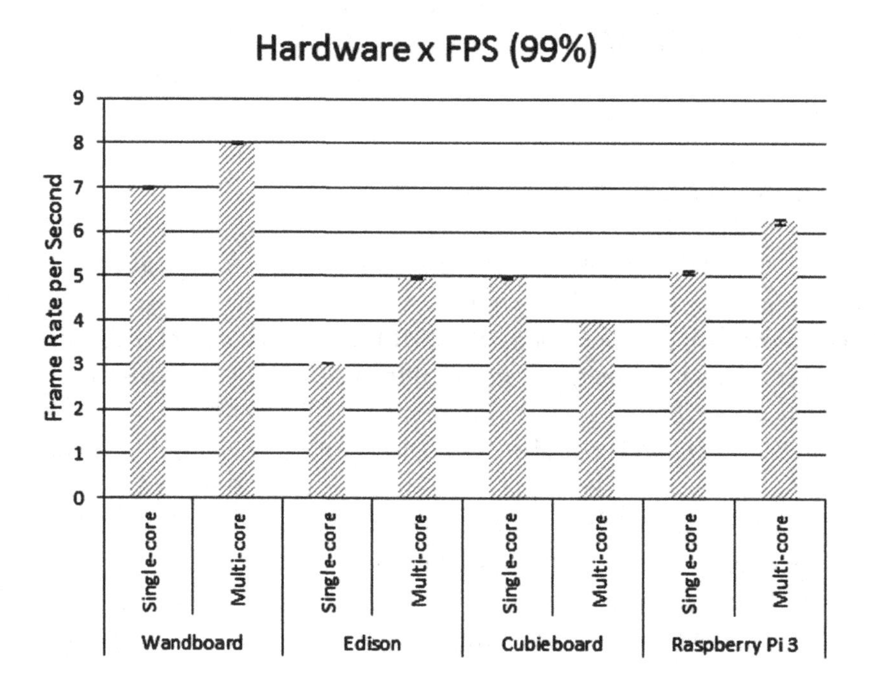

CONCLUSION

This work presented an overview about AR HMD's and eye center location methods. Also, an evaluation of an AR HMD prototype with eye center location capability is presented. The prototype was built with low-cost components, and its construction cost is less than US$ 100. An augmented reality display was developed using a beam splitter and an Android smartphone with Google Card Board API. A camera pointed to the user's eyes is used to estimate each eye center location through a digital image processing algorithm. The algorithm implemented for this function was evaluated in four different development boards. The Intel Edison development board had the best performance, considering both real-time and wearable requirements, while the Cubieboard presented the worst result, considering the same requirements. Therefore, the best development board for this project is the Intel Edison.

As future work, we plan to implement the algorithm using other programming languages, such as C or C++, and add new functionalities, which will be evaluated in the development boards. Our intention is to implement a gaze tracking algorithm to detect the user's point-of-regard and possible points of interest in the environment. The prototype can be applied for industrial applications. We also want to develop our own augmented reality application, independent of the Google Cardboard API, so that we can do a full integration of all prototype modules.

REFERENCES

Ackerman, E. (2015). 4-D Light Field Displays are exactly what Virtual Reality needs. *IEEE Spectrum*.

Al-Rahayfeh, A., & Faezipour, M. (2013). Enhanced frame rate for real-time eye tracking using circular Hough Transform. In *IEEE Long Island Systems, Applications and Technology Conference* (pp. 1–6). Long Island, NY: IEEE. doi:10.1109/LISAT.2013.6578214

Billinghurst, A. C. M., & Lee, G. (2014). A Survey of Augmented Reality. *Foundations and Trends in Human-Computer Interaction*, *8*(2), 73-272.

Bradski, G. R. (1998). Real time face and object tracking as a component of a perceptual user interface. In *Fourth IEEE Workshop on Applications of Computer Vision Proceedings* (pp. 214-219). Princeton, NJ: IEEE. doi:10.1109/ACV.1998.732882

D'Angelo, T., Delabrida, S., Oliveira, R. A. R., & Loureiro, A. A. (2016). Towards a Low-Cost Augmented Reality Head-Mounted Display with Real-Time Eye Center Location Capability. In *2016 Brazilian Symposium on Computing Systems Engineering* (pp. 24-31). João Pessoa, PB: IEEE. doi:10.1109/SBESC.2016.013

Delabrida, S., D'Angelo, T., Oliveira, R. A. R., & Loureiro, A. A. (2015). Towards a wearable device for monitoring ecological environments. In *2015 Brazilian Symposium on Computing Systems Engineering* (pp. 148-153). Foz do Iguaçu, PR: IEEE. doi:10.1109/SBESC.2015.35

Delabrida, S., D'Angelo, T., Oliveira, R. A. R., & Loureiro, A. A. (2016). Building wearables for geology: An operating system approach. *SIGOPS Operating Systems Review*, *50*(1), 31-45.

Fuchs, H., State, A., Dunn, D., & Keller, K. (2015). *Converting commodity head-mounted displays for optical see-through augmented reality. Project Report.* Department of Computer Science, University of North Carolina at Chapel Hill.

Gupta, K., Lee, G. A., & Billinghurst, M. (2016). Do you see what I see? The effect of gaze tracking on task space remote collaboration. *IEEE Transactions on Visualization and Computer Graphics, 22*(11), 2413-2422.

Hansen, D. W., & Ji, Q. (2010). In the eye of the beholder: A survey of models for eyes and gaze. *IEEE Transactions on Pattern Analysis and Machine Intelligence, 32*(3), 478-500.

Huang, F. C., Luebke, D., & Wetzstein, G. (2015). The light field stereoscope. In *Proceedings of ACM SIGGRAPH 2015 Emerging Technologies* (pp. 24-24). New York, NY: ACM.

Itoh, Y., & Klinker, G. (2014). Interaction-free calibration for optical see-through head-mounted displays based on 3d eye localization. In *IEEE Symposium on 3D User Interfaces* (pp. 75-82). Minneapolis, MN: IEEE.

Itoh, Y., Orlosky, J., Huber, M., Kiyokawa, K., & Klinker, G. (2016). OST Rift: Temporally consistent augmented reality with a consumer optical see-through head-mounted display. In *IEEE Virtual Reality Conference* (pp. 189-190). Greenville, SC: IEEE. doi:10.1109/VR.2016.7504717

Kiyokawa, K. (2015). Head-mounted display technologies for augmented reality. In W. Barfield (Ed.), *Fundamentals of Wearable Computers and Augmented Reality* (Vol. 1, pp. 59–84). Boca Raton, FL: CRC Press. doi:10.1201/b18703-7

Kramida, G. (2016). Resolving the vergence-accommodation conflict in head-mounted displays. *IEEE Transactions on Visualization and Computer Graphics, 22*(7), 1912-1931.

Kress, B. (2015). Optics for smart glasses, smart eyewear, augmented reality, and virtual reality headsets. In W. Barfield (Ed.), *Fundamentals of Wearable Computers and Augmented Reality* (Vol. 1, pp. 85–124). Boca Raton, FL: CRC Press. doi:10.1201/b18703-8

Lowe, D. G. (2004). Distinctive image features from scale invariant keypoints. *International Journal of Computer Vision, 60*(2), 91-110.

Mann, S. (2002). The eyetap principle: Effectively locating the camera inside the eye as an alternative to wearable camera systems. In *Intelligent Image Processing* (pp. 64–102). New York, NY: John Wiley & Sons, Inc.

Mann, S. (2013). Vision 2.0. *IEEE Spectrum, 50*(3), 64-102.

Microsoft. (n.d.). *Microsoft HoloLens*. Retrieved February 21, 2017, from https://www.microsoft.com/microsoft-hololens/en-us

Mirza, K., & Sarayeddine, K. (2015). *Key Challenges to Affordable See Through Wearable Displays: The Missing Link for Mobile AR Mass Deployment*. Retrieved February 21, 2017, from: http://www.optinvent.com/publications

Plopski, A., Itoh, Y., Nitschke, C., Kiyokawa, K., Klinker, G., & Takemura, H. (2015). Corneal-imaging calibration for optical see-through head-mounted displays. *IEEE Transactions on Visualization and Computer Graphics, 21*(4), 481-490.

Rolland, J. P., & Fuchs, H. (2000). Optical versus Video See-Through Head-Mounted Displays in Medical Visualization. *Presence, 9*(3), 287-309.

State, A., Keller, K. P., & Fuchs, H. (2005). Simulation-based design and rapid prototyping of a parallax-free, orthoscopic video see-through head-mounted display. In *Fourth IEEE and ACM International Symposium on Mixed and Augmented Reality Proceedings* (pp. 28-31). Vienna, Austria: IEEE. doi:10.1109/ISMAR.2005.52

Takagi, A., Yamazaki, S., Saito, Y., & Taniguchi, N. (2000). Development of a stereo video see-through HMD for AR systems. In *IEEE and ACM International Symposium on Augmented Reality Proceedings* (pp. 68-77). Munich, Germany: IEEE. doi:10.1109/ISAR.2000.880925

Valenti, R., & Gevers, T. (2008). Accurate eye center location and tracking using isophote curvature. In *IEEE Conference on Computer Vision and Pattern Recognition* (pp. 1-8). Anchorage, AK: IEEE. doi:10.1109/CVPR.2008.4587529

Valenti, R., & Gevers, T. (2012). Accurate eye center location through invariant isocentric patterns. *IEEE Transactions on Pattern Analysis and Machine Intelligence*, *34*(9), 1785-1798.

Valenti, R., Staiano, J., Sebe, N., & Gevers, T. (2009). Webcam-based visual gaze estimation. In *International Conference on Image Analysis and Processing Proceedings* (pp. 662-671). Vietri sul Mare, Italy: Springer.

Chapter 2
Wearable Internet of Things

Ranganathan Hariharan
Gojan School of Business and Technology, India

ABSTRACT

With the type of ailments increasing and with the methods of diagnosis improving day by day, wearable devices are increasing in number. Many times, it is found to be beneficial to have continuous diagnosis for certain type of ailments and for certain type of individuals. One will feel uncomfortable if a number of needles are protruding out of one's body for having continuous diagnosis. From this point of view, wearable diagnosis systems are preferable. With Internet of Things (IoT), it is possible to have a number of diagnostic sensors as wearable devices. In addition, for a continuous monitoring, the information from these wearable devices must be transferring information to some central location. IoT makes this possible. IoT brings full range of pervasive connectivity to wearable devices. IoT of wearable devices can include additional intelligence of location of the person wearing the device and also some biometric information identifying the wearer.

1. INTRODUCTION

Wearable technology is here to stay. The industry is poised for big leap in the years to come. Kosir (2015) reports that the wearable devices are becoming 'smart' with the help of innovative technologies. They are found to be very helpful in healthcare. This is due to the fact that the types of ailments are increasing and the diagnosis of each of them is getting more sophisticated. Some of the products that are catching up in the market place as of 2015 are Valedo back therapy, Intelligent Asthma Management by Healthcare Originals, Quell Relief, Helius by Proteus Digital Health, Google Smart Contact Lenses, Smart Stop by Chrono Therapeutics, Health Patch MD by

DOI: 10.4018/978-1-5225-3290-3.ch002

Vital Connect, iTBra by Cyrcadia Health, Diabetes care by Abbott, Vitaliti by Cloud DX, Ulcer sensor by Leaf Healthcare, Wearable ECG Monitor by Qardiocore and the like. All these products are available with corresponding applications on various mobile devices. Based on the information on the acceptance of these products and based on market survey, it is estimated the total market for wearable is likely to touch US $30 billion by the year 2020 on a CAGR of 20% between 2015 and 2020 (Markets and Markets, 2015). A closer analysis of some of these devices exhibits the type of information that is being generated and shared by these devices.

The accent by various healthcare communities is currently on early diagnosis of ailments, health status monitoring, maintaining healthy life-style and improving overall quality of life despite ailments. Lymberis and Dittmar (2007) say that the stress is also on delivering services through more economical and on faster response. The aim seems to be affordable healthcare to everyone whenever and wherever it is needed. Significant research has taken place through various efforts in Europe and in other parts of the world. It is reported that appreciable progress has been witnessed in integrating Information Communication Technologies (ICT) to use mobile technologies for personal health through Information Society Technologies (IST) activities of the Fifth Research and Development Framework Program (FP5) of European Commission (EC). According to Lymberis and Dittmar (2007), this progress is further enhanced in FP6 during the years 2002 – 2006 with more wearable systems such as smart textiles, body sensor networks and context aware sensor systems being used for psychological biochemical and physical monitoring of patients.

We have seen in various applications of wearable devices, the data is continuously obtained and transferred to a central location. There are many methods for achieving the transfer of data. Internet of Things (IoT) is one approach. As per Gubbi, Buyya, Marusic and Palaniswami (2013), IoT interconnects a number of physical objects to create domain-specific intelligence through the functions of pervasive sensing, data analytics and visualization of data with the help of cloud computing. The growth of IoT has been fuelled by proliferation of mobile devices, pervasive Internet and the technology of cloud computing. Coupled with such a growth of IoT, growth of wearable devices made it possible for the emergence of a new segment known as Wearable Internet of Things (WIoT). Hiremath, Yang and Mankodiya (2014) opine that WIoT fuses activities of sensing, computing and communication together and offers excellent solutions in various fields which were considered to be challenges earlier.

2. WEARABLE INTERNET OF THINGS

There are some factors that propel the use of wearable devices in healthcare. The factors are increasing cost of providing healthcare at a central location, increasing number of senior citizens, new and chronic diseases. These necessitate the healthcare to be provided at personally convenient location along with the disease monitoring and management, if possible continuously. The doctor is to interfere only when it is absolutely necessary. While these are the requirements of healthcare system, the simultaneous developments in the field of mobile devices such as smart phones and tablets have only accelerated the growth of wearable IoT. As per a report by Dolan (2015) of IMS, the market for wearable devices is likely to be US$ 6 Billion in 2016.

The wearable devices perform a number of functions such as collection of data on various body parameters using sensors, preprocessing the data so collected, storing of data for short periods and transfer of the data to nearby mobile devices such as smart phones etc. Through the smart phones, the data is normally transferred to remote servers through apps. Such wearable technology offers flexibility in healthcare for continuous monitoring of regular data and intervention by the medical professional as the need arises. Providing real time healthcare for senior citizens and babies by monitoring sleep and health conditions is the major driving force for the growth of wearable technology. The concept of wearable Internet of Things seeks to integrate seamlessly wearable sensors, smart phones and the Internet.

2.1 Internet of Things

The backbone of wearable technology for healthcare is the Internet of Things (IoT). It is one of the concepts of recent development. However, one has witnessed rapid rate of growth in this sector. The reasons for such an unprecedented growth are increasing processing power of various devices, steady progress in miniaturization of computer, communication devices and cameras, ever increasing storage capabilities built into various systems, emergence of big data and developments in the field of pervasive communication and networking capabilities. The important trend that one witnesses is the continued fall in prices. With the continuous price reductions, these are being integrated into various gadgets of everyday use allowing connectivity and ubiquitous computing. Cars, refrigerators, watches, eyeglasses, weighing scales and even apparels are becoming networked, integrating sensing, automation and communication functions.

IoT is considered to be ushering a 'third revolution in computing'. The word 'Internet of Things' was first introduced in 1999 by Kevin Ashton. He later gave the description of IoT as *If we had computers that knew everything, we would be able to track and count everything reducing waste and cost.* US National Intelligence Council defines IoT as *the general idea of things, everyday objects, that are readable, recognizable, locatable, addressable and controllable by Internet. It may be even possible to know when things needed replacing, repairing or recalling*. IoT makes the reach of Internet beyond desktops, laptops, servers, Smartphones etc. Swan (2012) reports that some of the area where IoT is currently in use are monitoring and controlling the performance of homes, buildings, automobiles, their movements and transportations, self tracking, environment monitoring and healthcare. This is possible with the computers with their own capability to see, hear or smell their ambience. Radio Frequency Identification Devices (RFID) coupled with sensors is able to make the computers to gather the information without having to depend on human to enter the data. These are able to operate around the world without being visible to the people using these technologies.

Gartner is a research and advisory firm based in US offering consultancy on IT related business. Orlando (2015) of Gartner reports that in 2020 mobile devices including wearable devices will be having maximum market penetration. GSMA is an association of mobile operators worldwide. GSMA forecasts huge business impact of IoT in two wide categories namely revenues and cost reduction and service improvements. They expect revenues connected devices and associated services to be of the order of US$ 2.5 trillion. From cost reduction and service improvements, the benefits to organizations are expected to be US$ 2 trillion, of which the benefit from global healthcare is to be US$ 660 billion.

IoT is a network of systems connecting real-world objects to the Internet with sensors. The layers of this network are being formed based on various functions of IoT. They are data acquisition, information creation, discerning meaning and action taking. These functions are carried out through Hardware sensor platform layer, Software processing layer, human – readable visualization layer and human – usable Action taken layer.

IoT is considered to be 'smart systems'. They bring in smartness to homes, buildings, appliances, healthcare, mobility and even cities. CEA predicts that the users can improve energy conservation, efficiency, productivity, public safety, health and education with the use of IoT. IoT is expected to not only expected to bring in good control of various aspects of lives, it is also expected let them have more free time by automating some of the regular repetitive work. Research studies have indicated that the economical impact of IoT will be around $6.7 trillion by 2025 having a compounded annual growth rate (CAGR) of 7.9%. Studies also indicate that the IoT's biggest impact can be expected in healthcare, energy, transportation and retail service.

2.2 Wearable Sensors

Wearable devices are sensors that measure the physical and chemical parameters of the body and the environment in an effort to derive information. The types of wearable sensors are Inertial Measurement Units (IMUs), bio potential sensors, chemical sensors, optical sensors, stretch and pressure sensors and the like. Accelerometers, gyroscopes, magnetometer and barometers are the IMUs. Other types of wearable sensors are optical heart rate monitors, Pulse plythesmography sensors, wearable electrodes, cameras, temperature sensors, microphones etc.

At the turn of the millennium, National Science Foundation of US predicted that wearable sensors and the computers are likely to increase one's knowledge about health, environment, pollutants etc. Now this prediction seems to be proving to be correct. Wearable smart bands showed 700% growth in the second half of 2013 compared to the first half. It is also indicated that wearable devices will have a CAGR of 78%. Many Smartphone companies such as Samsung, Apple are also gearing up to garner major portion of this explosive growth with health specific attractive apps using wearable devices. Another development in wearable area is 'sensing fabric', with which it is possible to incorporate wearable devices into the stylish apparels. Realizing the importance being gained by wearable technology and mobile phone applications proliferating healthcare, in 2013 FDA has issues definitions as to distinguish mobile health app can be called 'medical device' from other mobile health apps. It is also learnt that the health insurance companies are starting to provide wearable devices to customers with tailored plans and premiums. Some of the areas where wearable devices find increased use are surgery, emergency care, etc. Apart from healthcare, wearable finds use for enhancing personal convenience, such as automatically adjusting personal lighting levels and temperature in personal space. Some of the companies working in wearable technology and offering various healthcare solutions are discussed along with some brief description of the products. These are already available.

Valedo Back therapy (*https://www.valedotherapy.com*) uses a smart sensor attached to the person's back and the chest. These sensors continuously monitor even the tiniest real time movements in 3D. With these sensors, solution to back pain is given through a number of well designed exercises. The patients are expected to do exercises. The product is also useful for professionals for measuring various parameters in real time. An app helps to interface with the user.

The Asthma management device by Health Care Originals (*healthcareoriginals.com*) keeps tracking various symptoms such as cough count, respiration, wheezing and heart rate continuously. Based on the actual values of various factors, suitable alerts are transmitted. If immediate treatment is required, the treatment plans are displayed using an app.

Quell relief (*https://www.quellrelief.com*) is for pain relief without the drugs. It uses a brace to be worn under the knee and it works with the support of an app. It works round the clock, even while sleeping. It has an electrode which is placed inside the brace. On pressing a button, the electrode is activated and provides pain relief by blocking the pain signals to brain. The battery that comes with the device is guaranteed to provide pain relief for 40 hours after which the battery needs charging.

Proteus Digital Health (www.proteus.com) is the organization offering 'Digital Health Feedback System" in the form of a small patch "Helius" along with 1 mm sensor enabled pill. The system is ably supported by the cloud based service that gives information about the drug adherence of the patient in real time. The system is built around a sensor network that collects data continuously about the patient's heart rate, temperature and patters of activity and the rest of the patient. The pill is ingested and it transfers time stamped signals to the patch worn by the patient. The details are finally delivered to the patient's doctor.

Google has come out with smart contact lenses for the benefit of people suffering from diabetes. Google calls it the 'eye mountable device' to detect the glucose level with the help of sensors that takes the tears in the eye to measure the glucose level. Google has partnered with the Swiss Company Novartis for this. This device can be considered to be a contact lens along with the sensors to detect the analyte in the tear. Once the device is able to read glucose levels successfully over a longer term, there may be other analytes which can also be measured using this technology.

Chrono Therapeutics' (*chronothera.com*) "Smart Stop" is the answer to the smokers who are desirous of quitting the habit of smoking. As advertised by the company, to successfully quit the addiction of smoking, the cravings for nicotine must be understood, managed and overcome. Their 'Nicotine Replacement Therapy' (NRT) is delivered through the skin via the wearable pod without any needle. This simple, light weight and comfortable pod can be anywhere in the body. The pod has embedded sensors. With the help of the pod and a coaching app, the medicine usage by the smoker is monitored and reminder is sent to the user to stay on track of quitting program.

Health Patch MD by Vital Connect (www.vitalconnect.com/healthpatch-md) is a biosensor having 3 ECG electrodes and 3 axis accelerometer to keep track of body position (to detect if the person has fallen), heart rate, RR interval, Heart rate variability, steps, Respiratory rate and temperature. With its Bluetooth, the device can connect to any smart phone and transfer the real time data. It is said to be suitable for tracking vital parameters of in-patients, out-patients and home care patients. It is said to be capable of clinical grade measurements of vital parameters in continuous, configurable and non-obtrusive manner.

Cyrcadia Health *cyrcadiahealth.com*) has introduced iTBra for early detection of breast cancer through a comfortable, discrete and intelligent wearable technology. It is expected to provide monthly information about the wellness of the person wearing it. These breast patches are expected to detect even the tiny circadian temperature changes within the breast cells and to communicate the same to the central laboratory. It is set to detect normal circadian cellular baselines and also abnormal patterns as in cancer with the help of latest technology using neural networks and other algorithms.

Abbott Laboratories have introduced Free Style Libre (https://in.*abbott-diabetescare*.com/) wearable Glucose Monitoring System that makes the user to apply the sensor on the skin at a convenient place in the body, say, upper arm. The sensor can be scanned using a reader. The reader gets the reading. This type of glucose measurement does way with the system of pricking and using lancets and test strips. The water resistant sensor measures and records the glucose throughout the day and night and stores the readings. The reader can get the data by scanning the sensor. Abbott also has introduced handheld blood analyzer that can check 25 different blood markers such as Hemoglobin, hematocrit, glucose, potassium, calcium, pH, blood urea nitrogen (BUN), creatinine, and lactate. With this, it is possible to get lab quality results within minutes and the results can be transmitted to physician in real time. Even blood pathologies which were immeasurable earlier can now be measured. Papakostas, Shelton, Kinrys, Henry, Bakow, Lipkin, Pi, Thurmond and Bilello (2013) report that Ridge Diagnostics offers facility to test blood for depression by measuring serum levels of nine bio markers such as alpha1 antitrypsin, apolipoprotein CIII, brain-derived neurotrophic factor, cortisol, epidermal growth factor, myeloperoxidase, prolactin, resistin, and soluble tumour necrosis factor alpha receptor type II . Currently this test is expensive but it is likely that this test can predict the onset of depression in patients suffering from diabetes for over ten years.

Cloud DX has introduced Vitaliti (www.clouddx.com/vitaliti.htm) that can address a number of health conditions. It is a wearable device that can detect anemia, urinary tract infection, type-2 diabetes, atrial fibrillation, sleep apnea, COPD, pneumonia, otitis media, leukocytosis, etc. It can address the conditions such as hypertension, HIV, mononucleosis, tuberculosis, influenza, asthma, bronchitis, pneumonia, upper respiratory infection and respiratory syncytial virus. The system consists of four wireless devices including wearable vital sign monitor, wireless spirotoscope, in-virto diagnostic system and pulse wave health station connecting to an app. The wearable measures heart rate, oxygen saturation, respiration, core body temperature, blood pressure, movement, steps and posture.

Leaf Healthcare (*leafhealthcare.com*) has introduced pressure ulcer prevention wearable. It is a small wearable usually clip-on for patients in hospital beds. The wearable sensor monitors the movements of the patient and decides if the patient needs to be turned to prevent formation of pressure ulcers. The data obtained on various parameters is updated on to a central database which can be monitored periodically by the clinician. It can also alert nurses if the patient needs to be turned.

Qardiocore (https://www.*getqardio*.com/qardiocore-wearable-ecg-ekg-monitor-iphone) offers smart wearable ECG / EKG monitoring system free from any patches for continuous monitoring of ECG / EKG, body temperature, respiratory rate, activities, heart rate, Heart Rate Variability and stress. It can be useful for cardio training with its military technology. It offers real time data sharing with the doctor through mobile app.

SetPoint Medical (www.setpointmedical.com) has recently introduced an implantable nerve stimulator for rheumatoid – arthritis patients. It uses an iPad app to charge the battery of the device to transfer data.

Pillcam Colon (http://*ous.pillcamcolon*.com/) introduced a capsule which travels through the digestive system transmitting images to an external recorder. With this, colon cancer screening is expected to be less invasive but more accurate.

CardioMEMS HF system (https://www.*sjm*.com/en/sjm/*cardiomems*) has a device that is implanted in the pulmonary artery for transmitting various data related to the health of the patient to a medical through an external device. This system is expected to keep monitoring patient regularly by the medical team, reducing the necessities to hospitalize and the doctor's team will be ready with the required information in case of emergencies.

Another area that has been fascinating humans is the working of human brain. With the new sensor technology it is expected that a better understanding of brain may emerge. There are a number of consumer EEG rigs available such as 14-node EEG Emotive and single node Nurosky (http://*neurosky*.com/). In a study, Petersen, Stahlhut, Stopczynski, Larsen and Hansen (2011) found that the responses to the traditional EEG and Emotive were reported similar. There are also systems available for sleep analysis. Encouraged by these developments, researchers feel that wearable brain monitors will soon be available for continuous monitoring of neural activities. These data are likely to help associate the neural symptoms with emotion detection and intervention.

Continuous testing of body parameters with real time data transmission, real time feedback and personalized recommendations is a key deliverable in remote monitoring of patients. This sector alone is likely to provide a market of US$ 21 billion as compared to US$ 9 billion in 2011 (Lewis, N., 2012).

2.3 Mobile Applications

There are a number of sensors associated with the next generation personal tracking systems such as accelerometer, Galvanic Skin Response (GSR) sensor and temperature sensor. Coupled with such personal body based sensors, there are applications running on mobile devices and offering many solutions to a lot of specific problems. All the examples explained in section 2.2 are made useful to humans by means of specific applications developed for each of them. In fact these applications make the sensing devices more attractive to be used. Some of the apps are discussed.

Fitbit (https://www.fitbit.com) is a company that claims to "redefine fitness". They also claim to have the "Ultimate fitness app". They have complete set of fitness trackers in the form of various sensors. Coupled with the sensors, they have more than 30apps to support various aspects of fitness. FGor example, there is an app called "Fitstar personal trainer" which automatically synchronizes the statistics such as calories burnt etc during the workout sessions.

The company myZeo (*myzeo.com*) helps to identify the type of sleep one may have experienced each night. With the associated mobile head band, the app called Zeo Sleep manager tracks Rapid Eye movements (REM) and deep sleep one gets every night and suggests personalized methods of getting better sleep every night.

MapMyRun (www.mapmyrun.com) is an app for tracking and mapping every running exercise using GPS and tracker sensors. It will give regular feedback and tips to improve the performance. The company claims to have a membership in excess of 40 millions. The app tracks the route and saves the same. The data is made available on social networking sites.

RunKeeper (*https://runkeeper.com*) is a similar app to keep track of running exercise and other exercises such as walk. It allows the user to set goals, follow a plan and monitor progress on a plan.

MoodPanda (www.moodpanda.com) is a mood tracking app. The application tracks, graphs and stores the mood. It keeps the diary of moods of people. It is free to join.

Nike Fuelband is an activity tracking device coupled with an app. It tracks the wearer's physical activity, and the amount of energy burnt etc. The information collected through the wrist band is connected through the app and is integrated in to Nike+ online community. The product is discontinued now.

Apart from these products and apps, there are many companies who promote development of apps suitable for many wearable IoT devices. Some of the companies are Stanfy (https://stanfy.com), IBM (www.ibm.com/iot-developer), OpenXcell (https://www.openxcell.com/wearable-devices), Volansys *(volansys.com/connected-apps),* Fugenx *(fugenx.com)* and so on.

2.4 Wearable Technology

The wearable body sensors can function more meaningfully only if they are able to communicate through a central system. This is facilitated by combining data from wearable devices and the power of Internet of Things. The sensors need to transfer the data to the Internet so that the data can be accessed by all the stakeholders. The sensors do not have sufficient power to have long range communication. They need to send the data to nearby devices such as smartphones, tablets or stand alone computer systems. They make use of nearby Gateway devices with technology such as Bluetooth etc. The devices thereafter will use Wifi or GSM for further transfer of information over the Internet. A number of protocols are used for communicating the sensor data to the central server. For data transmission, protocols such as Wi-Fi, Bluetooth, ANT, ZigBee, USB, and 2G, 3G, and 4G are used. Since regular Bluetooth may consume considerable power, the latest low power protocol Bluetooth Low Energy (BTLE) is being used for conservation of battery power. The Gateway devices will also perform some preliminary preprocessing such as validating data etc. Zhang, Ren and Shi (2013) report applications wherein Smartphone works in synchronization with cloud server to report fall detection of elderly persons. To enhance the performance and the battery life, Smartphones are operated with less computational burden with the use of Mobile Cloud Computing (MCC) paradigm.

The amount of data generated by the 24 x 7 wearable sensors is so huge that latest Bigdata techniques are used and the data can be stored in Cloud (Somewhere in the Internet) taking care to address the concerns in areas of data privacy, security, ownership and access. Keeping the growth in this area in mind, there are a number of Cloud service vendors such as AT&T, Qualcomm developing healthcare specific services. Microsoft's Azure is an example of providing data collection and analytics using the public or private cloud such as Amazon Web services, Joyent and Rackspace.

There are organizations such as Sympho.me, Contiki, Cosm, Singly etc offering different types of solutions in WIoT. The WIoT has a number of operations such as sensor hardware platforms, sensor operating systems, software processing and development environments, and sensor data integration platforms. There are newer approaches to integrate the data from various WIoT devices, web services and social networking activity. In the near future, one may witness unified data platform and unified API platform with personal data based web services built on top of these unified platforms.

Three major areas of operations are discernible in WIoT. They are the set of body sensors, Interconnection to Gateways and Cloud and Bigdata support. The set of body sensors are called Wearable Body Area Sensors (WBAS) or Wearable Body Area Network (WBAN) or simply Body Area Network (BAN). Since the wearable body area network consists of a number of wearable sensors and since the sensors have to necessarily communicate with other devices through wireless protocol, Wearable Body Area Network (WBAN) can also be called wireless Body Area Network (WBAN).

The interconnection of various activities is as shown in Figure 1.

The sensors used in WIoT will have sensing hardware with an embedded processor. The sensing hardware will typically have capabilities for storage, power management and communication to nearby mobile devices. The power consumption rate and battery life are causes for concern. The success of wearable technology depends more on novel interface between the sensing element and the body. Some of the methods used for measuring various body parameters are ring sensor, chest worn ECG sensor and patches attached to body parts. Smart textiles are also used for measuring certain parameters such as autonomous nervous system responses. These sensors are available with global standards in quality for continuous and long term usage.

With the proliferation of wearable devices, it is expected that the cloud servers will be flooded with medical data of so many persons. To decipher information from mammoth volume of data is as critical as correctly measuring body parameters. The functions of data mining, machine learning and medical data analytics can be facilitated through the cloud computing infrastructure. Cloud assisted BAS (CaBAS) is being recognized as a cutting edge technology integrating MCC and WBAS to provide scalable, data driven pervasive healthcare. This, in turn, is expected to bring benefits to WIoT in the following areas:

Figure 1.

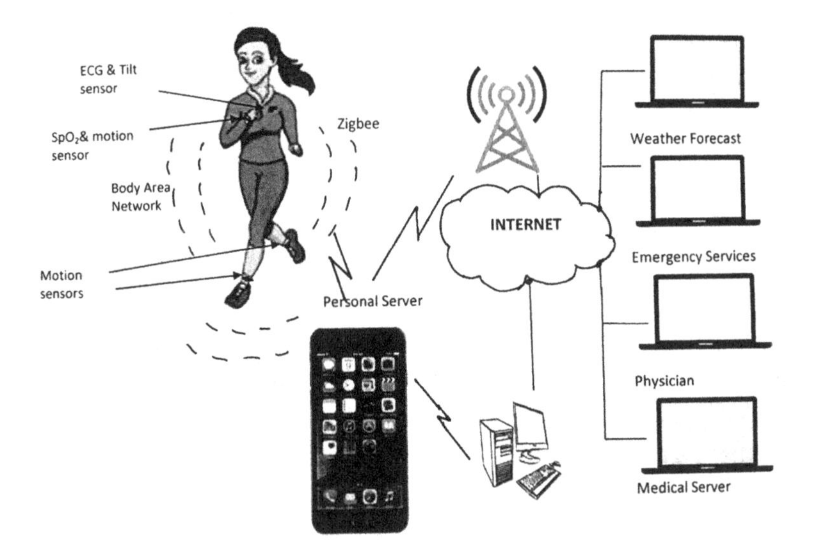

1. Connection of Smartphone and wearable sensors using energy efficient routing protocols.
2. Reduction of unnecessary data processing since event based data processing will be used.
3. Machine learning algorithms on Cloud enhancing the accuracy of data.
4. Preserving person-centric data for analysis.
5. Extension of decision support to physicians and patients through data visualizations.

Kartsakli, Lalos, Antonopoulos, Tennina, Di Renzo, Alonso and Verikoukis (2014) present a Wearable IoT operating in a three - tier architecture. The lower tier has body sensors in a network along with possible environment sensors. The data collected from body parts / environment is communicated through Bluetooth / Zigbee to the nearby devices such as Smartphone / Laptops / Desktops. Smartphone will use GPRS / 3G route to transfer the information to the Cloud. Laptop / Desktop computers can use Wi-Fi or broadband connections to transfer the information to the Cloud. The arrangement is shown in Figure 2.

Figure 3 shows the architecture for mHealth solutions with wireless M2M systems. The system has five major elements such as 2M wearable devices, M2M Body Area Network, M2M Gateway to a nearby wireless / wired network domain, M2M Access Communication Network to connect the Gateway to the Internet and M2M Application server.

3. ISSUES

We have considered the ways by which WIoT can be beneficial to mankind. Needless to say that such a beneficial technology is bound to see leaps and bounds growth. It is already predicted to be the most promising technology to provide meaningful healthcare. However, there are some issues associated with the use of this technology. Some of the issues are the battery life of the sensors, security issues as there is open exchange of data through wireless medium, legal issues, ethical issues and data protection issues. 'Biohacking' or hacking the system of body sensors is expected to grow and pose appreciable threats to healthcare. Encompassing all security and privacy concerns, in the US, FBI on 10 September 2015 has issued a Public Service Announcement vide Alert number I – 091015 – PSA under the heading "Internet of Things poses opportunities for cyber crime". The announcement clearly specifies the things that can be classified as IoT devices, how they operate and the risks associated with the IoT devices. The announcement also gives examples of risks and recommendations towards protecting consumer rights and defense.

Figure 2.

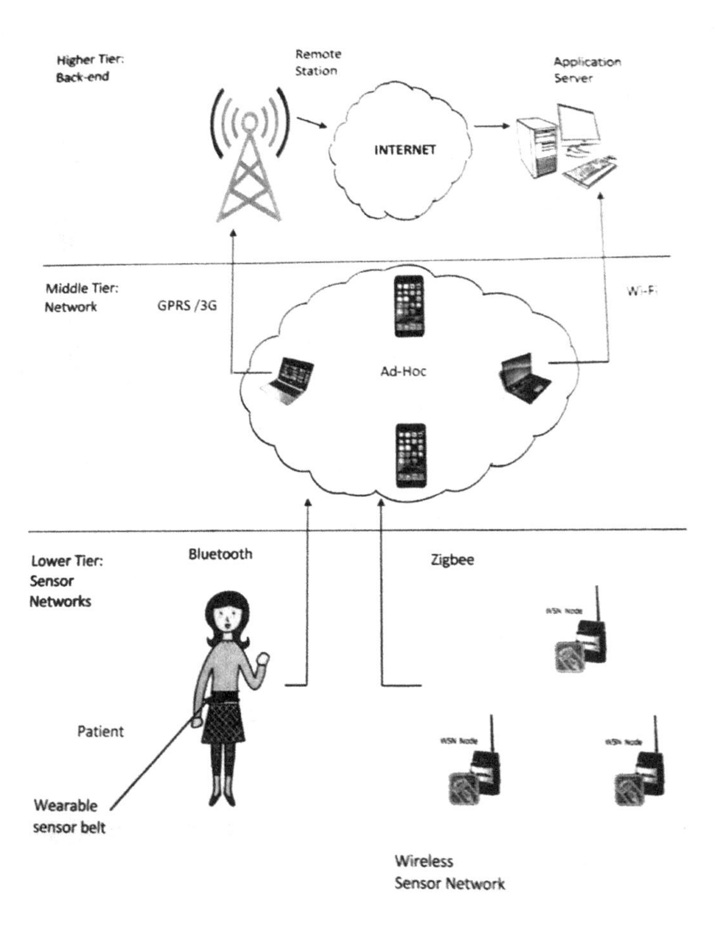

3.1 Battery Life of Sensors

Size and weight considerations pose serious restrictions on the batteries to be used in wearable devices. Wearable device has to sense, collect the data from sensing, perform some preprocessing operations on collected data, consolidate the information and transfer the data to the nearby wireless device, say, a Smartphone. These activities use the power from the batteries provided with the wearable devices. Battery lifetime is considered to be a major limitation for wireless sensor networks in general and wearable sensors in particular. There is progress in battery technology which promises higher power output batteries occupying smaller space. It must be borne in mind that the wearable sensors used in healthcare applications may not have the option of recharging on a regular basis. Not only the advancements in battery technology, but also the algorithms used for processing the data and the communication protocols

Figure 3.

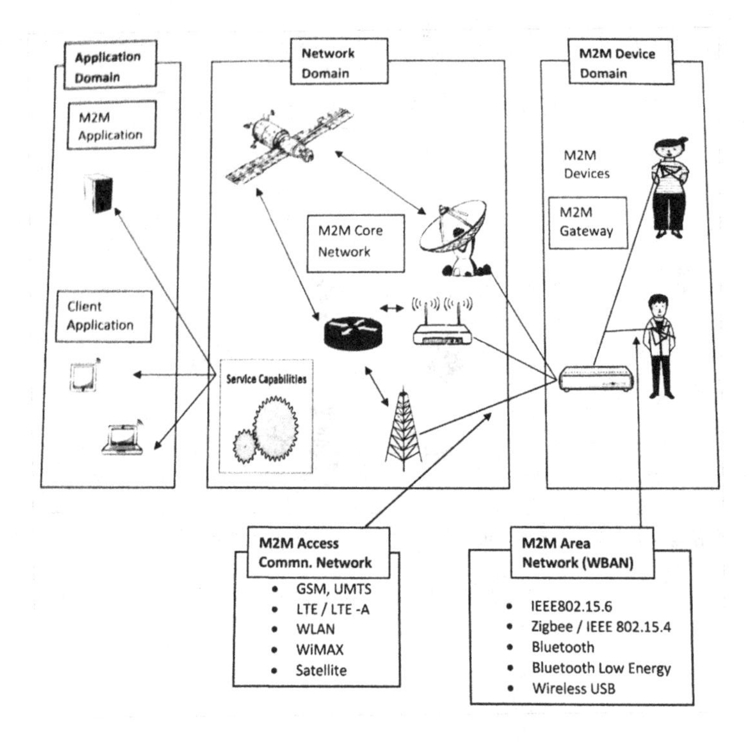

used for transfer of data are made in such a way as to enhance the lifetime of the battery. Large number of research papers is available for energy efficient operations of wireless sensor networks.

3.2 Data Security

Security is one of the major concerns in any wireless communication network. With pervasive deployment of sensors, wearable IoT is prone to attacks from eaves droppers. Moreover, the data is scheduled to be made available to various stakeholders through Cloud. Cloud technology is also known for being exposed to many security threats. A number of algorithms are being proposed by various researchers for enhancing the security in Wireless Sensor Networks and in Cloud Computing. If the eavesdropper is successful in hacking the wearable IoT, there will be large amount of personal intimate data about the person available on the wearable device and the same may get transferred to the eavesdropper's device. Hacking of personal laptop, tablet etc is quite common. The security breach with wearable devices can result in unprecedented consequences.

3.3 Legal Issues and Data Protection

Wearable technologies raise a lot of potential legal issues. Main areas of problems are about data protection and privacy. Potential legal problems include identity theft, profiling various persons, discrimination and stalking. The International Privacy Conference declared in October 2014 that the Bigdata derived from Wearable IoT should be considered to be personal data. The data is likely to be sensitive. Data collected through wearable technology must be covered under the Data Protection Laws. For example, a person wearing a wearable device is likely to have the heart rate monitored on a regular basis. This data may be available in the wearable device attached to the person's body. Supposing the person goes to school, the person may not want the details of heart rate etc to be known to Teacher in the school. So, there should not be an automatic transfer of data from the person's wearable device to the Smartphone of the teacher. However, the same person, when visiting the doctor, would like the data to be transferred to the Doctor's Smartphone immediately on arrival at the healthcare facility.

Ultimately the questions remain in the minds of the person who is using the wearable device: "Who owns the data generated through continuous monitoring of physical parameters? Who controls the flow of information? What are the safeguards in place to see that this data is shared only with the concerned parties? What are the safeguards available to make sure that this data is not sold in public domain either for marketing or advertising purposes?" Unconvincing responses and / or assurances may hinder users and companies to accept these technologies easily.

With wearable technology, in case of a legal problem it may become difficult to fix the liability on specific company / person. Any one of the stakeholder such as the end service provider of WIoT, or the device manufacturer, or the sensor designer or the software programmer or the Cloud server or the local Internet service provider or combination of them may be actually liable. It is possible that each one will blame the other in case of any mishap.

3.4 Ethical Issues

It is found that organizations use wearable technology to track employee behavior. Earlier, wearable technology was used to improve efficiency and safety of the employee. Current technology and gadgets combine the employee behavior and physiological parameters to gauge the employee performance. Green (2015) feels they go on to the extent of recommending a day off to the employee if the situation warrants. As per Connor (2015), some organizations have started tracking everything about their employees, gathering data on everything from sleep quality and heart rate to location and web browsing habits 24 hours a day. It is now possible to track employees' activities, stress levels and sleep through wearable devices. According to Thierer (2015), more and more companies are expected to adapt wearable technology for tracking their employee wellness. With wearable devices having memory, computing power and communication capabilities another ethical issue is if the Glasses with wearable devices can be worn inside the class or even during the exams. Such a situation allows the students to interact with each other during the exams, which may not be acceptable to many educational institutions. Usage of wearable device during the regular classes may be acceptable or encouraged. But use in exams is questionable. It is also possible for parents to keep monitoring the children's activities through their wearable devices. That is, parents will know what the children have browsed, mood swings etc. Parents may argue that this is essential to make sure that the children's safety is not in jeopardy.

With Wearable technology, one can expect an invisible and comprehensive surveillance of everyone's public and private life. One will feel that one is surrounded by things that can hear what is being spoken, see what is happening, sense some important parameters and report these to a central place. A day may not be very far when a customer enters a place, the 'customer profile' of the person would have already been relayed to the people concerned in that place. While a selected few well off clients may ward off such intelligence around them, vast majority will have to surrender some privacy in favour of increased productivity.

3.5 Mobile Application Issues

There are many issues associated with the development of suitable apps for better use of wearable technology. They are discussed below.

Fragmentation: Fragmentation happens due to mismatch between the devices and the apps and the user experience is reported to be inconsistent in many cases. This is mainly due to the fact that there are many operating systems and there is no cross-platform development of apps yet. Normally the developer develops the app without API and implementing each feature for each device separately. Such approach creates problems at a later date when all efforts are synchronized. An approach of API –first can help smoother development across devices. There is a set of new APIs being released by different organizations. Google has recently released Android new wearable software development kit API. NTT Docomo has released its Device Connect Web API.

Dearth of testing facilities: The companies offering testing services have not grown as fast as the growth of wearable devices. The available testing companies are very few and the companies offering wearable apps cannot test their software on their own. Making the wearable device to interact with others is becoming a challenge for testers. With many options such as Bluetooth, Wifi etc, the number of ways of interaction with wearable devices is large. Testing with emulator is not a good option. Considering all these points, there is a requirement for development of new methodologies for testing.

Localization issues: Many wearable devices and apps are aimed at Global market. But there are some specific points which are specific to some regions. For example, while majority of the people over the world are used to left to right way of reading, some people are used to the right to left way of reading. A product / app designed for one region will not be accepted in the other. Similarly, the fonts of different languages are so different that the uniform use of fonts is not possible.

Current technology: As indicated earlier, there are many options available for wearable devices with the nearby mobile devices for the associated apps to be run successfully. The interface can take any form such as Bluetooth, Wifi, Zigby, other hardware or wireless connections. However, in all the apps, it is expected that the wearable device will be interfaced with the mobile device through any one of the interfaces as stated earlier. But the main information transfer between the mobile device and the Internet is only through the existing wireless air interface. So, all the existing algorithm and protocols are expected to work with the same level of service. For example, if the mobile device goes out of range (say, in an app which tracks the runner) the information captured may not be transferred to the central database through Internet on real time. The data may be transferred at a later time when the connection is reestablished.

4. OVERCOMING ISSUES

The biggest dilemma that many organizations face today is whether to adapt and encourage further development of wearable technology or be bogged down with various issues as outlined above. The alternatives are between allowing innovation to take place and considering the concerns of legal, privacy and security issues. Better option would be to allow innovation with a creative handling of concerns employing various tactics such as self regulation, transparency and education of proper use of WIoT, pressure by public and watchdog mechanisms and promoting best practices in the industry. In the process of handling concerns, existing legal provisions and standards must be enforced. However, Thierer (2015) fears that handling concerns as detailed will result in reduction of pace of innovative development.

4.1 Battery Storage Issues

Sensors used in WIoT must be free from frequent replacement necessities, especially if they are used internal to the body. The sensor may not be active all the times. It may sense at regular intervals and remain idle during the intervals. If appropriate energy efficiency techniques are not employed, the battery may drain all the power within a few days requiring replacement frequently.

One major area of energy consumption in wireless sensor networks is routing. This is not a major issue in Wearable IoT especially in healthcare as the sensor normally connects to a Smartphone that may be available in short distance. The communication takes place through Bluetooth technology. Five possible approaches for energy conservation in Wireless Sensor Networks are proposed by Soua and Minet (2011). The approaches are Data reduction, Protocol overhead reduction, Energy efficient routing, Duty cycling and Topology control. Data reduction can be achieved by sampling only at specific intervals in place of continuous sampling as the parameter sampled may not vary regularly very dynamically. Variations, if any, are likely to be marginal and slow. Data aggregation and data compression can be considered as alternatives for improving energy efficiency. Protocol overhead reduction, energy efficient routing, duty cycling and topology control are techniques that may be applicable to Wireless Sensor Networks. In the WIoT communication is mainly between the sensor and the mobile device, these are not attractive energy efficiency improvement methods in WIoT.

Size of the wearable device (and so the size and energy capability of battery) and the necessity to encrypt data for protection are two important requirements in wearable devices. Reducing the size of device may necessitate reducing battery size with lesser battery life time and encryption for ensuring the security of the data transmitted may also reduce the battery life time. For getting a proper balance between the longer battery life time and smaller device with encryption algorithms, a suitable trade – off must be worked out. Huang Badam, Chandra. and Nightingale (2015) propose WearDrive as an energy efficient storage device for wearable devices providing the correct trade- off. They report that WearDrive improves the battery life by over 3.5 times. They claim to have demonstrated WearDrive with typical wearable applications. WearDriveis said to enhance performance and energy efficiency in wearable devices significantly.

While electronics devices are becoming smaller and the battery size is not keeping pace in size reduction, there is a growing need for better / prudent and efficient use of energy in the battery. As per the report from Shaw (2014) of Texas Instruments, energy harvested from body heat and motion and from light from the environment can be used to recharge battery in the wearable device Low power electronics both for computing and communication purposes is a very active research field. Wireless sensor networks use wireless sensors with limited hardware, memory and battery life. Kong, Ong, Ang and Seng (2012) expect that with design of Minimum Instruction Set Computer (MISC) processor in small area for encryption and error correction, energy efficiency can be improved in the sensor node.

4.2 Legal and Privacy Issues

Since the transfer of data takes place over open space, the data transfer is vulnerable for attacks from outside eaves droppers. Security is one of the major constraints that can hinder widespread acceptance of WIoT. Many users and stakeholders would want guarantees on appropriate confidentiality, authenticity and privacy of the information exchanged. Initially, the security issue of the IoT was to be handled in an ad hoc method. With many players and stakeholders in the fray now, there are a number of security challenges now. As per Miorandi, Sicari, De Pellegrini, and Chlamtac, (2012), three main areas of data security are in focus. They are data confidentiality, privacy and trust. The main points of focus remain to be trusted platforms, low complexity encryption (due to stringent power requirements), access control, data confidentiality, authentication, identity management and privacy.

Of various points of focus, data confidentiality is the important one. It is to ensure that the data is accessible for view or modification only to select authorized entities. For a successful operation of the scheme, there must be a clear identification scheme for each of the entity and a powerful access control mechanism. This is the primary concern in the acceptance of IoT. Due to a large volume of data being generated on a regular basis and due to very dynamic nature of data streams, conventional schemes for ensuring the data confidentiality may not be useful in WIoT. Role Based Access Control (RBAC) is a widely accepted practice in which the permission to access various objects is based on a person's role. The conventional RBAC considers static databases and static roles. The advantage of using RBAC is that the roles can be assigned dynamically. With real time data streams in WIoT, new forms of RBAC are to be used for better access control. Another aspect of security is the identity management. Though the concept of identity management is known for a long time, its role in smart objects throws up newer problems. For achieving better identity management, a clear agreement must be available on the concept of identity in general and identity of smart objects in particular.

Trust is another area that needs careful attention when it comes to data confidentiality. Trust is an aspect that cannot be measured in a network of systems as in WIoT. In the WIoT scenario, trust may be ensured by means of security policies which regulate the access to various resources and the credentials associated to satisfy such policies. Trust plays an important role in peer to peer communications and may involve repeated exchange of digital credentials between the communicating entities.

Knowledge can be shared only with trusted partners. TrustBuilder and KeyNote are possible solutions to this problem. However, these use static and computation intensive approaches. Considering the dynamic distributed nature of WIoT and the requirements to reduce complexity in computation as in WIoT, newer algorithms must be developed for building trust among the partners.

Privacy is being considered as fundamental requirement. Society can claim to have progressed only if the people can decide freely as per their beliefs and interests without fear or repression from others. Privacy can be considered to be a source of empowerment, dignity, regulating agent and utility. One possible approach to address the security concerns in use of wearable devices is to educate the users about the correct and incorrect uses of these technologies. It may be also advisable to educate users about the risks associated with the improper use of these technologies. According to Thierer (2015), possible steps in educating the users can be media literacy strategies, critical thinking skills and digital citizenship. Collectively these strategies are known as 'digital literacy' and are considered to be important skills to possess in the current century. The designers and developers will do well to follow best practices in building self regulating mechanisms to ensure safety and security of data in WIoT. However, cultural norms, public pressure and social sanctions can act as more powerful impediments that the self – regulatory mechanisms in ensuring security of the data. For example, there can be social sanction on persons using Google Glass in public places. These social norms and public pressures can be considered by the developers during the development stage and suitable designs proposed.

4.3 Data Security Issues

Many legal issues arise in WIoT due to misuse of information generated by wearable devices. Since a large amount of data is getting generated, transmitted, stored, shared and analyzed, the privacy of the person concerned must be maintained. Thierer (2015) concludes that more care should be taken especially when the information is taken from the person without the person being aware about the process. Legal and privacy safeguards related to WIoT are likely to figure in various fights involving, courts, lawmakers and regulators FTC launched an enquiry about these issues in November 2013 and submitted a report in January 2015. In May 2014 FTC also conducted a seminar on the topic "Consumer generated and controlled Health Data". This seminar explored various privacy concerns associated with websites and digital applications (including wearable devices) collecting personal information about health and fitness. These steps are expected to make Federal US Government to bring about regulations in using wearable technology. Many privacy related regulations are introduced in recent years in US, Europe and Asia. It is also found some users are willing to offer personal data in exchange of certain considerations such as 'self profiling' or continuous monitoring. Huge data collected from wearable devices constitute 'personal data' and appropriate data protection law (of the land) will be automatically applicable to such data. So, access to this data must be through notice and subsequent consent from the concerned parties wherein the parties are clear about the end use of the data. The consent must be obtained after providing all the details to all the stakeholders concerned in clear and no uncertain terms. Sweeping consent and large volume of fine print may not be acceptable.

There can be legal deterrents for violating norms including privacy norms. There can also be legal provisions against false claims made by service providers, breach of warranty, and liability for personal injuries sustained due to the use of wearable device. Separate rules must be governing functioning of wearable devices as compared to rules governing use of other consumer products. The companies involved in the business of wearable technology must be educated sufficiently about potential legal risks being faced inherently by the wearable technology companies. OPC Privacy Research Papers (2014) of the Canadian Government is of the opinion that while wearable technology exposes individuals to privacy risks, it also offers ample scope for improving privacy protection and user autonomy. They felt as of 2014 that the challenges to privacy were expected to intensify and the extent of privacy risk was difficult to predict.

4.4 Ethical Issues

The basic paradigm in wearable technology is that the computing and communicating elements recede in to background. This is seen more as an attempt by the technology to interfere with individuals' life without their knowledge possibly to circumvent resistance. Though there are some sinister theories being circulated about the real motives of such systems, the widespread acceptance of the wearable technology by many users is considered philosophical in which the users feel that the technology improves interaction with environment. According to Bohn, Coroamă, Langheinrich, Mattern, and Rohs (2004), the developers of these technologies are confident that the issues raised by many will not come in the way of acceptance as this technology is expected to simplify life, save time and relieve the responsibility of having to execute laborious tasks. Designers find that feasibility of the technology will overcome the doubts raised by others as it happened in the twentieth century 'smart' products. It is also expected that when man covers oneself with more such devices, the devices are treated more as supporting infrastructure than as prying one.

The other issues such as aversion to being tracked all time, being 'profiled', having personal information in public domain etc will continue to pose hindrance for acceptance of these technologies. Currently the scenario is not clear as to how much 'prying' will be allowed or accepted. It is also expected some standards will emerge ultimately to support ethical concerns of various stakeholders.

5. ROAD AHEAD

With specific issues associated with wearable technology, there is continuous research to address these issues. Under the circumstances, the question that remains in the minds of the stakeholders is if the wearable technology has a future. Current industry trends and projections as we have seen in various literatures predict healthy growth in this sector. The work is being carried out in brisk pace in areas of Bigdata, Cloud computing, Security in mobile computing, design of processors with less power consumption for computation and communication and algorithms to reduce power consumption in regular working of wearable IoT. It must be heartening to note that Article 29 Working Party is constituted and is working on the protection of individuals in respect of processing of personal data and free movement of such data. This party has declared that innovative thinking is required to meet the challenges of big data and to develop practical methods for implementing data protection principles.

So far the trend has been that the younger generation appreciates, adopts and depends on newer technology more as compared to the older generation. The acceptance of wearable technology by the younger generation is expected to be good as in earlier cases. The healthcare and fitness are important area of human interest. Wearable IoT addresses many issues in these areas and offers many innovative round the clock solutions. The benefits accrued from the use of such technology act as strong driving force for growth of WIoT. However, some types of devices are expected to be more popular as compared to other type of devices. For example, wrist based instruments are found to be preferred to ear based instruments.

There are some spin offs from wearable technology in fields other than healthcare and fitness. Advertisement seems to be a promising area with Google Glass. It offers excellent medium for advertisement. Pay Pal is developing payment options through wearable devices and Smartphones. Such innovative methods of interaction with customer are likely to improve customer relations and further business. Banking seems to open up to the idea of WIoT. Some immediate application areas are balance checking, alerts, real time information availability about receipts and payments and latest loan options. Stock markets and insurance companies can migrate to wearable technology to post their information online and customers can access them online. Travel industry, especially Airlines stand to benefit from the opportunities through wearable technology. Check – in and boarding experiences can be eased for better passenger comfort.

Another important driver for growth is the price. Currently the prices are such that the acceptance may not be forthcoming at an appreciable rate. With players such as Google, Samsung, Apple, Qualcomm and Microsoft already very much in the field, there is appreciable interest and business in the wearable technology. Based on current trends about new players entering the market, it is expected that the market place is likely to be crowded with many formidable players and such competition is expected to bring down the prices. If the price dynamics of PCs and Smartphones are to serve as indicators, one will see appreciable price reduction in wearable devices and then the acceptance rate will increase exponentially.

Rolin, Koch, Westphall, Werner, Fracalossi and Salvador (2010) propose a comprehensive approach to the problem of patient's data collection in healthcare using Cloud Computing paradigm. They present the practical advantages of such a system such as providing always – on real time data collection, eliminating possible data entry and resultant typographical errors and easy deployment process as the system works on the principle of wireless networks requiring no wires. They further add that such a system helps to improve the quality of delivery of medical services, especially for the needy communities. What may require assembly of a large number of specialists at a place of the community, can be handled easily with this type of a system. They claim that such an innovative system can be used by developing countries to offer cost effective modern healthcare services. This approach may pave the way for other healthcare institutions to use Software as a Service (SaaS).

Michahelles, Matter, Schmidr and Schiele (2003) present an interesting and life saving application of wearable technology. They proposed use of wearable sensors for rescue operations in an avalanche. They have explained how the rescuers in an avalanche situation can better plan the deployment of their resources during the rescue operation with their system using wearable sensors. They have stressed that basic sensor measurements may not be sufficient in such operations. They were able to incorporate 'urgency measure' associated with each victim in comparison with all the victims. They suggest that future work could include measurements regarding the depth of burial which can help to decide the number of members in the rescue team and also the possible approach for rescue. They suggest use of airbag which could be triggered by the suitable wearable sensor that can sense onset of avalanche. They argue such systems can be used in military rescue operations also.

The data generated by the wearable devices is huge and has a lot of potential to be used by marketing personnel, especially for targeted advertising addressing the needs of the individual. However, such a practice may not be acceptable to a large number of individuals. The individual might enforce the right o block the sharing of personal data. Instead of advertising based on analytics of the personal data, businesses can introduce anonymity by removing personalized information. Once the anonymity is introduced, the data is no longer personal data and can be used for advertising.

Apart from introducing anonymity, there can be other steps taken to handle the data issues from wearable devices. There are three critical points to iron out the issues. They are transparency, consent and security. By keeping the person informed about how the collected data is transferred or used or shared, the issue of transparency can be settled. After informing the wearer about the use of data, informed consent may be obtained from the wearer. Ensuring the security of data can be through contracts with various stakeholders.

Not all wearables perform similarly. The requirements for monitoring different body parts vary substantially. The way each sensor functions, placement of such sensor for each body part and their interactions are crucial points of consideration in making a wearable technology useful to different ailments. One may not be comfortable with a number of sensors at a number of locations in the body. Suitable design for each ailment / fitness regime is a complex and challenging process. With the growth of clothing based wearable technology, elite athletes will use products and services that will help them enhance their performance. For this, they have to be monitored while doing their practice / performance. Wearable technology offers best possible solutions towards this. One can see that smart workout clothing and undergarments gaining market share.

Due to increasing demand for healthcare, increasing costs for providing the care and vast improvements in treating previously untreated conditions make cost reduction in healthcare to be a major requirement for many countries. In many countries, the budget estimates for healthcare consume appreciable portion of GDP. It is possible to maintain patients' health records with accurate data from wearable devices. Designers of wearable technology products and services, healthcare providers (Public or private) and health insurers should come together and accept on certain standards and norms so that the cost of healthcare can come down and also result in overall well being of the society. Lymberis and Dittimar (2007) opine that the reduction in cost of healthcare is to be brought about by avoiding unnecessary hospitalization, ensuring deserving patients get treatment sooner and by reducing the medical errors through proper interaction between patient and healthcare giver.

Goncalves, Macedo, Nicolau, and Santos, (2013) propose another application of wearable IoT in healthcare for avoiding errors in intake of medicines. This assumes more significance in case of outpatients, especially elderly persons. The authors have proposed a new security protocol called M – Health Security Protocol which introduces well known SSL protocol for security in Application layer instead of transport layer. Though there is an increased possibility of 'denial of service' scenario with this protocol, the authors argue that such a situation will only introduce a small delay and so there is no major risk involved. This type of system may find acceptance widely especially in medication control.

Doukas, and Maglogiannis, (2011) suggest that the developers working in the area of wearable technology must make sure that the systems works seamlessly with the heterogeneous resources, sound data storage and data management principles and ubiquitous access issues. They have presented an open Cloud infrastructure that stores the motion and heartbeat data from open hardware and software. They propose that such system can be used for consistent surveillance and independent living of patients or elderly. Such systems can find better acceptance among users who require constant monitoring of patients, elderly or babies. Ted Leonsis, former CEO of AOL and currently the partner in Revolution Growth concurs with this view. He predicts that as the world population is aging, major advancements in the wearable technology will help to monitor the elderly as the elderly population will be more accommodative towards technology than the earlier elderly population.

Another area of utility with the wearable technology is from the availability of large volume of data. The data can be segregated segment wise, namely, healthcare, fitness etc. After taking care to introduce anonymity, the massive data can be the basis for data mining and deriving trends / categories / patterns etc. A number of physical and psychological parameters can be studied and a better insight may be gained from the Bigdata about various human behaviours / ailments.

To ensure the legal and the privacy issues are addressed to the satisfaction of the users and the stakeholders, various countries are enacting separate laws. So, many IoT companies are facing the problem of incorporating the safeguards for all the regulations. EU had Article 29 Working Party and its recommendations. The Government of US has released the IoT report prepared by Federal Trade Commission (FTC) containing three recommendations for IoT companies. The recommendations were in respect of data security, data consent and data minimization. Though the recommendations are in broad agreement with the Eu's Article 29 Working Party opinion, any IoT company must go through the fine print of all regulations and derive safeguards in their products and services.

Currently, Internet has reached far and wide. There is heavy traffic in the Internet. The traffic is regulated among the users based on various bandwidths available. Governments and Internet Service providers are expected to maintain 'Net Neutrality' ensuring that all users of Internet are treated equally and not to discriminate or charge differentially based on any criterion such as type of user or user location or website visited or equipment used. In a future date, when the traffic in Internet is set to grow exponentially due to exponential growth of WIoT devices, the Internet space may become so congested that some regulatory mechanism will become unavoidable then. It is a tricky situation and Governments and Internet service providers must deliberate on this issue (if not already done) and arrive at a suitable amicable working plan for smooth functioning of WIoT.

Many organizations in wearable technology are concentrating on meeting the technological challenges. Generally, little or no concern is shown towards the aesthetics of the product. Of late, companies have started showing interest in making their products more fashionable and visually appealing. Combining wearable technology with fashion and jewellery design, there can be new partners offering new products and services. The example of sportswear designed with wearable devices by Nike can be quoted as an example of such partnership.

6. CONCLUSION

Wearable technology is a promising technology offering innovative solutions in many fields. IoT is expected to propel the next information revolution. Since IoT allows control of various objects from remote locations and since it helps to integrate physical world and computers digitally, there is a lot of potential for automation in many fields. Many Governments are ushering the concept of smart cities with IoT. Its usefulness in healthcare and fitness is enormous. Realizing this, many organizations are working towards providing complete solutions using Wearable Internet of Things. Though this nascent technology is ridden with some issues, concentrated efforts are on for removing obstacles to make this technology acceptable to large population. Propelled by such extensive work, it is expected that the market for such products and services is likely to grow at a brisk CAGR of around 20% in the immediate years. The growth of wearable technology is related to the growth of Smartphone market. While Smartphone penetration is more than 50% in developed countries, it is lagging far behind in many developing countries such as India. However, with increased stress on healthy living and being fit, the wearable technology is likely to garner enhanced acceptance even in these countries.

REFERENCES

Bohn, J., Coroamă, V., Langheinrich, M., Mattern, F., & Rohs, M. (2004). Living in a world of smart everyday objects – Social, Economical and Ethical implications. *Human and Ecological Risk Assessment, 10*(5), 763–785. DOI: 10.1080/10807030490513793

Connor, S. O. (2015). *Wearables at work: the new frontier at Employee Surveillance*. Available online: http://www.ft.com/intl/cms/s/2/d7eee768-0b65-11e5-994d-00144feabdc0.html#axzz45rtn27V6

Dolan, B. (2012). *Wearable Devices, a $6B market by 2016*. Market Report by IMS. Available for online at: http://mobihealthnews.com/18194/report-wearable-devices-a-6b-market-by-2016

Doukas, C., & Maglogiannis, I. (2011). Managing wearable sensor data through Cloud Computing. *Proceedings of the Third International Conference on Cloud Computing Technology and Science*, 440 – 445. doi:10.1109/CloudCom.2011.65

Goncalves, F., Macedo, J., Nicolau, M. J., & Santos, A. (2013). Security architecture for Mobile E-Health applications in medication control. *Proceedings of the 21 International Conference on Software, Telecommunication and Computer Networks*, 1 – 8. doi:10.1109/SoftCOM.2013.6671901

Green, C. (2015). *Wearable technology: Latest devices allow employers to track behaviour of their workers*. Available online: http://www.independent.co.uk/life-style/gadgets-and-tech/news/wearable-technology-latest-devices-allow-employers-to-track-behaviour-of-their-workers-10454342.html

Gubbi, J., Buyya, R., Marusic, S., & Palaniswami, M. (2013). Internet of Things, A Vision, Architectural Elements and future directions. *Future Generation Computer Systems, 29*(7), 1645 – 1660. doi:10.1016/j.future.2013.01.010

Hiremath, S., Yang, G., & Mankodiya, K. (2014). Wearable Internet of Things. *Proceedings of the Fourth International Conference on Wireless Mobile Communication and Healthcare*, 304 – 307. DOI: doi:10.4108/icst.mobihealth.2014.257440

Huang, J., Badam, A., Chandra, R., & Nightingale, E. B. (2015). *WearDrive: Fast and Energy efficient storage for Wearables*. Available online: http://research.microsoft.com/pubs/244611/weardrive.pdf

Kartsakli, E., Lalos, A S., Antonopoulos, A., Tennina, S., Di Renzo, M., Alonso, L., & Verikoukis, C. (2014). A Survey on M2M Systems for mHealth: A wireless communication perspective. *Sensors, 14*(10), 18009 – 18052. DOI: 10.3390/s141018009

Kong, J. H., Ong, J. J., Ang, L. M., & Seng, K. P. (2012). Low Complexity processor design for Energy – Efficient security and error correction in wireless sensor networks. In *Wireless Sensor Networks and Energy Efficiency: Protocols, Routing and Management*. IGI Global. DOI: 10.4018/978-1-4666-0101-7.ch017

Kosir, S. (2015). *Wearables in Healthcare*. Available online at: https://www.wearable-technologies.com/2015/04/wearables-in-healthcare/

Lewis, N. (2012). Remote Patient Monitoring Market to Double By 2016. *InformationWeek Healthcare*. Available Online: http://www.informationweek.com/mobile/remote-patient-monitoring-market-to-double-by-2016/d/d-id/1105484

Lymberis, A., & Dittmar, A. (2007, May). Advanced Wearable Health Systems and Applications. IEEE Engineering in Medicine and Biology Magazine, 29 – 33.

Markets and Markets. (2015). *Reports*. Available online at: http://www.marketsandmarkets.com/Market-Reports/wearable-computing-market-125877882.html

Michahelles, F., Matter, P., Schmidr, A., & Schiele, B. (2003). Applying Wearable Sensors to Avalanche Rescue: First experiences with a Novel Avalanche Beacon. *Computers and Graphics, 27*(6), 839 – 847. DOI: 10.1016/j.cag.2003.08.008

Miorandi, D., Sicari, S., De Pellegrini, F., & Chlamtac, I. (2012). Internet of Things: Vision, applications and research Challenges. *Ad Hoc Networks, 10*(7), 1497 – 1516. DOI:10.1016/j.adhoc.2012.02.016

OPC Privacy Research Papers. (2014). Office of Privacy Commissioner of Canada. Available online: https://www.priv.gc.ca/information/research-recherche/2014/wc_201401_e.asp

Orlando, F. (2015). *Gartner identifies top 10 strategic technology trends for 2016*. Available online www.gartner.com/newsroom/id/3143521

Papakostas, G. I., Shelton, R. C., Kinrys, G., Henry, M. E., Bakow, B. R., Lipkin, S. H., & Bilello, J. A. et al. (2013). Assessment of a multi-assay, serum-based biological diagnostic test for major depressive disorder: A pilot and replication study. *Molecular Psychiatry, 18*(3), 332–339. doi:10.1038/mp.2011.166 PMID:22158016

Petersen, M. K., Stahlhut, C., Stopczynski, A., Larsen, J. E., & Hansen, L. K. (2011). Smartphones Get Emotional: Mind Reading Images and Reconstructing the Neural Sources. *Proceedings of Affective Computing and Intelligent Interaction Conference*, 578 - 587. Available online: http://www2.imm.dtu.dk/pubdb/views/edoc_download.php/6124/pdf/ imm6124.pdf

Rolim, C. O., Koch, F. L., Westphall, C. B., Werner, J., Fracalossi, A., & Salvador, G. S. (2010). A Cloud Computing solution for patient's data collection in healthcare institutions. *Proceedings of the International Conference on eHealth, Telemedicine and Social Medicine*, 95 – 99. DOI: doi:10.1109/eTELEMED.2010.19

Shaw, A. (2014). *Energy harvested from body, environment could power wearables, IoT devices*. Available online: http://www.pcworld.com/article/2463500/energy-harvested-from-body-environment-could-power-wearables-iot-devices.html

Soua, R., & Minet, P. (2011). A Survey of Energy Efficient techniques in Wireless Sensor Networks. *Proceedings of Wireless and Mobile Networking Conference*, 1 – 9. doi:10.1109/WMNC.2011.6097244

Swan, M. (2012). Sensor mania! The Internet of Things, wearable computing, objective metrics and quantified self 2.0. *Journal of Sensor and Actuator Networks*, *1*(3), 217–253. doi:10.3390/jsan1030217

Thierer, A. (2015). The Internet of Things and Wearable Technology: Addressing Privacy and Security concerns without derailing Innovation. *Richmond Journal of Law and Technology, 21*(2), 1 – 118. Available online: http://jolt.richmond.edu/v21i2/article6.pdf

Zhang, Q., Ren, L., & Shi, W. (2013). Honey: A multimodality fall detection and telecare system. *Telemedicine and E-Health, 19*(5), 415 – 429. DOI: 10.1089/tmj.2012.0109

KEY TERMS AND DEFINITIONS

eHealth: Healthcare supported by electronic processes such as sensing and communication including external mobile networks.

Fitness Application: Smartphone app that keeps tracks of activities, calories taken in, calories spent on each activity and helps the individual to stay fit.

Healthcare Applications: Smartphone app that offers healthcare from anywhere to anywhere through wireless sensor networks, Smartphones and Cloud Computing.

Internet of Things: Everyday objects, which are readable, recognizable, locatable, addressable and controllable by Internet.

Issues in Wearable IoT: Problems faced by popularization of Wearable Internet of Things for various everyday applications.

mHealth: mHealth is a component of eHelath in which medical and public health is practiced with the use of mobile devices such as Smartphones.

Wearable Sensors: Sensors that can be worn to sense different parameters with respect to various body parts continuously.

Chapter 3
When Wearable Computing Meets Smart Cities:
Assistive Technology Empowering Persons With Disabilities

João Soares de Oliveira Neto
Federal University of Reconcavo da Bahia, Brazil

André Luis Meneses Silva
Federal University of Sergipe, Brazil

Fábio Nakano
University of São Paulo, Brazil

José J. Pérez-Álcazar
University of São Paulo, Brazil

Sergio T. Kofuji
University of São Paulo, Brazil

ABSTRACT

In this chapter, wearables are presented as assistive technology to support persons with disabilities (PwD) to face the urban space in an autonomous and independently way. In the Inclusive Smart City (ISC), everyone has to be able to access visual and audible information that so far are available just for people that can perfectly see and listen. Several concepts and technologies – such as Accessibility and Universal Design, Pervasive Computing, Wearable Computing, Internet of Things, Artificial Intelligence, and Cloud Computing – are associated to achieve this aim. Also, this chapter discusses some examples of use of wearables in the context of Smart Cities, states the importance of these devices to the successful implementation of Inclusive Smart Cities, as well as presenting challenges and future research opportunities in the field of wearables in ISC.

DOI: 10.4018/978-1-5225-3290-3.ch003

INTRODUCTION

Cities have been the stage of most economical, societal, political and cultural human activities in the last centuries. Distinctively, almost as separated *silos*, these sectors have been profoundly influenced by the use of information and communication technology (ICT): the way we do business, communicate, interact with each other, the different alternatives we currently have to get informed about politics and politicians and so on. Yet only in the last years the urban space has become a subject of research and technological solutions.

New concepts, such as Urban Informatics and Urban Computing, have risen considering the employment of a wide range of *technologies* to problems that take *place* in the urban space (Foth, Choi, & Satchell, 2011). Due to ubiquitous and pervasive computing (Weiser, 1999), metropolitan areas are receiving more and more attention in order to have their problems addressed. Furthermore, cities can work as sources of information about their own *status* and about the location of their components – people, cars, parks, streets, avenues etc. – as well as what these components are performing.

Mobility is another key concept behind Ubiquitous Computing: everybody should be able to access every-piece of information, everywhere. For this reason, devices have to be reduced to the smallest size, weight and power consumption. This effort led to wearables: body-worn accessories that besides covering, adornment and protecting someone should be able to collect, to compute and to return information - armbands, anklets, bracelets, contact lenses, necklaces, glasses, gloves, jackets, rings, shoes, watches (Genaro Motti, Kohn, & Caine, 2014).

The rise of rapid prototyping boards and cheap yet powerful computational hardware motivated the development of a strong Dot-It-Yourself movement. The way Information and Communication Technology (ICT) solutions are prototyped, built and customized is changing fast and in favor of all people. Small computers built on a chip are now common and available to everyone, so are peripherals such as sensors, displays, actuators and so on. Everyone can be Gyro Gearloose. On the software side, tools to program these computers, libraries to use the peripherals, instructions to use the components and software are freely available to anyone willing to build "something". Also, people are more prepared and connected to deal with any challenge, providing conditions for fast device and tools development that can be directed towards inclusion of PwD as users or as developers of ICT solutions.

Due to built-in sensors, wearables can offer their users some extra powers - such as automatic face recognition provided by Google Glass – and directly impact the way we feel the reality around us. In fact, those devices have the power of improving the way we interact with our own body. Wearables can change the way we represent ourselves and can affect our cognitive abilities (Dunne & Smyth, 2007). For the purpose of being used in the urban space, wearables can support users to deal with the huge amount of data available in the city and can be seen as an innovative infrastructure to interact with the environment around.

In this chapter, the authors advocate wearables as data input/output interfaces for persons with disabilities (PwD) to benefit from applications available in the urban space. The authors do believe that Smart Citizens and Smart Administration are the pillars of successful Smart City projects. Technology can play a decisive role in leveraging citizens' smartness, since technology can self-adapt and provide customized interfaces/services to user needs, which implies having the required conditions to efficiently make decisions and to be responsive to feedback (C. Yin et al., 2015). The goals of this chapter are:

- Determining the challenges and innovative aspects of using wearable computing in the context of Inclusive Smart City Projects;
- Establishing associations between wearable computing and other concepts related to Inclusive Smart City such as Internet of Things, Multimodal interactions, Cloud Computing, and Machine learning;
- Analyzing the use of wearable devices by PwD in the city.

This chapter is organized as follows: section "Background" points out some contributions found in the literature covering the design principles of wearable devices, projects that use wearables in the context of Smart Cities, developments of Artificial Intelligence and wearables, and the underlying concepts of Inclusive Smart Cities; the next section, "Using wearables in the Inclusive Smart City", presents how wearables can strategically be used to make Inclusive Smart City projects more feasible; then, section "Future research directions" discusses some opportunities for the next years; and, finally, the "Conclusion" section brings some closing remarks.

BACKGROUND

This section presents some important definitions and summarizes the main contributions of research projects related to wearable devices, Artificial Intelligence, Smart Cities and Inclusive Smart Cities.

Wearables: Design Issues

The Wearable Computing market is expected to register a total of $70 million by 2017 related to the shipment of wearable devices (Moustafa, Kenn, Sayrafian, Scanlon, & Zhang, 2015). These devices have been used in a variety of sectors - from the entertainment industry to the healthcare services, from the military industry to sports and assistive technology – which realized that the future of computing is connected to ubiquity, to the movement, to the ability of changing in place/position rather than connect to a controlled ambient under the steady PC paradigm (Gemperle, Kasabach, Stivoric, Bauer, & Martin, 1998).

Personal Digital Assistants (PDA) can be considered the ancients of what we currently call wearables (Rawassizadeh, Price, & Petre, 2014). In the same way as pocket watches were replaced by wristwatches, mainly because information was accessible in a quicker manner, palms and PDAs were replaced by wearables, mainly because ubiquitous resources allowed small displays and electronic components to deliver information more promptly. Palm and PDA functionalities became accessible to human eyes in an instant blink, not requiring any extra movement of the hands and many times not even requiring to press buttons. In fact, Rawassizadeh et al. (2014) argue that the next move is the domain of Interaction Design and Cosumer Electronic Studies is the fact of wearables – specially smartwatches – to replace smartphones – which mostly requires at least one hand to manipulate the device.

In its essence, Wearable technology concurrently has to deal with two contradictory challenges: (i) device miniaturization or "disappearing computing" – from the ubiquity movement – and (ii) aesthetic appeal – from the product design field and jewelry sector. These devices must shape the support that will transport them – the human body – taking into account general intrinsic restrictions – such as low weight, not interfering with human gestures or disrupting body movements – and individual preferences – such as personal style or fashion trends. Briefly, a wearable product "should look good and be appropriate for the culture of the end-user, with garment functionality enhanced by embedding technologies such as electronics and computing into the clothing" (McCann, Hurford, & Martin, 2005).

Wearables can combine performance and convenience, since they connect the user with the right information, at the right place and at the right time (Rawassizadeh et al., 2014)(Siewiorek, 2002). Such handiness so expected in our days can in fact lead to a profusion of information causing distractions and information overload. Distractions can be minimized by *Context-aware computing*: the system is fed with user's information and preferences, as well as with data from the environment where the user is. This set of information is used to set the behavior of the system and to filter the information needed in a specific situation. Still in line with Siewiorek

(2002), the system can be adapted and customize the services and information it offers according to a range of attributes:

- Physical location;
- Physiological state: body temperature and heart rate, for example;
- Emotional state: such as angry, distraught, or calm;
- Personal history;
- Daily behavioral patterns, and so on.

On the other hand, wearables extract information from the environment regarding what is happening and/or what is available around the user. Associating user's attributes and environmental characteristics is the way in which wearables can improve the user's perception of his/her surroundings. Theses devices are capable of filtering the increasing amount of information and digital services, providing just what is really meaningful at a certain time, and, consequently, reducing the demands on human attention. Wearables can even proactively connect to other devices in the network for analyzing statements, making decisions, and anticipating user's needs – avoiding disturbing him/her with inappropriate messages, for instance.

The notion of *Wearability* is related to the interactional potential of the human body and wearable object (Gemperle et al., 1998), taking into account the dynamics of the human body, the physical shape of objects, comfort, amplitude and repetitiveness of movements, and even physiological aspects –Siewiorek (2002) remarks that the long-term use of wearables must be better investigated.

Wearables cannot obstruct daily life movements as they are fixed to the human body in a stable way, so they will be able to obtain accurate sensor data (Yamamoto, Terada, & Tsukamoto, 2011). Depending on the part of the body, some wearables are more or less recommended. Head mounted displays, caps and glasses, due to their position, are appropriated for the head. For hands, better wear a watch, a wristband, or a bracelet. Wrist-length gloves are difficult to wearing all the time. Knees move frequently, so it is difficult to fix wearables on them. For a foot, better choose shoes, socks, and sandals.

In addition, designers must consider the body positions users will engage in and if they are performing some activity while interacting with the wearable device. If the user is in a sitting position, his/her hands and feet can be used freely and precisely. Yet the designer must consider the possibility of the user holding a pencil or a cup with one hand, which can obstruct the interaction with the wearable. In a standing position, users can have both hands and feet free, but at least one foot needs to support the body weight for standing. However, when the user is on train, one or both hands can be unavailable. In walking situation, the user can be free. It is worth emphasizing that vibratory motions and interference with user's movement can affect the movement of the hands. Another restraint to be considered by designers when defining the interaction technique to operate the wearable device is the amplitude and acceleration of gestures and the parts of the body engaged in producing the gesture. Shaking hands and head, twisting arms and legs can be natural in a private environment, but can be deemed bizarre attitudes when performed in public and social spaces.

As Wearable Computing is a crossroad of distinct disciplines – such as Marketing, Computing, Ergonomics, human biology and textile/material/product/fashion design –the creation of common tools, models and guidelines is crucial to make the design process understood by all participating professionals. These tools have the challenge of being encoded as a unique and clear language, understood and spoken by all the members of the team, to guarantee the contributions from each discipline. For example, the sense of "usability" as a whole must be understood and implemented the right way by each discipline, even in different manners, for better user approval.

Gemperle et al. (1998) summarize a set of features that must be considered when developing wearable products:

- **Placement:** Where on the body it should go;
- **Form Language:** Defining the shape;
- **Human Movement:** Consider the dynamic structure;
- **Proxemics:** Human perception of space;
- **Sizing:** For body size diversity;
- **Attachment:** Fixing forms to the body;
- **Containment:** Considering what is inside the form;
- **Weight:** As it spreads over the human body;
- **Accessibility:** Physical access to the forms;
- **Sensory Interaction:** For passive or active input;
- **Thermal:** Issues of heat next to the body;
- **Aesthetics:** Perceptual appropriateness;
- **Long-Term Use:** Effects on the body and mind.

Ordered from the most simple to the most complex features, this list gives rise to design guidelines and principles that, early in the design process, can help designers to increase the functionality, usability and acceptability of wearable devices. In large terms, wearable design guidelines pursue adjustability and flexibility keeping in mind the variability of the human body forms and sizes: standards are not viable in a global scale – sometimes they are not adopted even in a local scale.

McCann et al (2005) suggest a wearable product design research and development process based on the "awareness of aesthetics, general creative design skills and an appropriate knowledge of garment and textile technology". The process flow is comprised of the following steps:

1. **Identification of End-User Needs:** Designers must trace 'Demands of the Body', 'Demands of the Activity/End-use', 'Demands of the Culture' and 'Aesthetic considerations':
   a. **Demands of the Body:** Basic understanding of human physiology concerning breathability, thermal regulation, movement, fit, agility, sensitivity and grip, for example, as a matter of assuring comfort to the user;
   b. **Demands of the Activity/End-Use:** Designers should look for answers to some questions related to the main activities the users are engaged in and about the environment where these activities take place, such as "Is the activity over very quickly or is it medium or long-duration?", " Is it uniform for daily use, leisure wear worn once a week or travel-wear worn for a couple of weeks per year?", "Is the activity seasonal, to be practiced indoors in a controlled environment or outdoors in an unknown terrain?", "Is it a contact or non-contact activity?", "Is there an extreme climate or a range of temperatures and degrees of humidity?", and so forth.
   c. **Demands of the Culture:** some cultural or even religious aspects may influence the requirements of wearable devices and can impact styling and color choice, for example; and
   d. **Aesthetic Considerations:** Designers must be aware of market trends as well as of new materials and specifications to provide the appropriate appearance to the end-user, not forgetting to balance functionality and comfort;
2. **Textile Development:** After understanding the user's needs, designers should analyze the best choice of material among those is available. Designers must consider the material specificities and must understand its characteristics, as well as to acknowledge generic fibers, microfibers, bi-component yarns, elastomerics and the range of knitted, woven, multi-layer and non-woven constructions, breathable membranes, coatings, insulations and water resistant finishes etc.;
3. **Garment Development:** In this phase, designers must be concerned about building the wearable device, joining its different pieces and mounting what is known as "layering system". This is the moment to define the cut and the sizing patterns of the wearable product;
4. **Integration of Smart and Wearable Technology:** At this step, designers blend aesthetic and functionality, textile/plastic and electronic components, art and intelligence. Designers must think carefully about incorporating sensors,

switches, conductive fibers, communication elements and circuit boards to the human body. Moreover, these components require power supply, which may interfere with the shape and final weight of the wearable product;

5. **Garment Manufacture:** The procedure used to manufacture wearable products can require specialist techniques. Designers must be aware of the capabilities, restrictions of the available and appropriate manufacturing methods;
6. **Distribution/Product Launch:** Selling and distribution language must be connected to the design process. Wearable products must be offered and addressed to the right customer – the one whose needs were elicited – creating an efficient branding strategy;
7. **End of Life Recycling:** The final step of the process is related to sustainability and ethics issues. Designers must be concerned about using biodegradable material and about the end of the non-biodegradable pieces used to create their wearable products. Indeed, sometimes, in some countries, it is a matter of legislation. Designers must thus take this into account.

This suggested process is very interesting because places the user's needs at the starting point, which makes it a user-centered design process, and engages the designer not only in the "creativity" phases but makes them intimately committed with all steps of the process. Depending on the user's needs, designers can prioritise functionality or appearance, can better choose materials and shapes and can be more confident during the decion-making process.

However, both the academy and the industry must address some issues if they want all the expectations, promisses and forecasts about the future of wearables come true. According to Rawassizadeh et al. (2014), restricted I/O and small hardware results are the two major wearablesobstacles. Those obstacles directly impact the way users interact with those systems, the overall user's experience and the need of required attention from professionals in the design process. Small displays can be a source of problems to input data, specially considering touch screens. Also, reading the information being processed by the device can be a serious problem since the user maybe moving, the amount of information may be much larger than the capacity of the display, or still the user is an ageing person. This is not an exclusive user concern, because designers have to be strenuously engaged with the task of fitting content, firstly conceived for PC displays, into smaller screens or even for non-display devices. This adaption sometimes changes the interaction techniques that the application formerly offered users.

The second issue, small hardware results, is related to how users have to deal with when the computational performance of the weareble device is below their expectation. This also affects the user's experiences because nowadays low computational performance means higher frustration, namely, bad user's experience. In addition, although mobile devices have advanced in the past years, battery life is still a source of trouble for users and digital service providers. Hardware designers have to deal with the need of smaller devices, but keeping high performance: batteries should be smaller but should also provide a longer lifespan; sensors should be smaller keeping the same eficiency and precision. The industry developed inteligent devices that switch services on and off as they are required by users. This solution works most of time but there are situations, such as location services, in which the mobile equipment must combine information from different sensors – such as GSM, Wi-Fi, and GPS – or must invoke several applications at a time, causing high power consumption.

Briefely, wearables' pros and cons must be balanced specially taking into account the users caracteristics and needs, as well as considering the overall environment where those artifacts will operate. The next section presents some projects that use wearables in the Smart City context.

Wearables in the Smart City

The rapid and disorganized growth of urban centers has challenged their managers. In 2008, half of the world population lived in urban areas. Some predictions indicate that this percentage will be 70% in 2050 (Streitz, 2011). Urban centers are more susceptible to natural disasters, industrial hazards, global warming, etc. Additionally, urban centers made little progress regarding the inclusion of people with disabilities. A possible solution to these issues are the smart cities, i.e., cities strongly founded on information and communication technologies that invest in human and social capital, to improve the quality of life of their citizens by fostering economic growth, participatory governance, wise management of resources, sustainability, and efficient mobility, whilst guaranteeing the privacy and security of the citizens (Pérez-Martínez, Martínez-Ballesté, & Solanas, 2013).

Wearables have contributed to the development of smart cities. Using wearables with modern sensing capabilities, it is possible to receive more significant information about the world around us, while making it more intelligent and flexible (Fletcher et al., 2010). Similarly, with wearable computing, we strive to enhance our own capabilities by allowing us to see/listen better, and to interact with surrounding technologies more easily (Klingeberg & Schilling, 2012).

Some smart city applications use the dataset sensed by the population. These applications use People as Sensors (Resch, 2013), i.e., they use anonymous data obtained from wearable devices. Examples for exploiting collective information include using data from mobile phone networks to sense and to influence urban dynamics (Reades, Calabrese, Sevtsuk, & Ratti, 2007), (Gonzalez, Hidalgo, & Barabasi, 2008), or leveraging Flickr photos to assess seasonal tourist behavior (Sagl, Resch, Hawelka, & Beinat, 2012). Solanas et al.(Solanas et al., others, 2014) present a new vision for wearables in the context of smart cities and health. The smart health (s-health) vision is the provision of health services by using the context-aware network and sensing infrastructure of smart cities.

Other approaches are concerned with the inclusion of people with disabilities in the context of smart cities. Martin et al. (Martin, Dancer, Rock, Zeleny, & Yelamarthi, 2009) present an obstacle detection system based on an ultrasonic sensor. The sensor is mounted on the RFID cane to extend its effective range and to perceive obstacles the cane alone would not be able to detect. A voice message played on the monitoring station alerts visually impaired persons when an obstacle is detected. The proposal developed by Sony tries to make the movie theater an inclusive space for the deaf. The Sony glasses (Paris & Miller, 2016) are similar to 3D glasses. Captions are projected onto the glasses and appear to float about 10 feet in front of the user. The glasses are attached to a receiver box, which allows the user to adjust the display distance (near, mid, far) and brightness (on a scale of 1 to 5). For individuals with enough residual hearing, the glasses provide volume control. Regal Cinemas adopt this device.

The BlindShopping project (López-de-Ipiña, Lorido, & López, 2011) enables accessible shopping for visually impaired people using mobile technologies. The solution has a navigation system and a product browsing mechanism. The navigation system guides blind people inside the market. The product location is found by combining RFID tags distributed throughout the corridors and a RFID reader attached to the white cane. The product recognition uses QR codes for retrieving information about products to the user. QR codes are recognized by the smartphone camera and the information about the products retrieved from the back-end. Verbal descriptions are generated so that the blind person can understand what is being described. This solution is easily extendable to public spaces in order to give place to a smarter, more inclusive city, both indoors and outdoors. For instance, tactile paving of some core parts in a city could be enriched with RFID tags that could then be read by an RFID reader embedded in white canes to guide the blind throughout the city and pinpoint information of their interest around them.

In order to increase the autonomy of blind people, the OrCam Company developed the MyEye solution. MyEye (Paris & Miller, 2016) is a device that promotes a significant improvement in the users' interaction with the environment surrounding them. It consists of a small camera clipped to the wearer's glasses using a magnet and connected by a thin cable to a portable computer designed to fit in the wearer's pocket. Additionally, it has a bone conduction speaker that bypasses the auditory channel. The MyEye system recognizes the text signaled by the user and plays it through the ear bones. It is able to recognize texts in newspaper articles and environment text such as bus numbers, billboard signs, and changing stoplights. The software also helps users to identify familiar faces in a crowd.

Since wearables are most of the time in constant contact with their owners, in a Smart City scenario, this device is able to recognize its owner's physical activities and location. This is one of the main features of the Smart City project: providing accurate information and services around the user. When users operate devices such as smartphones, their location is not precise, because there is a possibility of forgetting the devices in a different room or even laying them down on a shelf. Wearables are body mounted and, thus with more chances to provide the precise users' location –crutial for Smart City heath initiatives, as well as for Marketing initiatives.

Wearables and Artificial Intelligence

The evolution of wearable devices has attracted the interest of researchers in artificial intelligence for the development of new solutions that improve these devices. A common feature of wearables is sensing. The sensing activity generates a volume of data that can be useful to monitor and to detect patterns of the users. Machine learning techniques can be used for this purpose. Machine learning is a set of methods that can automatically detect patterns in data, and then use the uncovered patterns to predict future data, or to perform other kinds of decision-making under uncertainty (Murphy, 2012). Machine learning applications usually have two stages: training and evaluation. The training stage uses a dataset of measured attributes from the investigated object. The dataset undergoes a feature extraction step that generates relevant information from the raw signal. Then, the application uses learning methods to generate an activity recognition model from the dataset of extracted features. For testing, wearables should generate a new dataset that undergoes a feature extraction step. Finally, the recognition model evaluates this feature set.

Examples of measured attributes are the user's movement, environmental variables (e.g., temperature and humidity) and physiological signals (e.g., heart rate or electrocardiogram). It is quite common to use these attributes in Human Activity Recognition (HAR) systems. Some examples are HAR systems for detecting abnormal activities in patients with dementia (Yin, Yang, & Pan, 2008), behavior monitoring for personalized health care (Hanai, Nishimura, & Kuroda, 2009), movement analysis for sports assistants (He & Jin, 2008), etc. Approaches that adopt HAR systems to assist public health are also very common.

Falls are a serious public health problem. According to the Centers for Disease Control and Prevention, one out of three adult persons aged 65 and over falls each year. In 2015, falls among older adults cost the U.S. healthcare system over $34 billion dollars (CDC, n.d.). Özdemir *et al* (Özdemir & Barshan, 2014) present an automated fall detection system with wearable motion sensors. Each unit comprises three tri-axial devices (accelerometer, gyroscope, and magnetometer/compass). The approach uses six wireless sensor units tightly bound with special straps to the patient's head, chest, waist, right wrist, right thigh, and right ankle. They adopted six classifiers based on machine learning to distinguish between falls and activities of daily living and the classifiers achieve accuracies above 95%. Mazilu *et al* (Mazilu et al., 2012) propose the use of accelerometers and smartphones as wearable devices for Freezing of Gait (FoG) detection. FoG is a gait deficit associated with falls, interferes with daily life activities and impairs quality of life. This condition is common in patients with advanced Parkinson's disease. The approach uses a classifier for providing feedbacks to patients whenever a FoG event is detected. Ugulino *et al* adopt the decision tree technique to recognize the following activity classes: sitting, sitting down, standing, standing up, and walking. Data is gathered from four subjects wearing accelerometers mounted on the waist, left thigh, right arm, and right ankle of the user. In tests, the solution presented accuracy of about 99.4%.

Some works combine machine-learning techniques with intelligent agents. For example, the smart wearable device (MaRV) (Fraile, Bajo, Corchado, & Abraham, 2010) is an approach to automatically manage and to improve the assistance to patients in geriatric facilities by using smart wearable technology. This system uses smart wearable devices to obtain context-aware information, which allow identifying, locating, and gathering data for each patient. MaRv is a multi-agent system (MASs). The MaRv agents are in charge of managing, controlling and learning the information obtained by the system.

Another area that presents innovative solutions for wearables is Big Data (Jacobs, 2009). It is easy to understand the growth of big data applications for wearable devices. Considering the context of smart cities, where every inhabitant is a potential data generator, we have a large volume of complex and unstructured data. Ping Jiang *et al.* (Jiang et al., 2014) propose a system for using wearable sensors capable of continuously monitoring the elderly and forwarding the data to the big data systems. The system uses Hidden Markov Model for human behavior recognition for feeding meaningful data to the system. The approach aims to improve and customize the quality of care. In (Baimbetov, Khalil, Steinbauer, & Anderst-Kotsis, 2015), the authors describe a platform to extract insightful emotional information using cloud-based big data infrastructure. The architecture is capable of handling sensory data from users' devices, annotate it with users' context, process it with well-proven single-mode emotion recognizers, fuse the sensing results and interpret the results with respect to the users' context.

A problem of the applications presented so far is that they do not have elements to promote interoperability. The data generated by each of them, even for applications that deal with the same domain information, are incompatible. Some works have used semantics to deal with this issue. Mezghani *et al.* (Mezghani, Exposito, Drira, Da Silveira, & Pruski, 2015) propose a generic semantic big data architecture adopting the Knowledge as a Service (KaaS) approach to deal with data heterogeneity and system scalability challenges. In addition, they present WH_Ontology, an ontology to deal with the heterogeneity of wearable data to ensure semantic interoperability and to allow a more accurate knowledge about the patient such as detecting and/or predicting anomalies. The objectives of this architecture are (1) providing a scalable solution for storing the large volume of healthcare data generated by multiple sources, (2) supporting data sharing and integration for better decision making, and (3) extracting valuable information (knowledge) from heterogeneous data. Castro *et al.* (Castro, Normann, Hois, & Kutz, 2008) propose OASIS, a repository of ontologies for assistive systems. The approach extends Ontology Metadata Vocabulary (OMV) (Palma, Hartmann, & Haase, 2008), a standard that defines a set of descriptors that follows the principles of the Dublin Core. The aim is to promote the reuse of ontologies, flexibility, and interoperability.

Some approaches apply the view that a wearable device is an entity that provides services. They use the Service-Oriented Architectures' (SOA) view for promoting automatic service composition. The aim is to create new services from existing services using the user goals. Generally, they use Artificial Intelligence planning to identify individual services that, when compound, meet user demands. For example, Kaldeli *et al.* (Kaldeli, Warriach, Lazovik, & Aiello, 2013) propose automatic service composition in pervasive environments. The system selects and combines services in runtime. In addition, it performs different compositions for the same goal depending on the current state of the devices. This system uses a planner based on CSP (Constraint Satisfaction Problem).

Inclusive Smart Cities

Previous studies reinforce the hypothesis that cities still do not make very important information accessible – and that, in many cases, can be vital – to guide citizens in the urban space (Oliveira Neto & Kofuji, 2016). Some aspects observed in the central area of São Paulo (Brazil) determine that the urban space is more complex concerning PwD needs. On the other hand, these aspects establish that cities are a source of research opportunities regarding the improvement of assistive technology already installed and/or the development of new products and services driven to PwD to allow them to fully use the urban space.

As a large and permanent human settlement ("City," 2016), the city should be a territory of balanced forces, where everybody should have the same rights and duties. Instead being a segregated space that isolates different groups, cities should be better seen as places of associations, links and connections. Cities should have more bridges rather than walls. This democracy-based point-of-view conceives the *Inclusive Smart City.*

One of the key issues of contemporary societies, accessibility has brought together national and international organisms, governments and social movements around the needs of persons with disabilities. Concisely, the design (or redesign) of products, devices, services and environments ("Accessibility," 2015) means a wide range of advantages to PwD: the ability to access - products, places, information, systems and so on – that were only previously accessible to people without disabilities. Assistive technology is mostly developed making use of Universal Design principles, in order to guarantee access to the widest possible range of abilities. Designing taking into account accessibility principles is designing for everyone.

In their *Convention on the Rights of Persons with Disabilities,* the United Nations emphasize that accessibility has to "enable persons with disabilities to live independently and to fully participate in all aspects of life, States Parties shall take appropriate measures to ensure to persons with disabilities access, on an equal basis with others, to the physical environment, to transportation, to information and to communications, including information and communication technologies and systems, and to other facilities and services open or provided to the public, both in urban and in rural areas" (United Nations, 2006). This right to access concerns the physical layer (building, roads, and other indoor and outdoor facilities), as well as to the digital layer (information, communications and other services, including electronic services and emergency services).

Cities continue to be one of the greatest obstacles for PwD: wheelchair users must deal with potholes; impaired–hearing persons must count mostly on their vision to compensate for the lack of sound; people with limited walking abilities have to move over sidewalks with changes in level; visually impaired have do deal with the lack of appropriate signs regarding places and objects. Specific laws and governmental regulations have treated part of these problems associated to the physical infrastructure of the urban space. In Brazil, for instance, a number of municipalities have made it mandatory to adapt sidewalks equipping them with tactile floor indicators; buildings must be readapted with ramps, curb ramps and elevators – even old buildings are motivated to be refitted; parking lots must be reserved for elderly people and PwD; public spaces restrooms must fit PwD needs; buses must provide platform lifts, etc. As expected, those changes take time, but there have been advances and progress.

When the focus moves to the digital layer of the urban tissue (applications, systems, ICT services and electronic services), there is still a particular absence of digital services oriented to all diversity of citizens living in a city (including PwD, the elderly, children, pregnant women, foreign people that do not speak the local language etc.). When discussing Smart City initiatives, accessibility and assistive technology, we rarely find options of services. Most applications in this field (route tracers, maps, emergency systems, sharing economy apps, ride-sharing programs, point of interest maps, bus tracker apps, smart parking and others) are not at all capable to interact properly in a non-excluding manner. Public administrators have more and more provided free Wi-Fi zones in squares and public buildings, but unfortunately, most PwD are not able to fully benefit from this service.

As most of our activities are becoming digital, providing digital services in an inclusive manner is to allow every citizen to occupy a place in a digital society and in a moving democracy era, potentially assuring the mitigation of the digital divide caused by the lack of full access to both urban physical and digital layers.

In an Inclusive Smart City, all citizens benefit from the digital infrastructure to access useful data enabling them to obtain the information they need, when and where they need it. More than physical access, it is crucial to provide digital access to information in the variety of formats in which the information is recorded. Also, service delivery must be citizen-centered, meaning that citizens' needs must be in the forefront. City-administrators must collaborate with each other and share the information management in order to provide a coherent service and a consolidated view of data that, most of time, is spread over a multiplicity of silos. As already available to some citizens, in an Inclusive Smart City, an intelligent physical infrastructure (computers, sensors and other Internet of Things components) would be available to collect and to transport data supplying services and enabling those services to perform their tasks.

Another very important issue is transparency of outcomes/performance to feedback citizens/ enterprises with data collected from the city and to enable citizens to compare and to challenge performance. An Inclusive Smart City makes possible the engagement for PwD to play a consistent role as citizens in a broaden manner.

Hence, the Inclusive Smart City is an approach that deals with the observed lack of accessibility in the urban space and with improving the tools already offered to PwD in cities. Inclusive Smart City is a new citizen-centered approach that combines pervasive technologies (hardware and software) and the Universal Design methodology in order to: (i) provide mechanisms that allow people with disabilities to interact with the urban space and to access geolocalized information and services; (ii) use ICT to mitigate the segregation of people with disabilities, creating innovative solutions or adapting some of those already in use but not available to everyone.

The urban space is a rich source of visual, aural and spatial data. People with disabilities have to face the obstacle of not perceiving one (or several) of these channels of information. A blind person, for instance, does not perceive what is drawn in a traffic sign a few meters away. If he/she does not know the place, he/she will probably not find where the public bathroom is (even though the place shows the appropriate sign). Thus, *the main feature of the Inclusive Smart City is the ability of identifying places and objects (or things) and making this information digitally available*. Once this information is available, it can be sent to devices that receive the information and personalize this information according to the disability of the user.

USING WEARABLES IN THE INCLUSIVE SMART CITY

Although in most of the cases PwD strive against all kinds of difficulties in their daily routine, the urban space remains a scary territory, the unknown, where all sort of (mainly bad) surprises take place. This section describes features of the routine of PwD or limitations faced by PwD in which wearables can be useful.

First of all, wearables can be seen as the key technology in the context of Inclusive Smart City once these devices provide a way for PwD to connect to the digital layer (digital services and information) as well to the physical layer (other intelligent objects) of the urban space. Namely, wearables are the interface between PwD, intelligent objects, and the digital world. They work as gateways that allow users to interact with the environment, then send users' requests to be processed in the Cloud, and finally supply users with new and accurate information. Wearables are the glue that, even virtually, keeps all those pieces together.

When moving, PwD (i) carry assistive technology, such as canes; (ii) hold bags and packets of all sizes; (iii) wear accessories, such as hats and sunglasses, and specific clothes, such as rainwear and/or winter coats; (iv) operate smartphones, wheel chairs and so on. Most of them are separated pieces and can pose some kind of trouble to PwD: they may fall down –a nightmare for a visually impaired person or for someone with physical motor limitation; some of these pieces may hinder the movements; others require the use of hands– such as canes and smartphones; still, other pieces can be the target of thieves, such as jewelry and electronic devices. The first contribution for PwD to employ wearables in the urban space is that they are body-mounted and can let the hands free to other tasks, such as holding grab bars, touching Braille signs and other tasks.

When dressing in all-in-one wearables, PwD are set free of managing several separated pieces of clothing, accessories, electronic devices and assistive technology. Handling all this paraphernalia is an attention and cognitive-effort consuming task. Moreover, PwD have to keep the corporal balance even when walking through the crowd. When trapped to the body, wearables can support users to perform several tasks, concurrently offering the stability and the level of unawareness that pervasive devices can bring to ordinary life.

It is possible to associate intelligent built-in solutions in wearables to monitor PwD and warn them about imminent risks. In the context of Inclusive Smart Cities, machine-learning algorithms can be useful both to avoid accidents, as well as to identify their occurrence and automatically request assistance to the public health service or even to the user's relatives, or any other contact. Semantics and logical reasoning may be applied to customize wearables to the individual needs of PwD and hence promote a seamless integration between the wearable and the PwD.

Machine Learning relies on gathering and processing data. The framed application involves personal data such as location, id, and physical traits either of the user or of people in their surroundings, giving rise to privacy issues. There are technical ways of dealing with it, such as obfuscating data, extracting the features before any data transmission. In any case, a social/governmental agreement on what is private and how it should be treated seems unavoidable for its wide adoption.

For people who employ wearables while moving around the city, designers should conceive modular wearable devices. Hence, users would be able to increase the overall functionality of the device easily connecting blocks of components – each responsible for performing a specific task. A device as flexible as that presented here provides PwD with the capacity of adapting the device based on their needs and context. The user can prioritize the functionalities and select which of them he/she would pack in the wearable device to carry in his/her daily journey in the city. The heavier and the larger the equipment, the more difficult it is to travel having it on. The weight of a wearable should not hinder the body movement or balance.

Another important issue about how flexible the wearable device is that sometimes a single shape will not fit the body of all PwD, especially persons with motor disabilities, whose bodies sometimes present contortions and non-regular shapes. In these cases, flexible wearable devices can follow the format of the PwD's body and can be adapted to follow the body movement. As a design principle, wearables must be stable, solid, and comfortably attached to the human body when we consider its use in the context of the Inclusive Smart City.

PwD can benefit from the fact that wearables are body-mounted pieces and, consequently, these devices can perform their jobs in a (truly expected) smooth and non-obstructive way. Therefore, while the PwD are performing their daily tasks, wearables by themselves are collecting, computing and (when needed) spreading information without the awareness of the user. This functionality is very important, for example, to automatically keep the family informed about the PwD precise location. In addition, context-aware applications can run appropriate procedures, particularly in the environment situations providing information tailored to the user's needs. Also, wearables can communicate with other smart objects surrounding them to enrich the offering of services and information. All of this without the user's direct participation.

Our brain can perceive an aura around our body, or "body schema" (Head & Holmes, 1911). Dunne and Smyth argue that:

... psychologically, humans generate a dynamic understanding of the size, shape, and physics of their bodies, and it allows us to navigate physical spaces and manipulate objects. The space designated as part of the body schema is surrounded by a spatial area known as peripersonal space. Peripersonal space is generally differentiated

from extrapersonal space by our reaching abilities. Objects that can be reached without locomotion are within peripersonal space, objects out of grasping distance are in extrapersonal space. (Dunne & Smyth, 2007)

Due to impairments, limitations or even due to the ageing process, human beings can have this ability weakened. Even though the human brain rearranges itself using other senses to replace one sense when it is lost, the whole body perception is negatively affected. Wearables built with a combination of different sensors and other electronic components distributed over specific parts of the body can mimic the human sensorial system sometimes to simulate the original *stimuli* in the human brain. PwD need perceptual extra information when they can count on no one else or even when they are in a crowd of unknown people. The information captured by wearables can be processed and provided in the more appropriate format to the user, i.e., in the modality that best fits his/her needs. As a metaphor, wearable devices can be seen as the super power: the more powerful the sense(s) that PwD have trying to counterbalance the absence of (an) other sense(s).

Moreover, wearables can provide new interaction techniques to PwD. For instance, when outfitted with accelerometers, the application can use de variation in position of the wearable device covering the user's hand to infer the movements the user is making which will be understood as commands. Interacting using gesture movements is known as natural interfaces because this uses elements of the interaction between humans rather than traditional input devices, such as mouses, keyboards, and screens. Also, wearables can explore other parts of the body – such as ears, neck, feet and so on – as well as other sensations as interface modalities, such as touching and pressing.

Even though the use of wearables in Inclusive Smart City Projects seems to be advantageous and innovative, their achievement depends on several elements, such as:

- **Multidisciplinary Teams:** Professionals must be aware that they will work on projects that rely on the synergy of several domains. They must constantly exercise interchanging ideas, seeking for inspiration in other domains, keeping the mind open and willing to learn (and to teach) new tools and resources. Some fields (and professionals) are specially expected to be represented in these teams because they have the potential to contribute largely to the success of those projects, such as:
    - **Human-Computer Interaction (HCI):** This field provides a range of tools and techniques to map the user's needs applying Universal Design techniques, defining the best way to communicate with users – the available modalities to provide users –, understanding the context in which the user is, exploring the environment and how it can impact the

interaction with the system, evaluating the system before it is released, and so on. Another important contribution from the HCI area is a variety of tools to improve the system usability HCI professionals try to develop solutions that are easy to learn, efficient to use, easy to remember, and which lead to an overall user's satisfaction. However, physical aspects are not the sole concern of HCI. Cognitive load – a psychological aspect related to the amount of mental effort while, for example, using systems – is more and more common nowadays, and is an issue addressed by HCI professionals;

- **Hardware:** Area that defines electronic components and associate them forming the physical part of the system. Professionals from this area choose strategies to keep the device working the longer possible – such as high-capacity batteries, low-power devices, lighter communication protocols – and to minimize some inconvenient consequences of electronic components – such as heating. Communication is a key feature of pervasive computing. In the context of Inclusive Smart Cities, Wearables must be able to communicate to other objects surrounding it, but they must be able to communicate with Cloud-based services and remote servers. Also, they can calibrate sensors and define which sensor is more appropriate to a specific situation;
- **Artificial Intelligence (AI):** This field is responsible for the "smart" piece of the project. More than just collecting and providing information, wearable devices used by PwD has to be able to work as a useful assistive technology, discovering user's patterns, analyzing user's features, limitations and preferences in order to provide tailored information and services at the appropriate time. Another key issue is the influence of the social group the user is part of. AI algorithms are able to take into account the preferences of a group of users to infer the next choices and attitudes of a particular user. PwD must trust the functionality of wearable devices, namely, AI algorithms, and be able to outsource to those algorithms the constant attention the urban space requires;
- **Design:** This domains contributes by providing tools and techniques to define the material that fits de requirements of the wearable device, Ergonomy issues– i.e., studying the usage of the device for the place where it will be used, trying to minimize fatigue and discomfort –, aesthetic elements – such as colors and shape – and how to put all those items together in a market-driven product. Prototyping is a very important tool related to Design that helps designers and users to anticipate the final version of the product, and thus allows discovering

still in the early phase of the product development some flaws to be corrected;
    - **Software Engineering (SE):** As software is very relevant to the system, this field is important as it provides a range of tools that make the software development process more efficient, reusable, rapid and systematic. The quality of the final product depends on detecting and correcting flaws before the product is delivered to the client. SE professionals pursue the quality of the artifacts they produce. For this, they have a range of models, which standardize the work done by a distributed team, and guarantee the desired integration of separated modules. Also, software engineers deal with project management tools, which are primordial when working in silos and these tools are necessary to keep track of everyone's responsibilities, of tasks deadlines and the impact of not-achieved tasks.
- **Government:** This is the role that coordinates forces to improve citizens' quality of life. Also, Government balances the interests of social-economical-political sectors, including in the local agenda some initiatives that benefit minorities and groups that have little power or little representation within the society, such as PwD, ageing people, refugees and so on. Some of those initiatives cause disagreements and controversies, mainly because they are imposed by laws and those refusing to obey them must pay fines and even go to prison. The Federal Brazilian Government, for instance, has implemented a successful initiative of transforming cities into more physically accessible places. It is motivating other administrative political levels (states and municipalities) to create and to stipulate as obligatory by law and administrative rules that buildings, streets, sidewalks and other city equipment is prepared to be used by PwD and other persons in a situation of dependence. As a result, it is quite common to find tactile floors, accessible pedestrian signs, traffic lights with warning sounds and others. In the case of Inclusive Smart City Projects, a similar strategy must be applied: the government must first provide digital local information of public buildings and public spaces – i.e., tagging public buildings with accurate information on what the building is, what kind of services are offered there and making this information available in a format that can be processed by PwD wearables and other devices – and then making other sectors in the city provide the same facility. Municipalities that already have an Accessibility City Council must include this element in the decisions and conduction of the Inclusive Smart City policy.
- **Policy:** Once several technologies and stakeholders are engaged in a typical Inclusive Smart City project, defining a set of standards and rules that will work as the compass for everyone involved in the project and guarantee

devices' interoperation. Wearable devices manufacturers must follow the regulations and rules of communication protocols, the same followed by digital content and service producers. In fact, more than regulations and rules, innovation is also led by actions, norms and behaviors that people accept as good or take for granted.

- **Industry:** This is the element committed to innovation and has the power to perceive users' needs and transform them into new ideas and products. In fact, the Industry can create new demands, which makes this a crucial sector for the success of wearables in the Inclusive Smart City. Other applications of wearables supporting PwD can be envisioned, able to open new research and investment pathways. Also, an Inclusive Smart City relies on a variety of computing and communication technologies that are critical for achieving accessibility objectives and to be developed by the Industry.
- **Academy:** Systematic research can lead to more useful and efficient products and processes, which is one of the reasons that make it crucial to count with the support of universities and research centers. Important development principles, such as Universal Design and Accessibility studies, derive from academic researches. Moreover, the Academy is responsible for forming professionals that will be in charge of understanding the use and of proposing new uses for wearables in the context of Inclusive Smart Cities. Another important point is that the academy can propitiate the convergence of results from different fields. Hence, the academy can contribute by providing a holistic view of the problem to be solved and how to structure and to evaluate solutions.
- **Community:** People with disabilities form one of the most organized and articulated communities. They have political representation, despite having few representatives, and are spported by several Non-Governmental Organizations (NGO). Since their demands are particularly appealing, they can count on the support of a broad segment of society. PwD can thus use the engagement of their community to press the Government and the Industry for innovative assistive technologies that will be able to increase their participation in the public life, especially in the urban space. Moreover, PwD can contribute to the Academy by providing real-life needs/problems as sources for academic research and can congregate teams in a participative way, often as solution developers.

However, the most important is the association of two or more of those stakeholders. Everyone wins, especially the user community, when they are able to collaborate to solve problems of PwD interaction with the urban space. Some initiatives are remarkable, such as the Government creating public investment calls to motivate collaboration between the Industry (and its professional) and the Academy (and its researchers and community) to develop assistive technology – including explicitly wearable devices – for Smart Cities projects, to make them Inclusive Smart Cities. This type of initiative is suitable and is already implemented in the City of São Paulo, Brazil.

FUTURE RESEARCH DIRECTIONS

Studies concerning the use of wearable devices by PwD in the urban space are still in the very beginning. Despite existing research and advances in wearables in the context of Smart Cities, there are a number of open research issues related to their use particularly in the context of Inclusive Smart Cities. The authors believe this lack of studies and researches can be seen as opportunities, as listed below:

- **New Materials:** Wearable devices must work under the non-controlled urban environment conditions because they can be exposed to rain, heat, pollution, and communication interference zones. Also, since this kind of devices can be in constant contact with the skin, they must prevent skin allergies and other kinds of injuries to the human body;
- **New Interaction Techniques:** Researchers must take advantage of the embedment of PwD to develop new interaction techniques that use touchpads over the human body, as well as the combination of sensors, such as touch intensity, vibration, pressure, and force sensors. This can lead to the use of new input/output interaction modalities. In addition, context-aware and gesture recognition applications can explore spatial and temporal awareness, which impact the relative and absolute position and orientation of a user. Moreover, PwD can also take advantage of wearable devices to access social applications, simulating a contact that is commonly strong in the physical world;
- **Expanding PwD Surroundings:** Certainly, there is a set of other assistive technologies and aids that can also be considered and integrated in a whole solution besides the wearable devices used by PwD. One example is the development of wearable devices for guide dogs that can assist visually impaired people more efficiently.
- **Adaptation to User's Body and Needs:** Disabilities are not well specified and not well framed. In other words, there is a large scale of blindness, as well as there is a wide range of motor impairments. Wearables must therefore

be flexible and adaptable to fit users' needs and shapes. Also, some diseases, such as diabetes, can cause loss of sensitivity in some parts of the body, such as the extremity of fingers and toes. Designers must be aware that this can happens and anticipate an alternative way to guarantee the interaction between user and wearable device;

- **Avoiding Imbalance in the Urban Space:** PwD always complain about suffering imbalance in the urban space, mainly because they have to take care of accessories, assistive technology and have to deal with the urban noise, the crowd and abnormal conditions, such as rain, animals, emergency situations and so on. Wearables could concentrate some functionalities that have so far been distributed in different accessories and assistive technology. Also, wearables could anticipate abnormal situations or help emergency teams to contact PwD;
- **Batteries and Power Supply:** As PwD will depend more and more on wearables to walk around the city; these devices must have a long lifespan;
- **Trustful Communication Channels:** As in the case of batteries, communication channels must be stable to guarantee the access to digital information and to services on demand;
- **IoT Devices Integration:** IoT platforms are still defining standards and intercommunication protocols. Hence, teams that work with wearable devices must be aware of the decisions that have been made, because they will affect the facility or not integrate their wearable solutions to other devices.

CONCLUSION

A new arena of wearable device research and industry is emerging: the use of this technology in the context of Inclusive Smart Cities. Entrepreneurs and researchers are likely to be attracted by conceiving and developing wearable solutions that will work as bridges between the physical and the digital layer of the urban space. In the Inclusive Smart City, physical objects commonly found in the city are connected to digital content and services that improve the quality of life of PwD as a whole.

In this chapter, the authors discuss the principles behind the development of wearable solutions and provide several examples of projects that use wearables in the context of Smart Cities and support PwD in their daily life tasks. Wearables are then associated to a variety of technologies as a crucial element for achieving the Inclusive Smart Cities. It is necessary to reinforce that besides the technology itself, a multidisciplinary team must be engaged in such projects counting on the participations of the PwD community.

We finished this chapter by discussing some future research directions, such as *new materials, new interaction techniques, expanding PwD surroundings, adaptation to user's body and needs, batteries and power supply, and integration of IoT devices.* We hope this chapter is an inspiration for researchers, professionals and PwD to join forces to the conception, development and evaluation of wearable solutions in a very innovative way to make our society more equal end more inclusive.

REFERENCES

Accessibility. (2015, December 5). In *Wikipedia, the free encyclopedia.* Retrieved from https://en.wikipedia.org/w/index.php?title=Accessibility&oldid=693827449

Baimbetov, Y., Khalil, I., Steinbauer, M., & Anderst-Kotsis, G. (2015). Using Big Data for Emotionally Intelligent Mobile Services through Multi-Modal Emotion Recognition. In *Inclusive Smart Cities and e-Health* (pp. 127–138). Springer. doi:10.1007/978-3-319-19312-0_11

Castro, A. G., Normann, I., Hois, J., & Kutz, O. (2008). Ontologizing Metadata for Assistive Technologies-The OASIS Repository. In *Ontologies in Interactive Systems, 2008. ONTORACT'08. First International Workshop on* (pp. 57–62). IEEE. Retrieved from http://ieeexplore.ieee.org/xpls/abs_all.jsp?arnumber=4756196

CDC. (n.d.). *Important Facts about Falls | Home and Recreational Safety | CDC Injury Center.* Retrieved May 9, 2016, from http://www.cdc.gov/HomeandRecreationalSafety/Falls/adultfalls.html

City. (2016, June 12). In *Wikipedia, the free encyclopedia.* Retrieved from https://en.wikipedia.org/w/index.php?title=City&oldid=724986747

de Oliveira Neto, J. S., & Kofuji, S. T. (2016). Inclusive Smart City: an exploratory study. In M. Antona & C. Stephanidis (Eds.), *Universal Access in Human-Computer Interaction. Access to Learning, Health and Well-Being.* Springer International Publishing. doi:10.1007/978-3-319-40244-4_44

Dunne, L. E., & Smyth, B. (2007). Psychophysical Elements of Wearability. In *Proceedings of the SIGCHI Conference on Human Factors in Computing Systems* (pp. 299–302). New York, NY: ACM. http://doi.org/ doi:10.1145/1240624.1240674

Fletcher, R. R., Dobson, K., Goodwin, M. S., Eydgahi, H., Wilder-Smith, O., Fernholz, D., & Picard, R. W. et al. (2010). iCalm: Wearable sensor and network architecture for wirelessly communicating and logging autonomic activity. *Information Technology in Biomedicine. IEEE Transactions on, 14*(2), 215–223.

Foth, M., Choi, J. H., & Satchell, C. (2011). Urban Informatics. In *Proceedings of the ACM 2011 Conference on Computer Supported Cooperative Work* (pp. 1–8). New York, NY: ACM. http://doi.org/ doi:<ALIGNMENT.qj></ALIGNMENT>10.1145/1958824.1958826

Fraile, J. A., Bajo, J., Corchado, J. M., & Abraham, A. (2010). Applying wearable solutions in dependent environments. *Information Technology in Biomedicine. IEEE Transactions on, 14*(6), 1459–1467.

Gemperle, F., Kasabach, C., Stivoric, J., Bauer, M., & Martin, R. (1998). Design for wearability. In *Second International Symposium on Wearable Computers, 1998. Digest of Papers* (pp. 116–122). doi:10.1109/ISWC.1998.729537

Gonzalez, M. C., Hidalgo, C. A., & Barabasi, A.-L. (2008). Understanding individual human mobility patterns. *Nature, 453*(7196), 779–782. doi:10.1038/nature06958 PMID:18528393

Hanai, Y., Nishimura, J., & Kuroda, T. (2009). Haar-like filtering for human activity recognition using 3d accelerometer. In *Digital Signal Processing Workshop and 5th IEEE Signal Processing Education Workshop, 2009. DSP/SPE 2009. IEEE 13th* (pp. 675–678). IEEE. Retrieved from http://ieeexplore.ieee.org/xpls/abs_all.jsp?arnumber=4786008

He, Z.-Y., & Jin, L.-W. (2008). Activity recognition from acceleration data using AR model representation and SVM. In *Machine Learning and Cybernetics, 2008 International Conference on* (Vol. 4, pp. 2245–2250). IEEE. Retrieved from http://ieeexplore.ieee.org/xpls/abs_all.jsp?arnumber=4620779

Head, H., & Holmes, G. (1911). Sensory Disturbances from Cerebral Lesions. *Brain, 34*(2-3), 102–254. doi:10.1093/brain/34.2-3.102

Jacobs, A. (2009). The pathologies of big data. *Communications of the ACM, 52*(8), 36–44. doi:10.1145/1536616.1536632

Jiang, P., Winkley, J., Zhao, C., Munnoch, R., Min, G., & Yang, L. T. (2014). *An intelligent information forwarder for healthcare big data systems with distributed wearable sensors*. Retrieved from http://ieeexplore.ieee.org/xpls/abs_all.jsp?arnumber=6775278

Kaldeli, E., Warriach, E. U., Lazovik, A., & Aiello, M. (2013). Coordinating the web of services for a smart home. *ACM Transactions on the Web, 7*(2), 10. doi:10.1145/2460383.2460389

Klingeberg, T., & Schilling, M. (2012). Mobile wearable device for long term monitoring of vital signs. *Computer Methods and Programs in Biomedicine, 106*(2), 89–96. doi:10.1016/j.cmpb.2011.12.009 PMID:22285459

López-de-Ipiña, D., Lorido, T., & López, U. (2011). Blindshopping: enabling accessible shopping for visually impaired people through mobile technologies. In *Toward Useful Services for Elderly and People with Disabilities* (pp. 266–270). Springer. Retrieved from http://link-springer-com.ez67.periodicos.capes.gov.br/chapter/10.1007/978-3-642-21535-3_39

Martin, W., Dancer, K., Rock, K., Zeleny, C., & Yelamarthi, K. (2009). The smart cane: an electrical engineering design project. In *ASEE North Central Section Conference*. Retrieved from http://people.cst.cmich.edu/yelam1k/CASE/Publications_files/Yelamarthi_ASEE_NCS_2009.pdf

Mazilu, S., Hardegger, M., Zhu, Z., Roggen, D., Troster, G., Plotnik, M., & Hausdorff, J. M. (2012). Online detection of freezing of gait with smartphones and machine learning techniques. In *Pervasive Computing Technologies for Healthcare (PervasiveHealth), 2012 6th International Conference on* (pp. 123–130). IEEE. Retrieved from http://ieeexplore.ieee.org/xpls/abs_all.jsp?arnumber=6240371

McCann, J., Hurford, R., & Martin, A. (2005). A design process for the development of innovative smart clothing that addresses end-user needs from technical, functional, aesthetic and cultural view points. In *Ninth IEEE International Symposium on Wearable Computers (ISWC'05)* (pp. 70–77). http://doi.org/ doi:10.1109/ISWC.2005.3

Mezghani, E., Exposito, E., Drira, K., Da Silveira, M., & Pruski, C. (2015). A Semantic Big Data Platform for Integrating Heterogeneous Wearable Data in Healthcare. *Journal of Medical Systems, 39*(12), 1–8. doi:10.1007/s10916-015-0344-x PMID:26490143

Moustafa, H., Kenn, H., Sayrafian, K., Scanlon, W., & Zhang, Y. (2015). Mobile wearable communications [Guest Editorial]. *IEEE Wireless Communications, 22*(1), 10–11. doi:10.1109/MWC.2015.7054713

Murphy, K. P. (2012). *Machine learning: a probabilistic perspective*. MIT Press. Retrieved from https://books.google.com.br/books?hl=pt-BR&lr=&id=NZP6AQAAQBAJ&oi=fnd&pg=PR7&dq=Machine+Learning+A+Probabilistic+Perspective&ots=KQTfz1Blgu&sig=IU4rkH5LF1lnicHaA1Ta0kvYccM

Özdemir, A. T., & Barshan, B. (2014). Detecting falls with wearable sensors using machine learning techniques. *Sensors (Basel, Switzerland)*, *14*(6), 10691–10708. doi:10.3390/s140610691 PMID:24945676

Palma, R., Hartmann, J., & Haase, P. (2008). *Ontology Metadata Vocabulary for the Semantic Web*. OMV Consortium.

Paris, D. G., & Miller, K. R. (2016). Wearables and People with Disabilities: Socio-Cultural and Vocational Implications. *Wearable Technology and Mobile Innovations for Next-Generation Education*, 167.

Pérez-Martínez, P. A., Martínez-Ballesté, A., & Solanas, A. (2013). *Privacy in Smart Cities-A Case Study of Smart Public Parking* (pp. 55–59). PECCS.

Rawassizadeh, R., Price, B. A., & Petre, M. (2014). Wearables: Has the Age of Smartwatches Finally Arrived? *Communications of the ACM*, *58*(1), 45–47. doi:10.1145/2629633

Reades, J., Calabrese, F., Sevtsuk, A., & Ratti, C. (2007). Cellular census: Explorations in urban data collection. *IEEE Pervasive Computing / IEEE Computer Society [and] IEEE Communications Society*, *6*(3), 30–38. doi:10.1109/MPRV.2007.53

Resch, B. (2013). People as sensors and collective sensing-contextual observations complementing geo-sensor network measurements. In *Progress in Location-Based Services* (pp. 391–406). Springer. doi:10.1007/978-3-642-34203-5_22

Sagl, G., Resch, B., Hawelka, B., & Beinat, E. (2012). From social sensor data to collective human behaviour patterns: Analysing and visualising spatio-temporal dynamics in urban environments. In *Proceedings of the GI-Forum* (pp. 54–63). Retrieved from http://gispoint.de/fileadmin/user_upload/paper_gis_open/537521043.pdf

Solanas, A., Patsakis, C., Conti, M., Vlachos, I., Ramos, V., Falcone, F., … Perrea, D. (2014). Smart health: A context-aware health paradigm within smart cities. *Communications Magazine, IEEE, 52*(8), 74–81.

Streitz, N. A. (2011). Smart cities, ambient intelligence and universal access. In *Universal Access in Human-Computer Interaction. Context Diversity* (pp. 425–432). Springer. Retrieved from http://link.springer.com/10.1007%2F978-3-642-21666-4_47

United Nations. (2006). *Convention on the rights of persons with disabilities.* Retrieved February 25, 2016, from http://www.un.org/disabilities/convention/conventionfull.shtml

Weiser, M. (1999). The Computer for the 21st Century. *SIGMOBILE Mob. Comput. Commun. Rev.*, *3*(3), 3–11. doi:10.1145/329124.329126

Yamamoto, T., Terada, T., & Tsukamoto, M. (2011). Designing Gestures for Hands and Feet in Daily Life. In *Proceedings of the 9th International Conference on Advances in Mobile Computing and Multimedia* (pp. 285–288). New York, NY: ACM. doi:10.1145/2095697.2095757

Yin, J., Yang, Q., & Pan, J. J. (2008). Sensor-based abnormal human-activity detection. *Knowledge and Data Engineering. IEEE Transactions on*, *20*(8), 1082–1090.

Chapter 4
Wearables Operating Systems:
A Comparison Based on Relevant Constraints

Vicente J. Peixoto Amorim
Federal University of Ouro Preto, Brazil

Saul Delabrida
Federal University of Ouro Preto, Brazil

Ricardo A. R. Oliveira
Federal University of Ouro Preto, Brazil

ABSTRACT

Wearable devices have increasingly become popular in recent years. Devices attached to users body remotely monitor his daily activities/health. Although some of these devices are pretty simple, others make use of an operating system to manage memory, resources, tasks, and any user interaction. Some of them were not initially designed and developed for this purpose, having a poor performance requiring the use of more resources or better hardware. This chapter presents a characterization of wearable devices considering the operating systems area. Some constraints of this context were designed to analyze the operating system's execution when inserted into a wearable device. Data presented at the end shows that there is a lack of performance in specific areas, letting to conclude that improvements should be made.

DOI: 10.4018/978-1-5225-3290-3.ch004

INTRODUCTION

Nowadays, wearable computing has become one of the most trending topics on market and research areas. The possibility to have a computer attached to the body gives the user a superior level of context information inside an interactive environment. Although wearable computing is not a new concept (Sutherland, 1968; Mann, 1996; Mann, 1997), a recent advance in electronic components miniaturization has supported this type of device evolution. Watches (smartwatches), fitness trackers, glasses, augmented reality (AR) / virtual reality (VR) equipment and others have been increasingly gaining space on industry and user daily activities.

Regarding the hardware scope, the wearable devices can vary from simple ones to powerful machines fully integrated to the users' needs, or even connected to other local equipment/services, such as smartphones and tablets. Some of them are just vital signals monitors while others may have a visual interactive display where a real-time processing is made and feedback is provided, letting the user be immersed in a virtual environment.

Despite the number of different wearable devices on the market today, until now, there is no operating system (OS) used as a reference for this environment. Most of the existent proposals start from a desktop or mobile OS distribution adapting it to deploy on a final wearable product. Current solutions that can be applied to wearables vary from real-time approaches (L. Foundation Zephyr, 2016; Pebble, 2016) to mobile devices operating systems adaptations (Google Brillo, 2016; Google Android, 2016). There are also specific proposals aiming this environment (Apple, 2016; Samsung, 2016) although they were not strictly developed for this purpose.

This chapter presents a characterization between different types of wearable devices, separating them according to its main functionalities and applicability. The characterization provided here helps to understand the difference between wearable devices and Internet of Things (IoT) devices. In addition to that, a comparison between operating systems applied on wearable devices is also made. The main contributions depicted here are a better comprehension of wearables environment organization and operating systems needs, besides the evaluation of current solutions on the market. Operating systems selected and listed here were specifically designed to wearable computing or have a close relationship with this context. In the end, tests results figured out, considering peculiar environment constraints, which is the operating system that closes reach the wearables environment requirements. Furthermore, the authors expect that this discussion raises the understanding about this environment essential characteristics as a whole, once hardware and software aspects were covered.

The work presented here is based on an earlier work: A Constraint-Driven Assessment of Operating Systems for Wearable Devices (Amorim, Delabrida, & Oliveira, 2016).

RELATED WORKS

Operating systems and wearable devices can be analyzed in a separate way: the first one concerns about software specificities that control and manage the attached hardware. The second one is related to the context where the solution is applied. In this way, evaluate operating systems performance may be considered separately to the wearables context. However, it will lose precision, once the main area of applicability is not considered. Here the authors depict main related works that evaluate operating systems used by different fields. Conversely, in a second part, particularities related to wearable computing are also analyzed as a way to understand the main desired features when using an operating system inside this context.

A simple conclusion can be made when investigating these related works: There is no solution covering both worlds, and it can impact the selection of a right or wrong operating system running on a wearable device/solution. This chapter exposes the main relationship between these two areas while provides empirical data results to sustain the results.

Operating Systems Performance

When considering operating systems context as a whole, several works evaluate its performance. From desktops to real-time devices, each operating system is considered according to the environment in which it is inserted.

Work depicted by (Acquaviva, Benini, & Riccó, 2001) presents an assessment methodology for real-time operating systems inside a wearable device. Some energy consumption key parameters were raised. However, no comparison was made against other operating system taking wearables requirements into account. Similarly, (Cho, & Lee, 2010; Cho, Lim, & Lee, 2009) describes the development of a low power Real-Time Operating System (RTOS) focused on wearable devices. Despite the good memory footprint results, proposed solution does not take another relevant metric into account. In fact, even energy consumption was directly compared to the competitors.

In (Tan, & Nguyen, 2009) the authors evaluate a specific operating systems subset. This considered subset encloses real-time operating systems (RTOS) running on small microcontrollers. A well-done contextualization was made, making clear the applicability of an RTOS in comparison to a conventional operating system. Although it presents a functional evaluation between considered operating systems, no relationship was established with the wearable computing environment.

Wearable Computing Assessment

Commonly, conventional computing systems (e.g. desktop computers) do not have resources restrictions at the same level as the embedded world. As one solution gains in mobility, it also increases its resources dependability and constraints. Mobile and wearable computers are usually categorized in this class, with heavy restrictions regarding energy availability and use, processing power, and so on. Indeed, the wearables environment is even more restricted when compared to the mobile scenario. These limitations led to studies that try to figure out how to increase devices autonomy, maintaining or increasing its processing power.

Works depicted by (Benini, Bruni, Macii, Macii, & Poncino, 2003) and (Shi, Yang, Huang, & Hui, 2015) focus on energy optimization. The first one presents a computational offloading guideline aimed at an AR application on a wearable device. The second one outlines a new battery lifetime maximization policy taking the wearables environment into account. Despite the fact that both proposals have a good performance, these works have been focused only on energy autonomy maximization with no performance conclusion being made regarding operating systems part.

Conversely, (Park, & Jayaraman, 2003) raise challenges and solutions for wearable devices inside medicine area. A set of performance metrics is depicted considering a smart shirt system. Although it is valid, no conclusion is made regarding how these metrics can be related to operating systems installed on wearable devices.

The research outlined in (Liu, & Lin, 2016) analyzes Android Wear operating system (Google Android, 2016, September). A functional characterization is presented considering variables such as CPU usage, thread-level parallelism, and general behavior, leading to conclude that this OS has some inefficiencies. Despite its Android Wear in-depth analysis, it still limited to just one operating system, with no comparison to other related wearable operating system, what creates a data lack once there is no solution to be used as a reference.

WEARABLE DEVICES CONTEXT

Bellow is described the most important aspects related to wearable devices, taking into account a classification regarding its features set.

Wearable computing was initially defined as a head-mounted display that provides virtual information to the user in a real world (Sutherland, 1968). Evolutions on integrated circuits allowed wearable computing to be applied to several other applications and areas (Merkouris, & Chorianopoulos, 2015; Nassani, Bai, Lee, & Billinghurst, 2015; Ye, Malu, Oh, & Findlater, 2014). Indeed, to support many different applications, wearable devices received complex hardware assistance that, most of the time, requires a significant processing power.

On the software side, operating systems are now being used to provide a more efficient and robust control and management of resources. Commonly, operating systems attached to a wearable equipment does not have all the features as in its desktop version. As presented by (Delabrida, D'Angelo, Oliveira, & Loureiro, 2016), a wearable operating system must optimize specific variables, such as efficient process scheduling and energy use. Additionally, a wearable device that provides a visual user interface must consider optimizations on graphics processing. Conversely, a device without a visual user interface/display may focus on sensors input/output performance.

Main Characterization

This section aims to classify the wearable devices into pre-defined classes according to its functionalities and hardware requirements. Nowadays, the market has a significant number of products of this class used in different areas. There are devices to monitor fitness activities, overall sports, health, gaming, rescue, activity recognition and so on. Although each equipment may have its requirements and features, many hardware solutions share a common internal organization and architecture. Extending the operating systems characterization presented in (Delabrida, D'Angelo, Oliveira, & Loureiro, 2016), wearable devices may also be classified according to its hardware components and performed functions.

- **User Interface (UI) Dependent:** Devices commonly high dependent on graphical interfaces operations, providing high-quality visual feedback to the end-users. Focused on a visual display to perform its core tasks. As an instance, AR/VR solutions require enriched graphical layers and sometimes 3D objects drawing (Microsoft, 2016, September; Google Glass, 2016, September);
- **Input/Output (I/O) Dependent:** Devices focused on I/O operations. Sensed data commonly must be retrieved and delivered to a cloud server considering real-time constraints. Indeed, sensors attached to the hardware may require a good data transfer throughput and low latency. This fact lets these devices be appropriated to create Internet of Things (IoT) solutions, once its focus is on

user/environment sensing, providing a quick response to condition changes. Examples are devices like remote activity monitors (Nguyen, ChenLuo, Yeo, & Duh, 2011) and context awareness solutions (Randell, & Muller, 2000);
- **Hybrid:** Devices that commonly take into account aspects from both categories above mentioned. While it considers a minimum level of visual feedback, it also senses the environment in a way to provide end-user feedbacks or even send gathered data to a remote server. Smartwatches (Pebble, 2016, September) and fitness trackers (Fitbit, 2016, September) are in some way examples of this type. It is expected that an equipment characterized inside this category should equally give priority to user interface and input/output operations, once they commonly retrieve and process user bio-data, periodically providing a visual feedback through small displays.

Figure 1 depicts presented classes visual organization. In fact, instead of a discrete definition, wearable devices are closer to have a continuous characterization along the three grades presented above.

The first of three classes presented above can be extended to a more detailed description. It is possible because a visual interface provides more or less interaction with the users, forcing the hardware requirements to be increased or decreased. For its turn, Input/Output (I/O) requirements commonly focus on sensor's data gathering/monitoring and transmission, a fact that frequently does not require much user interaction. However, both worlds when contextualized with energy requirements may use a high amount of power to sustain its functionalities: To power a rich colored display or even to maintain the sensor's data up-to-date on a cloud server. In this way, taking the interactivity level into account, a more accurate characterization can be given considering the following layers/levels, from the most basic one (Level 0) to the most complex one (Level 4):

- **Layer 0:** Devices that does not have a visual display and are mainly focused on users' data gathering and transmission. Health monitors and fitness trackers

Figure 1. Wearable devices distribution according to performed activities and dependency level

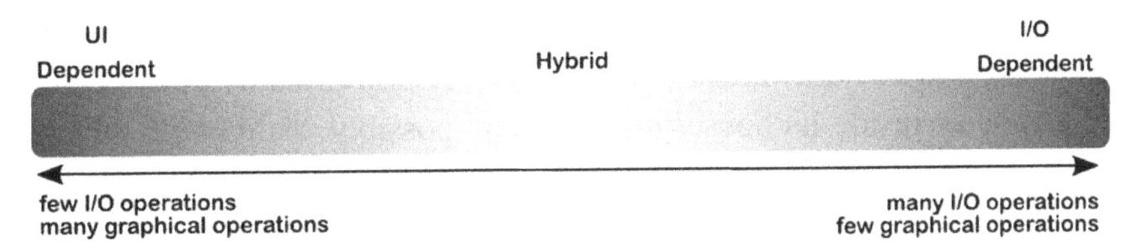

are included in this category that is the closest point to simple Internet of Things (IoT) devices;

- **Layer 1:** Wearable equipment that uses a simple display to show a layer of text and 2D graphics (squares, circles, icons, and pre-processed images) drawn on the screen to represent physical places or interesting points;
- **Layer 2:** Wearable equipment with 3D-rendering capabilities that may (or may not) be able to process user interaction. 3D graphics layer can be used by Augmented Reality (AR) and Virtual Reality (VR) applications;
- **Layer 3:** Addition of interaction to the features provided by Layer 2. Users can now manipulate graphics on screen overlaying the real world. A certain level of human interaction must be considered to build better applications for this environment;
- **Layer 4:** Use of pattern recognition, machine learning and other AI techniques to extract and recognize real world objects. The GPU hardware may help graphics processing task to reach real-time execution.

Figure 2 presents a better organization of the five layers presented above. It is imperative to note that, the top layers may reuse functionalities from bottom layers, creating an extensible concept to be used from simple wearable devices to complex ones.

WEARABLES OPERATING SYSTEMS

The performance of a wearable device is directly related to the software it runs. Here the authors present a description of main operating systems that eventually can be used by some wearable solution. Furthermore, it is valid to highlight that some solutions will not even require an operating system to manage its resources.

Wearable vs. IoT?

IoT and wearables solutions are emerging as new top trending technologies. The advent of ubiquity, context-aware technologies, and mobile devices has increased personal-centered applications development. Today, these solutions are focusing on user micro-contexts, from the own user, its home or any other surrounding places that he/she stays in contact. As a consequence, these contexts knowledge are gathered by more hardware sensors and locally processed by devices attached to user's body or inside its house/office/car and possibly shared with other users through a common cloud solution.

Figure 2. Wearable devices layers organization according to its features

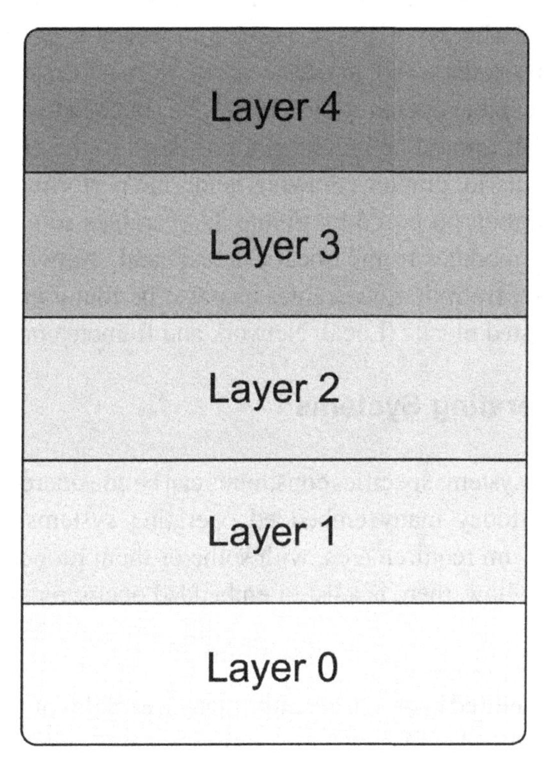

Until now, wearables and IoT concepts coverage were defined in different ways, sometimes being listed as distinct groups, sometimes as nested subsets. There are solutions (Hassanalieragh et. al., 2015) using wearables and IoT notions as complementary technologies, keeping the first one as an IoT subset. Conversely, other characterizations are classifying IoT solutions from the network perspective, once it is acting as wearable devices communication layer (Arias, Wurm, Hoang, & Jin, 2015; Azariadi, Tsoutsouras, Xydis, & Soudris, 2016).

Regardless these previous definitions, this research tends to accept the definition presented in (Hiremath, Yang, & Mankodiya, 2014) as the most appropriated one when taking operating systems into account. At that work, authors proposed a concept called "WIoT (Wearable Internet of Things)" making the intersection between both areas discussed here. Indeed, this research concludes that an extension of this paradigm is possible, once it is very common to have a network infrastructure independent wearable solution. This type of wearable device only retrieves and process obtained data, without any communication with a remote entity, once generated information is locally consumed.

Associating this latest context with existing operating systems helps to define what is a wearable OS and what is not. In some situations wearable operating systems can work retrieving sensors data, and, in that scenario, I/O performance is essential. On the other hand, a wearable operating system may be attached to a hardware solution where processed data should be graphically presented to the end-user using local displays. In that scenario, graphics drawing/rendering performance is capital.

Given the separation proposed by Figure 3, wearables solutions can be found implemented using modules from "Local Context" and "Network Context" with a limited range. Apart from that, wearables can also be found in IoT solutions that enclose the three listed blocks (Local, Network and Remote contexts).

Embedded Operating Systems

Wearable operating systems specifics constraints can be considered as the assessment metrics. However, today many embedded operating systems provide wearable environment minimum requirements, with some of them proposed specifically to these conditions. Bellow, there is a list of embedded operating systems considered by this work:

1. **Generic Embedded Linux:** Currently, many wearables products have focused on a generic embedded Linux as a solution. Despite its flexibility, embedded Linux have a generic build for a broad range of embedded devices. From the market perspective, build tools like Yocto (L. Foundation Yocto, 2016, September) and Buildroot (Buildroot, 2016, September) can ease the task of creating an embedded Linux distribution, allowing to customize system software components. Although it can reduce developing time, there is no wearable environment;

Figure 3. Contexts considered by wearables and IoT applications

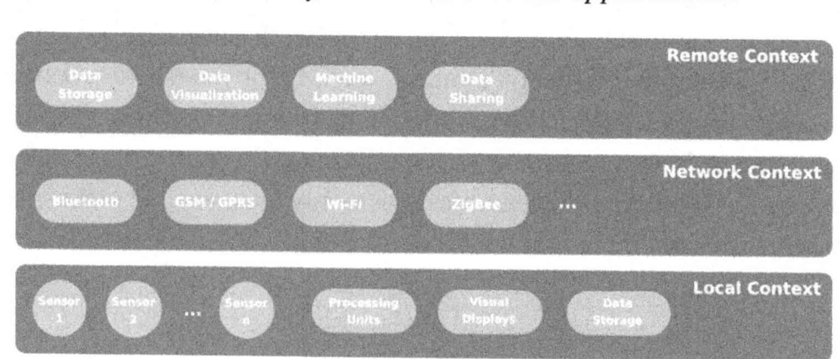

2. **Samsung Tizen:** Samsung's open source operating system encloses mobile, wearable, and In-Vehicle-Infotainment (IVI) devices (Samsung, 2016, September). The wearable version was specifically developed for this purpose and currently is embedded on smartwatches company's version. Furthermore, Tizen's latest release (3.0) delivers a common software base between all the environments. However, until now, there is no real-time constraints specification or more details regarding wearables context-specific variables that were optimized for this OS;
3. **Google Android Wear:** Created to support smartwatches devices, this Google's open-source operating system (Google Android, 2016) provides a development environment to build specific applications. Despite its wearable characteristics, most of Android Wear features are anchored in an interface with Android/iOS smartphones. Commonly, its operating system is used as a software base taken by other manufacturers to customize its devices and final products;
4. **Pebble OS:** Pebble OS is the Pebble smartwatches operating system, specially developed for wearable devices. It is based on FreeRTOS operating system (R. T. Engineers, 2016, September), and currently is being integrated with other mobile platforms, such as Android and iOS. Unlike their competitors, Pebble OS does not have an open-source license on its core parts;
5. **Linux Foundation Zephyr:** Designed for IoT environment, Zephyr (L. Foundation Zephyr, 2016, September) operating system has a small footprint, according to Linux Foundation. It has an open-source code and was developed as a real-time operating system (RTOS), providing a network stack, device management capabilities and support to hardware sensors. Besides nothing was said about its feasibility on wearable devices, Zephyr was taken into account by this work owing to its applicability on an environment with low resources;
6. **Google Brillo:** Developed for Internet of Things (IoT) area, Google's operating system is portable to many hardware architectures, such as ARM, Intel-x86 and MIPS-based (Google Brillo, 2016). Moreover, Google Brillo can support a communications platform specially developed for IoT, named Weave. Although its applicability in IoT area, Brillo may also be considered for wearable devices, due to both environments common characteristic, the energy consumption restriction;
7. **RIOT OS:** Focused on Internet of Things (IoT) solutions, RIOT OS (Riot OS, 2016, September) development was started in 2008. Provides an open-source operating system with real-time support and low memory footprint (RAM / ROM), running on most common architectures and boards in the market. Currently, it has native Linux and Mac OS X ports that aim to help during the development phase.

ASSESSMENT CONSIDERATIONS

This section explains operating systems selection step, the criteria and how their assessment was made. The work conducted by this research was separated into four sequential steps: Operating systems selection, environment configuration, test applications development and evaluation.

Operating Systems Selection

The following three operating systems, presented in subsection III-C were chosen to have a detailed evaluation considering its provided functionalities and the wearable computing requirements:

- **Generic Embedded Linux:** Selected due to its customization capabilities. For this research, Buildroot (Buildroot, 2016, September) tool was taken into account to generate a customized embedded Linux version. Only strictly necessary software modules were selected to be inside the kernel and root filesystem. For graphics side, Direct Frame Buffer library (version 1.7.1) (Direct Frame Buffer, 2016, September) was selected once it is a lightweight solution that provides low overhead on drawings;
- **Samsung's Tizen:** Provides a specific solution for the wearables environment and currently is embedded on market products, such as Samsung Gear S2 smartwatch. Here, tests were made using Tizen Wearable operating system version 2.3.1; and
- **Google's Android Wear:** Selected due to the fact it is a wearable environment specific solution. Wear version taken into account was based on Android 5.1 API 22.

Operating systems other than those listed above were not deeply examined due to particular reasons, such as limitations on functionalities provided by Software Development Kit (SDK), lack of an SDK (or its documentation), low images maximum resolution, and a limited number of colors. For instance, Pebble OS has limitations on images number of colors in addition to restrictions regarding files size in persistent storage.

Environment Configuration

Selected operating systems were configured inside three similar virtual QEMU machines with identical hardware arrangements, running on the same host. For every platform, no hardware acceleration or intervention was allowed, in a way that even graphics processing was made through software. Furthermore, network interfaces (Wi-fi, Bluetooth, and NFC) and cameras (front and rear) were disabled as well as any emulated sensor, such as gyroscope, accelerometer, proximity, heartbeat, and GPS. Indeed, all the virtual machines have been configured with an x86 architecture single core processor. The remaining configurations were: QEMU Virtual CPU version 2.0.0 with 2.4GHz, 4096KB of cache memory and 1024MB of RAM memory.

Test Applications Development

To properly evaluate selected operating systems, the authors have built four applications that figure out which is the best operating system for wearables environment, considering the classification depicted by section III. As a result, developed applications focuses on I/O and graphics drawing performance. Except for Embedded Linux operating system, which has used Direct Frame Buffer Library and GNU C Library (glibc) (GNU, 2016, September), SDK functions provided by the manufacturers were taken into account. Indeed, these four applications were re-implemented respecting each platform particularities: I/O latency: A test application considering I/O operations to the internal filesystem. The authors evaluate the time spent to write 30MB of raw data, separated in 512bytes-size packets, to a file; I/O throughput: Also consider I/O operations to the internal filesystem. The time spent to write 30MB of raw data, in a single packet, to a file was also evaluated; Image rendering time: Render time was verified through an application that draws an image to the display. Same 800x600 pixels image was considered for every platform; and Graphics drawing: Aims to check operating system performance when drawing vector figures. We evaluate the time needed to draw serially on the screen 200 squares of the same size.

I/O test applications were considered as a way to assess the operating system behavior when dealing with external components/ devices (e.g. wearable sensors or network interfaces). Data size (30MB) was estimated by this work as the worst case for a transfer. Conversely, graphics test applications aim to check OS behavior when drawing an image to the screen (e.g. AR/VR devices or smartwatches). Moreover, applications performance was measured through execution time calculation, considering only instructions and function calls associated with the related functionality. For instance, only routines related to drawing preparation and rendering were taken into account when evaluating "Image rendering time".

Evaluation

Resulting data was taken considering that every application was executed 30 times on each operating system. Applications were started/stopped between two subsequent executions to ensure that there was no OS optimization/cache. Figure 4 shows three snapshots of Tizen, Android Wear and Embedded Linux machines running the test application that draws an image on the screen.

RESULTS

This section present results obtained from test applications running on specific operating systems. Moreover, we also present results obtained from test applications running on specific operating systems.

Table 1 shows a comparison between all operating systems. It is possible to verify that, in most of the cases, an operating system has focus IoT area or in wearables area. In addition to that fact, the majority of operating systems are open-source and have an SDK available to build its applications. Still, Table 2 summarizes the results presented in next subsections regarding operating systems functionalities evaluation. This table also displays the time spent to execute each test application, with average

Figure 4. Test applications snapshot

Table 1. Operating Systems Classification Based on a Characteristics Set

OS / Characteristic	Graphical Support	User Interface	IoT Focus	Wearable Focus	RTOS	Open Source	SDK Available
Samsung Tizen OS	X	X		X		X	X
Linux Found. Zephyr OS			X		X	X	X
Google Brillo OS			X			X	
Embedded Linux	X	X				X	X
Pebble OS	X	X		X	X		X
Google Android Wear OS	X	X		X		X	X
Riot OS	X	X	X		X	X	X

Table 2. Performance results regarding tests executed on each operating system

OS / Functional Test	Latency		Throughput		Image Rendering		Graphics Drawing	
	Average (µs)	Std. Dev.	Average (µs)	Std. Dev.	Average (µs)	Std. Dev.	Average (µs)	Std. Dev.
Samsung Tizen OS	$111{,}69 \times 10^5$	$16{,}24 \times 10^5$	$11{,}18 \times 10^5$	$0{,}26 \times 10^5$	$1{,}26 \times 10^5$	$0{,}09 \times 10^5$	$0{,}01 \times 10^5$	$0{,}005 \times 10^5$
Embedded Linux	$42{,}21 \times 10^5$	$1{,}23 \times 10^5$	$3{,}02 \times 10^5$	$0{,}05 \times 10^5$	$11{,}55 \times 10^5$	$0{,}24 \times 10^5$	$0{,}44 \times 10^5$	$0{,}05 \times 10^5$
Google Android Wear OS	$16{,}05 \times 10^5$	$1{,}41 \times 10^5$	$2{,}50 \times 10^5$	$0{,}34 \times 10^5$	$1{,}43 \times 10^5$	$0{,}14 \times 10^5$	$0{,}02 \times 10^5$	$0{,}01 \times 10^5$

execution time values listed in microseconds (µs) and the standard deviation values considering a 95% confidence interval.

I/O Latency

Figure 5 shows the average latency graphical analysis. According to obtained results, Android Wear has the lowest latency between the three operating systems, followed by the Linux Embedded solution. Tizen OS reached the highest values for latency, with its average value greater than four times the value obtained during Android Wear experiments. Still, Tizen error rate is the highest one, letting us conclude that it also have a high jitter on this type of I/O.

Figure 5. Measured average latency when running test applications on selected operating systems

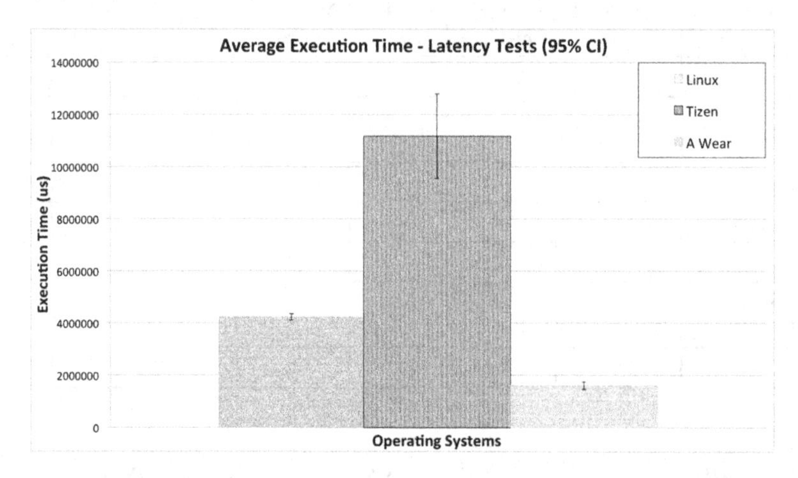

I/O Throughput

Figure 6 outlines the average throughput results measured in each operating system. Here, a 95% confidence level was also taken into account. As happened in latency tests, Tizen OS got the worst performance, with Android Wear and Embedded Linux achieving best results.

Figure 6. Measured average throughput when running test applications on selected operating systems

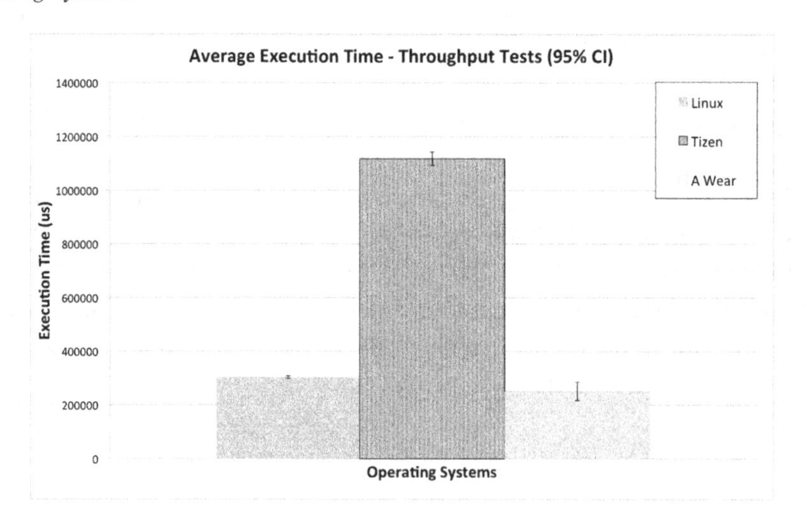

Image Rendering Time

Unlike previous results, Figure 7 shows that the average time spent to draw an image on the screen was statically equaled on Tizen and Android Wear operating systems. However, Embedded Linux with Direct Frame Buffer library has not performed well as in the other evaluations.

Graphics Drawing

Figure 8 shows the results regarding the graphics drawing tests execution. Embedded Linux got the worst average time, while Tizen and Android Wear presented statistically equivalent results. Graphics tests results may lead to conclude that Tizen and Android Wear have been optimized to prioritize tasks that draw on the device display.

Size on Storage Memory

Here we have measured the binaries size generated by each test application on every operating system. Table 3 outlines each test application size on storage memory. It is possible to verify that Embedded Linux has better performance, providing a smaller application binary in all the considered cases. Still, application packets generated by Android Wear requires 60 times more storage space than Tizen application binaries.

Figure 7. Measured average time to render an image using selected operating systems

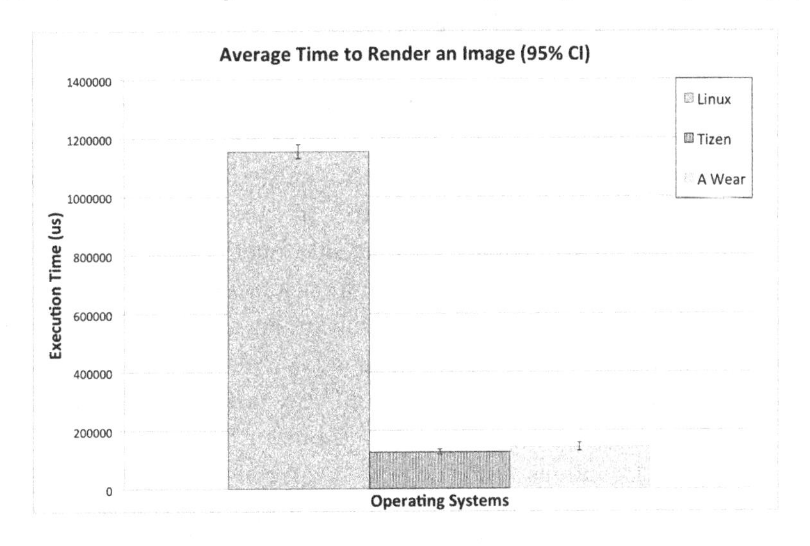

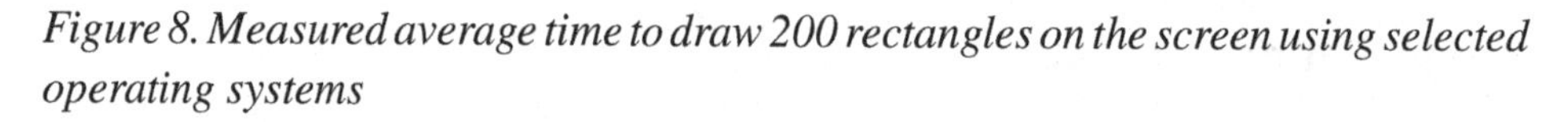

Figure 8. Measured average time to draw 200 rectangles on the screen using selected operating systems

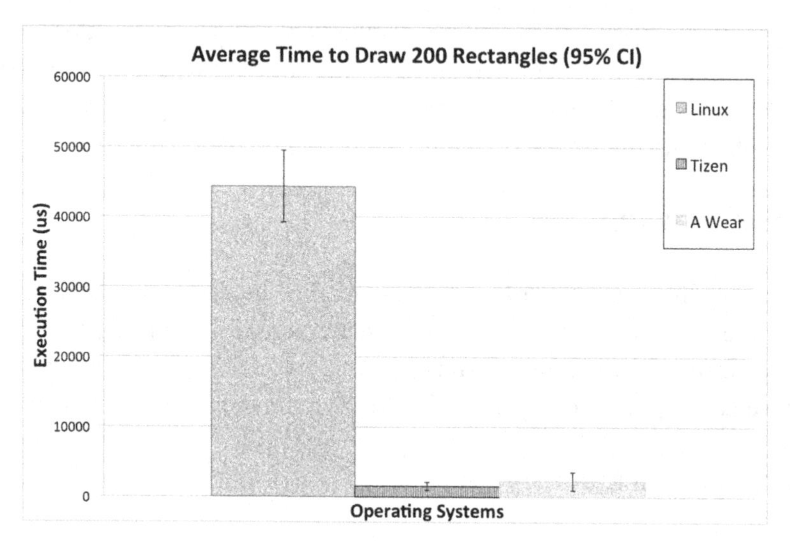

Table 3. Test applications binary size (in bytes) on storage memory

	Embedded Linux	Tizen	Android Wear
I/O Latency	6216	31959	1955908
I/O Throughput	6284	32013	1955905
Image Rendering	6124	31629	1981663
Graphics Rendering	6228	31432	1972293

CONCLUSION AND FUTURE WORKS

Wearable devices are increasingly becoming a hot topic on the market, thanks to recent evolutions on hardware miniaturization. From health monitors to smart glasses, affordable devices can be found in stores. Despite its popularity, the separation between wearable computing and IoT concepts still unclear, leading to different interpretations by the authors. On the software side, operating systems specifically proposed for wearables environment are commonly focused on just one segment of products, such as smartwatches. An operating system needs characterization can help developers to enhance final applications and deliver a higher quality product.

This work presented a functional comparison between different operating systems that can be used inside wearables context. Results present here allow concluding that none of the evaluated solutions performed well in all considered constraints. While

Embedded Linux has an excellent performance on I/O, Samsung Tizen and Google Android Wear obtained good results on graphics rendering tests. This information also points out where optimizations can be made on these platforms. As future works, a closer evaluation of RTOS operating systems should be done to figure out its feasibility to be applied on a generic wearable device. Moreover, a common hardware board may be used as a reference to provide results in a real environment.

REFERENCES

Acquaviva, A., Benini, L., & Riccó, B. (2001, December). Energy characterization of embedded real-time operating systems. *SIGARCH Comput. Archit. News*, *29*(5), 13–18. doi:10.1145/563647.563652

Amorim, V. J. P., Delabrida, S., & Oliveira, R. A. (2016). A Constraint-Driven Assessment of Operating Systems for Wearable Devices. *Proceedings of the 6th Brazilian Symposium on Computing Systems Engineering (SBESC)*, 150-155. doi:10.1109/SBESC.2016.030

Apple. (2016, September). *Watchos*. Retrieved September 19, 2016, from https://developer.apple.com/watchos/

Arias, O., Wurm, J., Hoang, K., & Jin, Y. (2015, April). Privacy and security in internet of things and wearable devices. *IEEE Transactions on Multi-Scale Computing Systems*, *1*(2), 99–109. doi:10.1109/TMSCS.2015.2498605

Azariadi, D., Tsoutsouras, V., Xydis, S., & Soudris, D. (2016). Ecg signal analysis and arrhythmia detection on iot wearable medical devices. *5th International Conference on Modern Circuits and Systems Technologies (MOCAST)*, 1–4. doi:10.1109/MOCAST.2016.7495143

Benini, L., Bruni, D., Macii, A., Macii, E., & Poncino, M. (2003). Discharge current steering for battery lifetime optimization. *IEEE Transactions on Computers*, *52*(8), 985–995. doi:10.1109/TC.2003.1223633

Buildroot. (2016, September). *Buildroot project*. Retrieved September 21, 2016 from https://buildroot.org/

Chen, J. B., Endo, Y., Chan, K., Mazières, D., Dias, A., Seltzer, M., & Smith, M. D. (1996, February). The measured performance of personal computer operating systems. *ACM Transactions on Computer Systems*, *14*(1), 3–40. doi:10.1145/225535.225536

Cho, M. H., & Lee, C. H. (2010, August). A low-power real-time operating system for arc (actual remote control) wearable device. *IEEE Transactions on Consumer Electronics*, *56*(3), 1602–1609. doi:10.1109/TCE.2010.5606303

Cho, M. H., Lim, J. S., & Lee, C. H. (2009). ertos: The low-power realtime operating system for wearable computers. *IEEE 13th International Symposium on Consumer Electronics*, 1015–1019. doi:10.1109/ISCE.2009.5156939

Delabrida, S., DAngelo, T., Oliveira, R. A., & Loureiro, A. A. (2016, March). Building wearables for geology: An operating system approach. *SIGOPS Oper. Syst. Rev.*, *50*(1), 31–45. doi:10.1145/2903267.2903275

Direct Frame Buffer. (2016, September). *Direct frame buffer library*. Retrieved September 21, 2016 from http://elinux.org/DirectFB

Engineers, R. T. (2016, September). *Freertos*. Retrieved September 21, 2016 from http://www.freertos.org/

Fitbit. (2016, September). *Fitbit activity trackers*. Retrieved September 20, 2016 from https://www.fitbit.com/

GNU. (2016, September). *The gnu c library (glibc)*. Retrieved September 21, 2016 from https://www.gnu.org/software/libc/

Google Android. (2016, September). *Android wear*. Retrieved September 19, 2016, from https://www.android.com/wear/

Google Brillo. (2016, September). *Brillo project*. Retrieved September 19, 2016, from https://developers.google.com/brillo/

Google Glass. (2016, September). *Glass*. Retrieved September 20, 2016 from https://www.google.com/glass/start/

Hassanalieragh, M., Page, A., Soyata, T., Sharma, G., Aktas, M., Mateos, G., . . . Andreescu, S. (2015). Health monitoring and management using internet-of-things (iot) sensing with cloud-based processing: Opportunities and challenges. *Services Computing (SCC), 2015 IEEE International Conference on*, 285–292. doi:10.1109/SCC.2015.47

Hiremath, S., Yang, G., & Mankodiya, K. (2014). Wearable internet of things: Concept, architectural components and promises for person-centered healthcare. *Wireless Mobile Communication and Healthcare (Mobihealth), EAI 4th International Conference on*, 304–307. doi:10.4108/icst.mobihealth.2014.257440

L. Foundation Yocto. (2016, September). *Yocto project*. Retrieved September 21, 2016 from https://www.yoctoproject.org/

L. Foundation Zephyr. (2016, September). *Zephyr project*. Retrieved September 19, 2016, from https://www.zephyrproject.org/

Liu, R., & Lin, F. X. (2016). Understanding the characteristics of android wear os. *Proceedings of the 14th Annual International Conference on Mobile Systems, Applications, and Services, ser. MobiSys '16*. New York, NY: ACM. doi:10.1145/2906388.2906398

Lynch, W. C. (1972, July). Operating system performance. *Communications of the ACM, 15*(7), 579–585. doi:10.1145/361454.361476

Mann, S. (1996, August). Smart clothing: The shift to wearable computing. *Communications of the ACM, 39*(8), 23–24. doi:10.1145/232014.232021

Mann, S. (1997, February). Wearable computing: A first step toward personal imaging. *Computer, 30*(2), 25–32. doi:10.1109/2.566147

Merkouris, A., & Chorianopoulos, K. (2015). *Introducing Computer Programming to Children through Robotic and Wearable Devices*. Paper presented at WiPSCE 2015: The 10th Workshop in Primary and Secondary Computing Education. doi:10.1145/2818314.2818342

Microsoft. (2016, September). *Hololens*. Retrieved September 20, 2016 from https://www.microsoft.com/microsoft-hololens/en-us

Nassani, A., Bai, H., Lee, G., & Billinghurst, M. (2015). Tag It! AR Annotation Using Wearable Sensors. *Proceedings of SIGGRAPH Asia 2015 Mobile Graphics and Interactive Applications*, 12:1—-12:4. doi:10.1145/2818427.2818438

Nguyen, K. D., Chen, I. M., Luo, Z., Yeo, S. H., & Duh, H. B. L. (2011). A wearable sensing system for tracking and monitoring of functional arm movement. *IEEE/ASME Transactions on Mechatronics, 16*(2), 213–220. doi:10.1109/TMECH.2009.2039222

Park, S., & Jayaraman, S. (2003). Enhancing the Quality of Life Through Wearable Technology. *IEEE Engineering in Medicine and Biology Magazine, 22*(June), 41–48. doi:10.1109/MEMB.2003.1213625 PMID:12845818

Pebble. (2016, September). *Pebble developer*. Retrieved September 19, 2016, from https://developer.pebble.com/

Randell, C., & Muller, H. (2000). Context awareness by analysing accelerometer data. *Digest of Papers. Fourth International Symposium on Wearable Computers*, 175–176. doi:10.1109/ISWC.2000.888488

Riot, O. S. (2016, September). *Operating system for the internet of things*. Retrieved September 21, 2016 from https://riot-os.org/

Samsung. (2016, September). *Tizen project*. Retrieved September 19, 2016, from https://www.tizen.org/

Shi, B., Yang, J., Huang, Z., & Hui, P. (2015). Offloading Guidelines for Augmented Reality Applications on Wearable Devices. *Proceedings of the 23rd ACM International Conference on Multimedia*, 1271–1274. doi:10.1145/2733373.2806402

Sutherland, I. E. (1968). A head-mounted three dimensional display. In *Proceedings of the Fall Joint Computer Conference*. New York, NY: ACM.

Tan, S. L., & Nguyen, B. A. T. (2009). Survey and performance evaluation of real-time operating systems (rtos) for small microcontrollers. IEEE Micro.

Ye, H., Malu, M., Oh, U., & Findlater, L. (2014). Current and future mobile and wearable device use by people with visual impairments. *In Proceedings of the SIGCHI Conference on Human Factors in Computing Systems*. New York, NY: ACM. doi:10.1145/2556288.2557085

Chapter 5
Human Context Detection From Kinetic Energy Harvesting Wearables

Sara Khalifa
Data61, CSIRO, Australia & University of New South Wales, Australia

Mahbub Hassan
University of New South Wales, Australia & Data61, CSIRO, Australia

Guohao Lan
University of New South Wales, Australia & Data61, CSIRO, Australia

Wen Hu
University of New South Wales, Australia & Data61, CSIRO, Australia

Aruna Seneviratne
University of New South Wales, Australia & Data61, CSIRO, Australia

ABSTRACT

Advances in energy harvesting hardware have created an opportunity for realizing self-powered wearables for continuous and pervasive Human Context Detection (HCD). Unfortunately, the power consumption of the continuous context sensing using accelerometer is relatively high compared to the amount of power that can be harvested practically, which limits the usefulness of energy harvesting. This chapter employs and infers HCD directly from the Kinetic Energy Harvesting (KEH) patterns generated from a wearable device that harvests kinetic energy to power itself. This proposal eliminates the need for accelerometer, making HCD practical for self-powered devices. The authors discuss in more details the use of KEH patterns as an energy efficient source of information for five main applications, human activity recognition, step detection, calorie expenditure estimation, hotword detection, and transport mode detection. This confirms the potential sensing capabilities of KEH for a wide range of wearable applications, moving us closer towards self-powered autonomous wearables.

DOI: 10.4018/978-1-5225-3290-3.ch005

INTRODUCTION

Recent advancements in wearable devices enable a wide era of human context-aware services in various domains, including healthcare (Osmani et al., 2008; Chipara et al.,2010), indoor positioning (Altun & Barshan, 2012; Khalifa et al., 2013), and fitness management (Albinali et al., 2010). Particularly, wearable sensors-based Human Context Detection (HCD) has recently become the focus of intense research and development, thus producing a wealth of tools and algorithms to accurately detect human context from data collected by the wearables (He et al., 2012). For example, a wearable sensor attached to the patient body can enable health care authorities to continuously monitor the current status of a patient from a remote centre. HCD then is expected to play a key role in reducing hospital costs by reducing the need for hospital admissions. Similarly, HCD can help individuals in monitoring their fitness level and having a better well-being by recognising various ambulation activities, such as walking, running, sitting, jogging, and so on. It has been confirmed that wearable technology coupled with HCD algorithms have the potential to improve the user's experience and quality of life.

The market of wearable devices is large, it has been found to be $20 billion in 2015 and expected to grow and reach $70 billion by 2025 (Harrop et al., 2015). Healthcare is considered the dominant sector of the wearable market, which combines medical, fitness, and wellness. It has big names such as apple, Fitbit, Google, Samsung, Nike, and Adidas. According to the International Data Corporation (IDC) Worldwide Quarterly Wearable Device Tracker report in 2016 the top leaders of the wearable market are Fitbit, Apple, Xiaomi, Samsung, and Garmin. A total of 78.1 million wearable units have been shipped in 2015, with 171.6% increase over 2014.

Almost all existing wearable products are powered by batteries. While battery technology has improved over the years, battery-powered devices cannot provide sustained operation without frequent charging. To achieve sustained operation, we either need to instrument the wearables with large batteries or be prepared to manually replenish the batteries when they die. Neither of these options is desirable because large batteries make the wearables heavy and less convenient to wear, while manual replacement is inconvenient and not a practical option for many elderly users, who may have to critically depend on such systems.

Over the past few years, a research trend in Energy Harvesting (EH) has emerged and gained the attention of the research community (Hamilton, 2012; Elvin & Erturk 2013). EH is commonly defined as the conversion of ambient energy such as vibrations, heat, wind, light, etc into electrical energy. EH devices can eliminate the need for battery replacement and significantly enhance the versatility of consumer electronics. In fact, significant advancements have been recently made in the EH hardware technology leading to many off-the-shelf products available at low cost. These developments point to future mobile devices that will be equipped with EH hardware to ease the dependence on batteries (Lee et al., 2013).

This means that it is conceptually possible to replace the battery of a wearable sensor with an EH unit to achieve perpetual sensing in many applications including HCD. Of all the ambient energy options, kinetic energy harvesting (KEH) is the most relevant option used for HCD because it can generate power directly from human motion and context. Advances in KEH hardware have motivated us to consider the concept of self-powered wearables for continuous and pervasive HCD, where numerous wearable tiny devices continue to sense and monitor the human on a permanent basis.

However, there is a caveat. KEH generally suffers from low power output (Bickerstaffe, 2015), which may challenge the power requirement of the wearable sensor's components, such as the accelerometer used for sampling human motion. Given that the sensor will also have to turn on its radio for occasional communications with a nearby sink, the power generated from energy harvesting is clearly too small to simply port the existing battery-powered wearables into energy-harvesting wearables. In fact, using energy harvesting to provide self-powered wearables is a very challenging problem that requires innovative sensing and communication solutions.

This chapter discusses a novel paradigm that may potentially overcome the power limitation of KEH, towards self-powered autonomous wearables. Although the primary purpose of KEH is to convert ambient vibrations into electric power, in principle, it could also be used as a potential sensor to detect or identify the source of the vibration. The ability to detect the vibration source can lead to many potential applications for the KEH hardware beyond its primary use of energy harvesting. More specifically, this novel approach employs KEH and infers information directly from the KEH patterns without using any other sensors such as accelerometers which need continuous power to operate. The underlying idea lies in the fact that different ambient vibrations generate energy in a different way producing different energy generation patterns in the KEH circuit. Because no actual sensor such as accelerometer is needed, a significant percentage of the limited harvested energy can be saved.

In this chapter, we discuss in more details the use of KEH patterns as a source of information for five main applications, human activity recognition, step detection, calorie expenditure estimation, hotword detection, and transport mode detection. This confirms KEH as a novel energy efficient source of information for a wide range of wearable applications, moving us closer towards self-powered autonomous wearables.

BACKGROUND

Human Context Detection

Human context has been initially perceived by the computer science community as a matter of the user location. However, in the last few years this notion has been generalised to all related aspects of the user (Orsi & Tanca, 2010). For example, a context-aware system may know the current physical activity of the user (walking, running, sitting, ...etc), each step the user has taken, the daily calorie expenditure, and even what kind of transportation mode the user is using. In fact, Human Context Detection (HCD) is increasingly being used for a wide range of applications including healthcare (Osmani et al., 2008; Chipara et al.,2010) and fitness monitoring (Albinali et al., 2010), smart living, and localization (Altun & Barshan, 2012; Khalifa et al., 2013). Context-aware systems involves two basic processes: the acquisition of user's context using sensors and understanding of user's context by context modeling

There are two fundamentally different approaches to acquire the user's context, using infrastructure sensors (Wongpatikaseree et al., 2012; Singla et al., 2010) and wearable sensors (He et al., 2012). In the former, the sensors are installed at fixed locations to detect human context (e.g. physical activities) when a user visits these locations and interacts with the sensors. For example, cameras installed at fixed locations can be used to detect user activity whenever the user comes within their vicinity (Bodor et al., 2003; Poppe et al., 2010). However, deployment and maintenance of infrastructure sensors are costly. On the other hand, wearable sensors provide an alternative option by placing various types of sensors on the human body. For example, a wearable device in a wristband can help identify user's context by simply collecting and analysing data from the wearable. In existing wearable devices, accelerometer is the dominantly used sensor to acquire the user's context. Typically a triaxial accelerometer is used to measure the acceleration of the user in three dimensions. Machine learning techniques can be used then to model the context of the user from the acquired accelerometer data; this is called accelerometer-based human context detection. Consequently, wearable device can help achieving pervasive HCD without the need to deploy infrastructure sensors.

However, the major challenge of wearable devices is the battery lifetime. A typical wearable device will need power for sensing, processing and communication which can quickly drain the battery life of the wearable. Accelerometer is widely used for sensing human motion and context. There are several types of accelerometers; however, the type that is used most in wearable and mobile devices is the capacitive accelerometer. In a capacitive accelerometer, a capacitor is formed by a "stationary" plate (the housing which moves with the base acceleration) and a "moving" plate attached to the seismic mass. The distance between these plates determines the capacitance which can be monitored to infer acceleration (change in capacitance related to acceleration). Bsching et al., (2012) tested the power consumption of six commonly used capacitive accelerometers when a 3.3v power supply and a 50 HZ sampling rate were used. Their results showed that accelerometers consume hundreds of microwatts at only 50 Hz sampling rate.

Moreover, the datasheets of the three widely used capacitive accelerometers ADXL150 (used in wearable sensors), SMB380 (used in Samsung Galaxy smartphones), LIS302DL (used in IPhone smartphones) showed that the average power consumption of the accelerometer is a linear function of the sampling rate (Yan et al., 2012). For example, Weinberg (2002) showed that the ADXL150 accelerometer consumes about 5 μW on average per Hz, which means that it would require 250 μW if a sampling rate of 50 Hz were required for a given activity set. The required sampling rate depends on the set of activities monitored and typically ranges from 1-50Hz (Ravi et al., 2005; Wang et al., 2005; Kwapisz et al., 2011; Khan et al., 2008). This means the battery must supply 5-250 μW to the accelerometer. This is simple for battery-powered wearable devices. However, it is an issue for energy harvesting wearable devices.

While it is possible to extend the battery lifetime by providing more energy-efficient solutions (Yan et al, 2012; Qi et al, 2013; Khalifa et al., 2013; Zappi et al., 2008), battery-powered sensors cannot provide sustained HAR without the need for frequent charging or battery replacement. This motivates us to explore Energy Harvesting (EH) solutions. EH is commonly referred to the conversion of ambient energy such as solar, kinetic, vibration, etc, into electrical energy. EH eliminates the need for battery replacement and significantly enhances the versatility of consumer electronics.

Kinetic Energy Harvesting Overview

In theory, electrical energy can be obtained from many types of energy, including kinetic (vibration) (Vocca & Cottone, 2014; Mitcheson et al., 2008), thermal (Xu et al., 2013) and radio frequency (Zungeru et al., 2012; Nintanavongsa et al., 2012). Table 1 shows the power density estimates of typical ambient energy sources from Texas Instruments (Raju, 2008). Of all ambient energy options, kinetic energy harvesting (KEH) is the most relevant for wearables because it can power the wearable directly from human motion. Kinetic energy also produces 4 times as much energy as RF (as shown in Table 1) and is more abundant. A brief review of KEH is presented in this chapter.

Kinetic energy harvesting (KEH) is a process of converting environmental vibrations into electrical energy. Kinetic EH and vibration EH are synonyms, environment around us is full of sources of kinetic or vibration energy such as natural

Table 1. Power Density Estimates of typical ambient energy sources

Energy Source	Characteristics	Harvested Power Density
Vibration	Human Machine	4 $\mu W / cm^2$ 100 $\mu W / cm^2$
Light	Indoor (illuminated office) Outdoor (direct sun)	10 $\mu W / cm^2$ 10 mW / cm^2
Thermal (Heat)	Human Industrial	25 $\mu W / cm^2$ 1-10 mW / cm^2
Radio Frequency	GSN WIFI	0.1 $\mu W / cm^2$ 1 $\mu W / cm^2$

Source: Texas Instruments, Energy Harvesting White Paper 2008 (Raju, 2008).

seismic vibration (e.g. earthquakes), wind movement, sea waves, vehicular traffic, machinery vibration and human motion. In this chapter, we discuss the system architecture of a KEH-based device, the transduction mechanisms, the commercially available products implementing KEH, and the possible applications of KEH.

System Architecture

Figure 1 shows a block diagram of a KEH-based device. KEH-based Hardware typically comprises three parts: a transducer to convert vibration into electrical energy, an AC/DC converter to convert the AC generated from the transducer into regulated

Figure 1. A block diagram of a KEH-based sensor

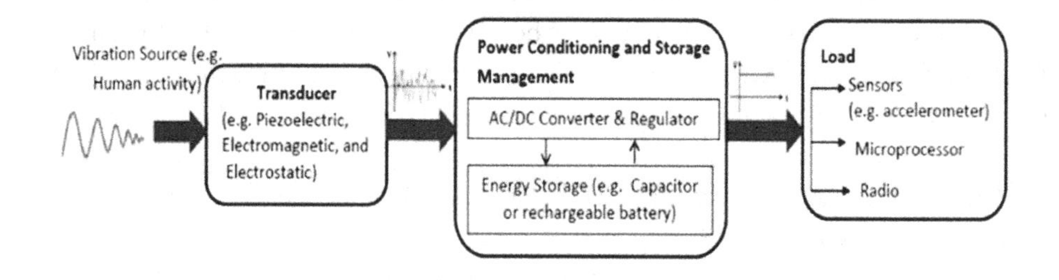

DC, and a battery or capacitor to store the harvested energy and provide a constant power flow to the load. The load normally consists of sensors (e.g. accelerometer), microprocessor, and Radio Frequency Transceiver.

Transduction Mechanisms

From a hardware point of view, there are three main transduction mechanisms for converting vibration energy to electric power (Rao et al., 2013): piezoelectric, electromagnetic (capacitive), and electrostatic (inductive). Depending on the mechanism used, the operating principle differs.

- Piezoelectric harvesters make use of certain piezoelectric materials such as PZT and MFC, which have the ability to generate an electrical potential when subjected to a mechanical strain (Sodano et al., 2005; Kim et al., 2011). The resulting strain on the material will result in an output of alternating current which is converted into power.
- Electromagnetic harvesters make use of an oscillating mass (magnet) which traverses across a fixed coil, creating a varying amount of magnetic flux, inducing an alternating current that is converted to power (Chae et al., 2013).
- Electrostatic (capacitive) harvesters are based on separating the plates of an initially charged variable capacitor (varactor) using vibrations and converting mechanical energy into electrical energy (Boisseau et al., 2012). Electrostatic harvesters are widely used though they are not as popular as piezoelectric or electromagnetic transducers since Electrostatic harvesters need a polarization source to work and to convert mechanical energy from vibrations into electricity.

Table 2 summarises the advantages and disadvantages of the three transduction mechanisms. Generally speaking, piezoelectric and electrostatic systems are well suited to micro-scale (small scale) applications, while electromagnetic systems are preferable for macro-scale (medium scale) devices. Piezoelectric transducers are the most favorable due to their simplicity and compatibility with MEMS (Lefeuvre et al., 2006). Electromagnetic-based energy harvesters are usually bulky in size and difficult to integrate with MEMS. Moreover, electrostatic transducers need external voltage to operate. Many kinetic or vibration EH models have been recently developed (Gorlatova et al., 2014; Biswas & Quwaider, 2013; Yun et al., 2008). The main focus of these models is to optimise the parameters of the harvester to maximise the output harvested power. To maximise the output power, the harvester is mechanically tuned to an optimized resonant frequency present in the application environment.

Table 2. Transduction mechanisms of VEH

Type	Advantage	Disadvantage
Piezoelectric	No need for smart material Compatible with MEMS Compact configuration	Depolarization brittleness in PZT charge leakage
Electromagnetic	No need for smart material No need for external voltage source	Bulky size Difficult to integrate with MEMS
Electrostatic	No need for smart material Compatible with MEMS	External voltage source (or charger) is needed Mechanical constraints are needed

Commercially Available KEH/VEH Devices

Several kinetic or vibration energy harvesters are commercially available. The prevalent commercial VEH devices are based on the piezoelectric and electromagnetic transduction mechanisms. Table 3 provides a list of the commercially available VEH devices. Perpetuum and Ferro Solutions produce electromagnetic-based VEHs,

Table 3. Commercially available KEH/VEH devices

Manufacturer (Country)	Product	Material&	Dimensions (in) $L \times W \times H$	Weight (grams)	Output (in voltage)
Perpetuum (UK)	PMG FSH	Electromagnetic	3.4×2.6	1075	DC (5 V and 8 V)
Ferro Solutions (USA)	VEH 460	Electromagnetic	-	430	DC (3.3V)
LORD MicroStrain (USA)	PVEH&	Piezoelectric	1.87×1.75	185	DC (3.2 V)
	&MVEH&	Electromagnetic	2.25×2.56	216	DC (3.2 V)
MicroGen (USA)	BoLT PZEH	Piezoelectric	$1.18 \times 1.04 \times 0.69$	10	DC (3.3 V)
MID'E(USA)	Volture V25W	Piezoelectric	$2.00 \times 1.50 \times 0.03$	8	AC
PI Ceramic GmbH (Germany)	P-876.A11 DuraAct	Piezoelectric&	$2.4 \times 1.38 \times 0.02$	-	AC
Smart Material (USA	MFC M2503-P1	Piezoelectric	$1.81 \times 0.93 \times 0.01$	-	AC
OMRON and Holst Centre/ imec	Still under testing	Electrostatic	1.96×2.36	15.4	DC

however, MID'E, MicroGen, PI Ceramic GmbH, and Smart Material produce piezoelectric-based VEHs. MicroStrain produces both electrodynamic generators (MVEH™ Harvester) and piezoelectric materials (PVEH™ Harvester). Recently, OMRON and Holst Centre/imec unveiled a prototype of an extremely compact electrostatic-based VEH. Figure 2 shows some of the commercially available VEHs. Piezoelectric transducers are simple and compatible with MEMS. The characteristic of the products show that electromagnetic-based energy harvesters are usually bulky and not compatible with MEMS as mentioned previously. Moreover, the only electrostatic transducer is still under testing and not commercially available.

Most VEH devices are available as packaged systems, including the transducer, power conditioning circuit, and local storage. They provide a constant (regulated) DC voltage which is suitable to power multi-sensor nodes, controllers, peripherals, memory, etc; however, the intermediate outputs such as the AC voltage, or the unregulated DC, cannot be accessed. Some companies (such as MID\'E) make these intermediate outputs accessible by offering customizable energy harvesting evaluation kits, which provide modular components for power conversion and storage that afford plug-and-play compatibility with their transducers.

Figure 2. Commercial kinetic or vibration energy harvesters (a) Perpetuum, (b) Ferro Solution (VED 460), (c) MicroStrain MVEH, (d) MicroStrain PVEH, (e) Mide Volture, (f) MicroGen, (g) PI Ceramic, (h) Smart Material (MFC), and (i) OMRON and Holst Centr

(a) Perpetuum *(b) Ferro Solution* *(c) MVEH MicroStrain* *(d) PVEH MicroStrain*

(e) MIDE Volture *(f) MicroGen* *(g) PI Ceramic* *(h) Smart Material (MFC* *(i) OMRON and Holst Centre/imec (under testing)*

KEH Applications

KEH has a wide area of applications, such as medical implants, consumer applications, building technologies, vehicles and aerospace. A brief summary of how KEH can be used for each of these applications is presented below.

- **Medical Implants:** KEH can use a patient's own body movement and heartbeat to provide power for medical devices deployed inside the body, and which are vital to the life and well being of the patient.
- **Consumer Electronics:** KEH is suitable for many low-power consumer electronics, used as a sole power source or as a means to extend battery life.
- **Building Technologies:** KEH is suitable for building technology applications such as infrastructure sensing system battery and safety systems for buildings in the event of a power loss.
- **Vehicles and Aerospace:** KEH provides safe, reliable, cost effective solutions to those applications in which traditional power sources are not reliable or preferred, e.g. supplying power to tyre air pressure sensors (where batteries are difficult to change and hard-wiring is impossible), supplying power to sensors mounted inside an aircraft which monitor in-flight mechanical loads on the airframe.

KEH Limitation

KEH is mostly used for harvesting energy from machine vibrations because machines vibrate at high frequency, hundreds of Hz, which produces a reasonable amount of energy to sense and transmit data. However, KEH performance drops when harvesting energy from human motion because human motion has lower frequency (in the order of tens of Hz). The amount of power that can be practically harvested from human motion is too small to power all necessary functions of a wearable device. KEH from human activities can produce only limited power (measured in μW), which is not sufficient to simultaneously power all components in a wearable device including the accelerometer used to acquire user context.

Table 4 shows the power that could be generated using a commercial kinetic energy harvester for different activities (Olivares et al., 2010). It shows that some activities generate only a few μW which is much lower than what is required to sample the accelerometer at a sufficiently high rate for accurate context detection. Clearly, this will force the device to reduce the power to the accelerometer, i.e., use a lower sampling rate and accept a lower context detection accuracy, each time the user switches to one of the activities that produce small amount of power. Even if

Table 4. Average Harvested Power for different activities when the device is attached to the shank

Activity	Average Harvested Power (μW)
Walking	10.30
Running	28.74
Cycling	0.36
Sitting	0.02
Lying	0.36

the harvested power is enough to operate the accelerometer at the required sampling rate, it reduces the amount that could be accumulated in the capacitor for future radio communications. Insufficient stored energy in the capacitor will force more aggressive duty cycling of the radio or more drastic reduction in the transmission power. In summary, when the power supply is limited by energy harvesting, powering the accelerometer trades off the quality of radio communication. In fact, using KEH to provide a self-powered HAR is a challenging problem that requires innovative sensing and communication solutions.

KEH-BASED HUMAN CONTEXT DETECTION

KEH-based Human Context Detection (HCD) aims at providing a self-powered HCD which does not need batteries to operate. It allows continuous and permanent monitoring of human activities, which will improve the user's experience and quality of life. KEH -based HCD is an alternative approach to HCD that does not use an accelerometer, which can have relatively high power requirements on relatively low-power energy harvesting wearables, but instead uses the generated KEH signal for HCD.

Since the wearables that rely on KEH to self-power themselves are still in their early stage of development, we built a KEH wearable prototype to evaluate the performance of the proposed KEH-based HCD. It is basically a data logger which records the generated signals of a commercially available piezoelectric KEH transducer, called Volture from MIDE (see www.mide.com). It provides AC voltage as its output. We also added a three-axis accelerometer (MMA7361LC) to the design for comparison purposes. We used an Arduino Uno as a microcontroller device for sampling the data from both the Volture and the accelerometer. We used a sampling rate of 1 kHz for data collection. We saved the sampled data on an 8-Gbyte microSD card, which we equipped to the Arduino using microSD shield. A 9V battery was used to power the Arduino. The data logger also includes two switches, one to switch on/off the device and the other to control the start and stop of data logging. Figure 3 shows the external appearance of our data logger, a user holding the data logger during the data collection process, and the internal appearance of the data logger including the details of its components.

This data logger is used to collect KEH patterns for five main applications, human activity recognition, step detection, calorie expenditure estimation, hotword detection, and transport mode detection. In this chapter, we show the performance of using KEH patterns as a source of information for those five applications including the used algorithms in each application.

Activity Recognition

Human Activity Recognition (HAR) is becoming critical in many applications, including aged health care, fitness monitoring, and indoor positioning. Accelerometer has been widely used for human activity recognition as it is considered low-power electronics drawing only about a few μW per sample per second (Hz). However, we showed that accelerometer power requirements is considered relatively high when used in KEH powered devices (Khalifa et al., 2015a). Using experimental data, we showed that the power requirement of accelerometer for HAR ranges

Figure 3. KEH data logger

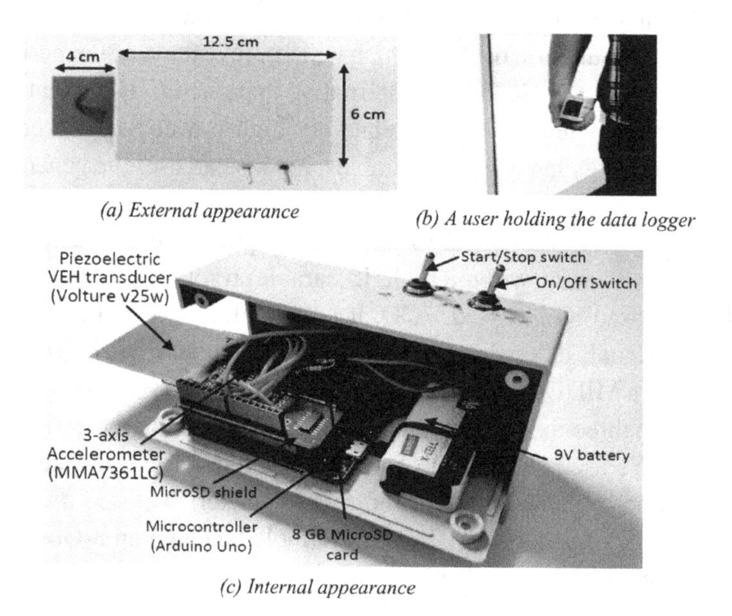

(a) External appearance

(b) A user holding the data logger

(c) Internal appearance

between 35-515% of the harvested kinetic power. We also demonstrated that down scaling power supply to the accelerometer reduces HAR accuracy exponentially. These results indicate that although accelerometers are considered low-power electronics in general, they can be the bottleneck of self-powered pervasive HAR. To address this challenge, we proposed the use of KEH patterns as a new source of realising HAR in a kinetic-powered device. Figure 4 shows our proposal of using KEH patterns for HAR compared to the conventional use of acceleration patterns in a kinetic-powered device. Our proposal eliminates the need for accelerometer, making HAR practical for self-powered devices.

Initially, We used a well know mass spring damping model to estimate KEH patterns form motion data due to the absence of commercially available kinetic energy harvesting portable devices that could be used to collect energy traces from users. By applying information theoretic measures on the estimated KEH patterns, we confirmed that KEH patterns contain rich information for discriminating typical

Figure 4. KEH-based HAR compared to accelerometer-based HAR in a kinetic-powered devices

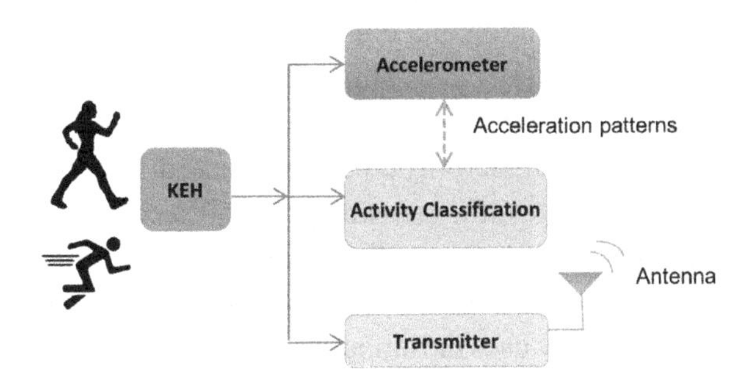

(a) Conventional accelerometer-based HAR in KEH powered devices

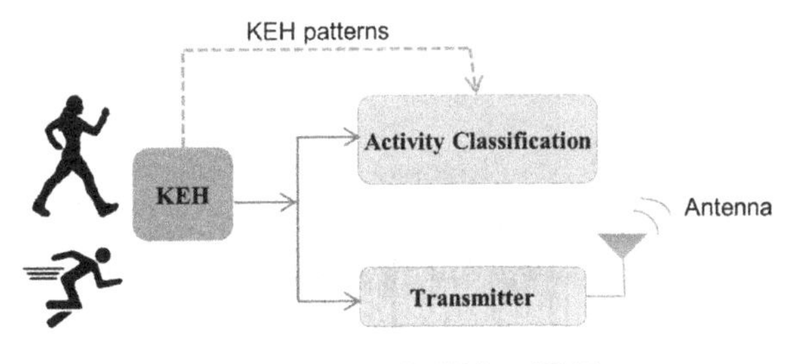

(b) Proposed KEH-based HAR

activities of our daily life. We evaluated our proposal using 14 different sets of common activities each containing between 2-10 different activities to be classified. We showed an average accuracy of 83%, which is within 13% of what could be achieved with an accelerometer without any power constraints. However, the used KEH patterns were an approximation of the real data.

These results motivated us to build a data logger (whose components have been presented previously) to collect real KEH data and investigate whether the generated patterns by a real KEH hardware contain information about human activity as reported in our previous study from estimated KEH power patterns. We collected data for three different activities from ten different subjects holding the datalogger in hand.

Figure 5 shows the KEH patterns for three different activities: standing, walking, and running. This shows that the generated signal of a piezoelectric KEH transducer switches to clearly distinguishable patterns as the user changes her activities. Our experimental analysis showed that KEH-based HAR can achieve 98% accuracy for distinguishing three basic activities: standing, walking, and running, which as accurately as accelerometer-based HAR. We have also done some energy analysis which showed that KEH-based HAR consumes 72% less energy compared to the conventional accelerometer-based HAR (Khalifa et al., 2015b).

Following the power savings of not using accelerometer for HAR, we further reduced both on-node classification and communication overhead by proposing a new method that guarantees energy neutrality (Khalifa et al., 2016a). In this study, we used the kinetic energy accumulated in a fixed-length time window to transmit an unmodulated signal, called an "activity pulse". Because different human activities generate power at different rates, the transmission and receiving signal strengths are different among different activities. Thus, those signal strengths can be used to classify the activities. Energy neutrality is guaranteed because the transmission power of the activity pulse only uses the amount of energy harnessed in the last time window, and no additional energy is required to power any sensing or classification components in the wearable device.

Figure 5. KEH patterns for three basic activities: walking, running, and standing

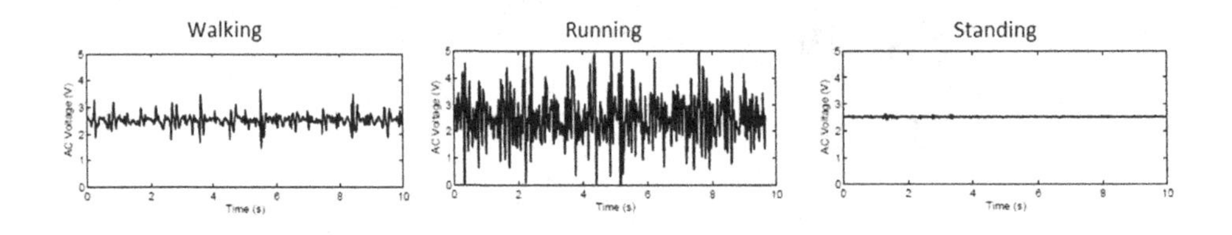

Figure 6 shows our proposed architecture for energy neutral KEH-basedHAR. The KEH component in the architecture utilizes a capacitor to store the energy harvested for a given time window; and then uses all the stored energy to transmit an unmodulated signal, called an *activity pulse*. Because different activities generate power at different rates (Gorlatova et al., 2014), the receiving signal strengths are different among different activities. Thus, the signal strengths can be used to classify the activities. Assuming that the distributions of the received signal strengths of the activities are known, the classification can be done at the receiver side using Bayesian decision theory. Our proposed architecture guarantees the energy neutrality because we only use the accumulated energy in the last time window to transmit the activity pulse; no additional energy is required to power any sensing or classification components in the wearable device (Khalifa et al., 2016a).

We evaluated the performance of our proposed idea of transmitting an "activity pulse" by collecting a real dataset from piezoelectric KEH prototype coupled with a Bluetooth prototype. We achieved an overall accuracy of 91% when the distance between the transmitter and the receiver is set to 30 cm. We also pointed out that the overall accuracy goes down to 85% and 65% when the distance is increased to 60 cm and 100 cm, respectively (Khalifa et al., 2016a).

Step Counting

Step detecting wearable devices are increasingly being used for health monitoring and indoor positioning applications. In the study shown in (Khalifa et al., 2015c), we conducted the first experimental study to validate the concept of step detection from the generated patterns of KEH wearables. Figure 7 shows the raw output patterns of a piezoelectric KEH from a wearable device attached to the waist of a subject walking along straight walkway for 11 steps.

Figure 7 shows that KEH patterns exhibit distinctive peaks for steps, which can be detected accurately using widely used peak detection algorithms. We collected data from four different subjects under different walking scenarios, including walk along straight and turning paths as well as descending and ascending stairs, covering a total

Figure 6. A proposed energy neutral KEH-based HAR

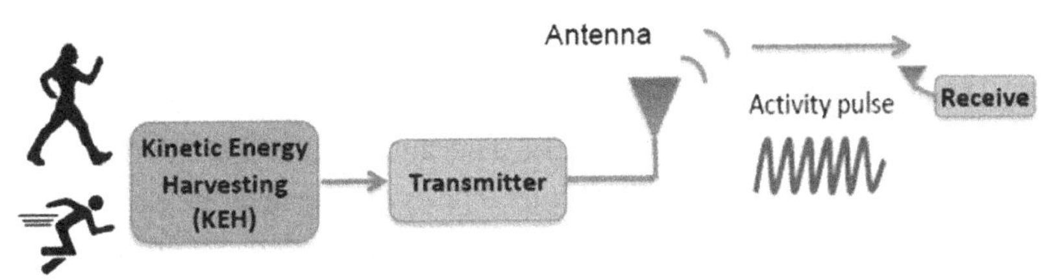

Figure 7. The raw output patterns of a piezoelectric KEH harvester from a wearable device attached to the waist of a subject walking along straight

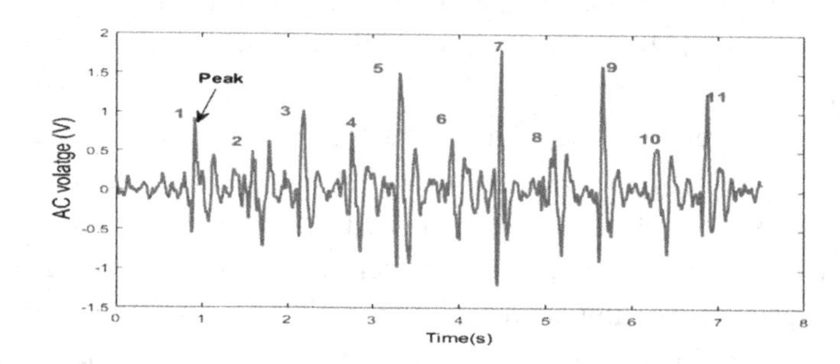

of 570 steps. Our analysis showed that PEH-based step detection can be achieved with 99.08% and 100% accuracy for straight and turning walkways, respectively. However, the accuracies for ascending and descending stairs scenarios are 92.97% and 93.42%, respectively. In total, over all subjects and all walking scenarios, 550 steps out of 570 have been successfully detected achieving 96% step detection accuracy when PEH patterns are used, compared to 100% accuracy when the accelerometer is used. All of our results in this study (Khalifa et al., 2015c) were based on a waist placement of the KEH hardware on the subjects' body. Therefore, more experimentation is still needed to study the effect of different device placements on the results.

Calorie Expenditure Estimation (CEE)

Calorie expenditure estimation (CEE) is valuable in monitoring many health problems, such as obesity, an epidemic which is predicted to be the most preventive health problem in the future.

Unlike the conventional works that highly rely on accelerometers for CEE, we conducted the first experimental study in (Lan et al., 2015) to assess the suitability of using KEH data for accurate CEE. We used KEH prototype to collect real data from ten different subjects for two different activities, walking and running. Figure 8 shows the instantaneous estimation results of KEH-based and ACC-based CEE for both walking and running activities.

Figure 8 shows that although the instantaneous estimations of the proposed KEH-based method are different from that of the accelerometer-based, the averages of the KEH-based CEE over a period of time (one second or longer), are very close to that of the accelerometer-based CEE. The authors used a standard statistical regression model to drive the results. Figure 9 plots the mean of the estimated calorie expenditure over one second of the KEH-based and accelerometer-based methods

Figure 8. The instantaneous estimation results of KEH-based and ACC-based CEE for both walking and running activities

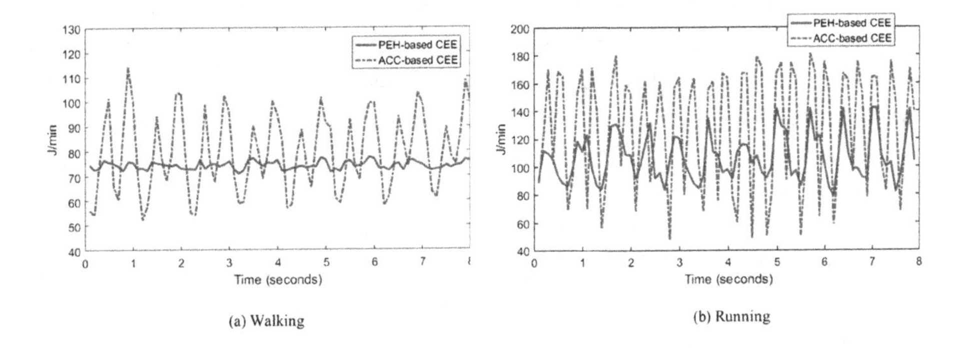

Figure 9. The mean of the estimated calorie expenditure over one second of the KEH-based and accelerometer-based methods for both walking and running activities

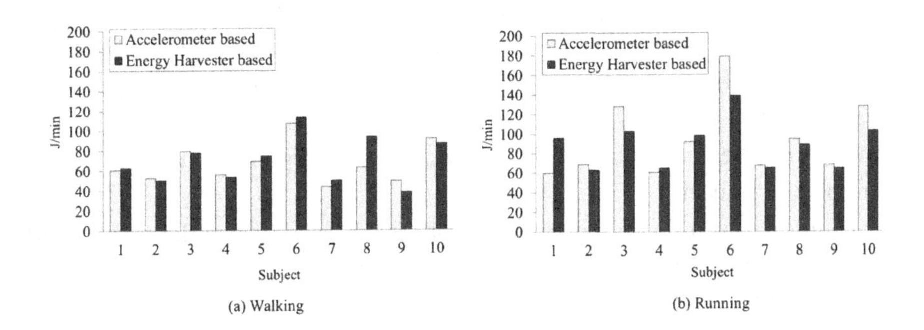

for both walking and running activities. The results show that for most subjects, the calorie estimations obtained from KEH patterns are very close to those obtained from a 3-axial accelerometer (Lan et al., 2015).

Hotword Detection

Detecting hotwords, such as ``OK Google", is a recent method used by voice control applications to allow verbal interaction with devices by delineating user commands from background conversations. Pervasive hotword detection requires continuous sensing of audio signals, which results in significant energy consumption when a microphone is used as an audio sensor.

We conducted the first experimental study to validate the feasibility of using the vibration energy harvested (VEH) patterns generated from human speech as a potential new source of information for detecting hotwords, such as ``OK Google" (Khalifa et al., 2016b). Figure 10 shows the architecture used in our study for VEH-based hotword detection. The generated AC voltage data is continuously fed to a trained binary classifier, which classifies the input signal into either hotword or non-hotword. No actions will be taken during the normal conversation (speech contains no hotword), but if hotword is detected, the system will switch to the command mode.

We conducted a comprehensive experimental study involving 8 subjects using our KEH datalogeer. We chose the phrase "OK Google" as a repetitive of the hotwork category and three choices of a non-hotword phrases "fine, thank you", "good morning", and "how are you". Figure 11 shows the VEH patterns for silence and when the four phrases are spoken by one of the subjects involved in the study.

We see that the voltage produced by silence is significantly lower than those produced by voice. We also notice that silence has a more periodic voltage pattern, which captures the background (noise) vibrations, while the voltage is markedly biased in the positive direction when phrases are spoken. This is expected because, in this scenario, sound waves continuously hit directly on one surface of the piezoelectric beam causing it to vibrate asymmetrically around the neutral position. The experiments involved the analysis of two types of hotword detection, speaker-independent, which does not require speaker-specific training, and speaker-dependent, which relies on speaker-specific training.

Our results showed that a simple Decision Tree classifier can detect hotwords from KEH signals with accuracies of 73% and 85%, respectively, for speaker-independent and speaker-dependent detections (Khalifa et al., 2016b). We further demonstrate that these accuracies are comparable to what could be achieved with an accelerometer sampled at 200 Hz.

Figure 10. VEH-based hotword detection

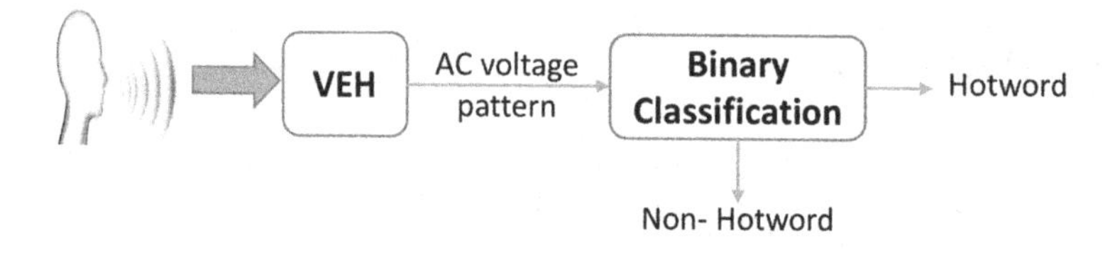

Figure 11. VEH patterns for silence and when the four phrases are spoken

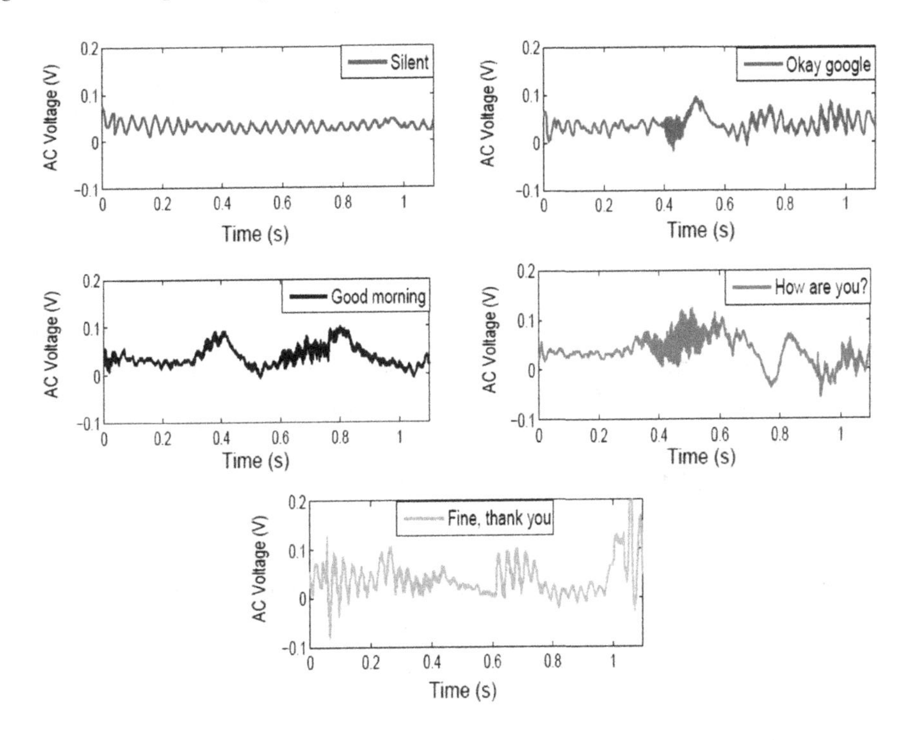

Transportation Mode Detection

Detecting the transportation mode of an individual's everyday travel provides useful information in urban design, real-time journey planning, and activity monitoring. In existing systems, accelerometer and GPS are the dominantly used signal sources which quickly drain the limited battery life of the wearable devices. However, we investigated the feasibility of using the output voltage from the KEH device as the signal source to achieve transportation mode detection (Lan et al., 2016). Figure 12 gives the high-level overview of the KEH-based transportation mode detection system. Instead of relying on any accelerometer or GPS signal, the proposed system exploits the AC voltage generated from the KEH wearable devices as the signal to achieve transportation mode detection. The proposed idea is based on the intuition that the vibrations experienced by the passenger during motoring of different transportation modes are different. Thus, voltage generated by the energy harvesting devices should contain distinctive features to distinguish different transportation modes.

Figure 12. Overview of KEH-based transportation mode detection

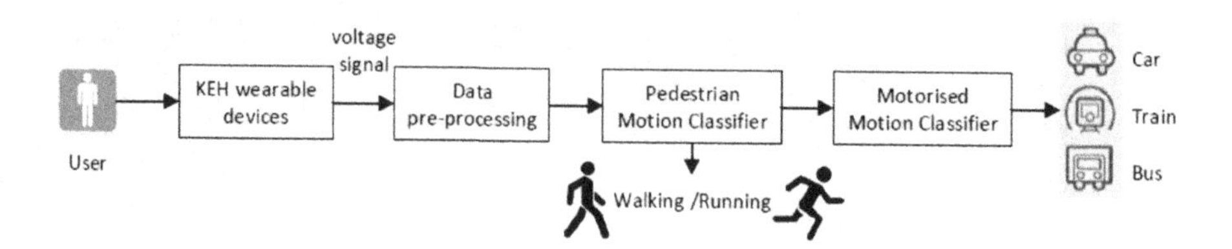

The system decomposes the overall detection task into three subtasks. First, the raw voltage signal from the KEH device is going through the data pre-processing which applies a lowpass filter to eliminate possible noise. In addition, we have designed a stop-detection algorithm to classify and filter the stop/pause data out from the voltage signal profile. Figure 13 shows a trace of the VEH voltage signal recorded during a train trip with an illustration of stop/pause periods of the train. Then, in the second level of classification, a pedestrian motion classifier is applied on the processed voltage signal; the classifier determines whether the person is traveling via walking/running. When the pedestrian motion classifier determines the ongoing traveling as non-pedestrian mode, the process progresses to the motorized motion classifier which determines whether the user is in a motorized transport, and what kind of vehicle is used.

Figure 13. Illustration of stop/pause periods of a vehicle (train)

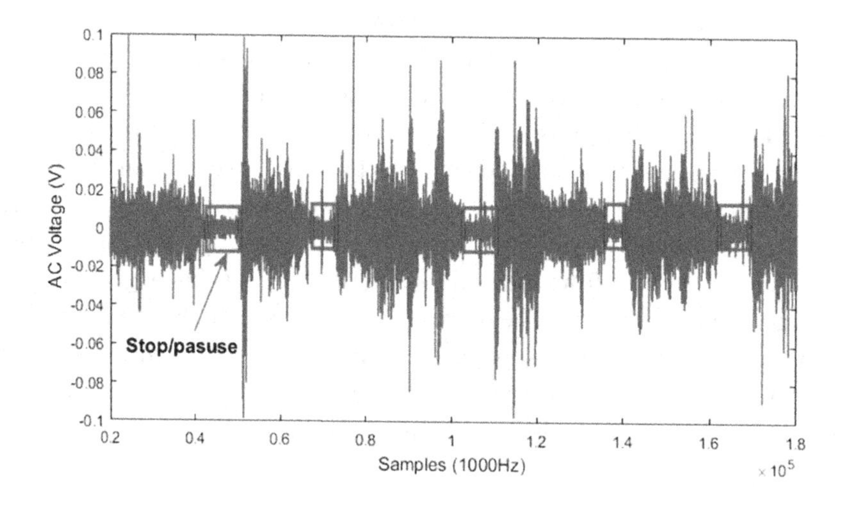

To develop and evaluate the performance of our approach, we collected over 3 hours of data trace using our KEH data logger. We collected four traces of data by 4 volunteers for pedestrian motions (walking/ running) and three motorized modes (bus, driving, and train) for different traveling routes across Sydney on different days. Figure 14 compares the voltage generated by pedestrian motions (walking and running) and motorized motions (bus, car, and train). We can observe that the amplitudes of voltage from the pedestrian motions are much higher than that from the motorized motions. Intuitively, this is because when traveling by vehicles, people's motion is relatively stationary (assuming the user is sitting or standing in the vehicles during the trip), thus, the output voltage from the KEH device is quite low. On the contrary, when people are walking/running, the KEH device experiences considerably heavier vibration, and thus, generates higher voltage. Our results show that an accuracy of 98.84% can be achieved in determining whether the user is traveling by pedestrian or motorized modes, using threshold-based classification algorithm. However, in a fine-grained classification of three different motorized modes (car, bus, and train), an overall accuracy of 85% is achieved using voltage peak based learning algorithm (Lan et al., 2016).

In this chapter, we present our studies of using KEH patterns as a source of information for five main applications, human activity recognition, step detection, calorie expenditure estimation, hotword detection, and transport mode detection. Table 5 summaries the data collection details, the algorithms used, and the accuracy reported in each study, confirming KEH as an efficient source of information for a wide range of wearable applications.

Figure 14. A comparison of KEH voltage signal from different transportation modes

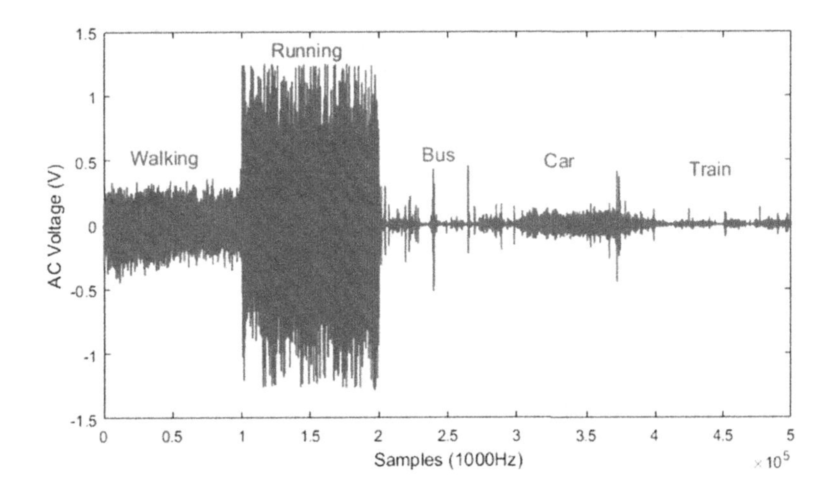

Table 5. A summary of our studies of using KEH patterns as a source of information for different applications and the corresponding data collection campaign, the algorithms used, and the accuracy reported for each application.

Application	Data Collection	Proposed Algorithm	Accuracy
Human Activity Recognition	10 Subjects (6F, 4 M) 5 different activities 2 holding positions	K-nearest neighbour algorithm	81% for hand 87% for waist
Step Counting	4 subjects Different walking scenarios, including walk along straight and turning paths as well as descending and ascending stairs, covering a total of 570 steps.	Peak detection algorithm	96%
Calorie Expenditure Estimation	10 subjects 2 different activites walking, running	Standard statistical regression	88% for walking and 84% for running
Hotword Detection	8 subjects (4 F, 4 M) 6 instances (30 hotwords, 30 non-hotwords) per subject Hotword: OK Google Non-hotwords Good Morning, How are you? Fine, thank you	Decision tree classifier	73% for speaker-independent 85% speaker-dependent
Transport Mode Detection	3 hours of data trace for three motorized modes (bus, driving, and train) Different traveling routes across Sydney different days.	Voltage peak based learning algorithm	85%

CONCLUSION

This chapter shows the potential use of KEH as a novel source of information for a wide range of wearable applications including, human activity recognition, step detection, calorie expenditure estimation, hotword detection, and transport mode detection. Unlike existing sensors, like microphones, or accelerometers, KEH does not require any power supply to operate, offering a unique power saving opportunity if used as a sensor for these applications.

REFERENCES

Albinali, F., Intille, S., Haskell, W., & Rosenberger, M. (2010). Using wearable activity type detection to improve physical activity energy expenditure estimation. *Proceedings of the 12th ACM International Conference on Ubiquitous Computing, Ubicomp* '10. doi:10.1145/1864349.1864396

Altun K. & Barshan B. (2012). Pedestrian dead reckoning employing simultaneous activity recognition cues. *Measurement Science and Technology, 23*(2), 1-20.

Bickerstaffe, J. (2015). *Energy harvesting*. Sagentia. Retrieved June 11, 2016, from http://www.sagentia.com/resources/white-papers/2011/energy-harvesting.aspx

Biswas, S., & Quwaider, M. (2013). Modeling energy harvesting sensors using accelerometer in body sensor networks. *8th International Conference on Body Area Networks (BODYNETS)*. doi:10.4108/icst.bodynets.2013.253588

Bodor, R., Jackson, B., & Papanikolopoulos, N. (2003). Vision-based human tracking and activity recognition. *Proceedings of the 11th Mediterranean Conference on Control and Automation.*

Boisseau, S., Despesse, G., & Seddik, B. A. (2012). *Electrostatic Conversion for Vibration Energy Harvesting, Small-Scale Energy Harvesting*. Intech.

Bsching, F., Kulau, U., Gietzelt, M., & Wolf, L. (2012). Comparison and validation of capacitive accelerometers for health care applications. *Computer Methods and Programs in Biomedicine, 106*(2), 79–88. doi:10.1016/j.cmpb.2011.10.009 PMID:22153570

Chae, S. H., Ju, S., Choi, Y., Jun, S., Park, S. M., Lee, S., & Ji, C. H. et al. (2013). Electromagnetic vibration energy harvester using springless proof mass and ferrofluid as a lubricant. *Journal of Physics: Conference Series, 476*(1).

Chipara, O., Lu, C., Bailey, T. C., & Roman, G. (2010). Reliable clinical monitoring using wireless sensor networks: Experiences in a step-down hospital unit. *Proceedings of the 8th ACM Conference on Embedded Networked Sensor Systems, SenSys '10.* doi:10.1145/1869983.1869999

Elvin, N., & Erturk, A. (2013). *Advances in Energy Harvesting Methods.* Springer Link. doi:10.1007/978-1-4614-5705-3

Gorlatova, M., Sarik, J., Grebla, G., Cong, M., Kymissis, I., & Zussman, G. (2014). Movers and shakers: Kinetic energy harvesting for the internet of things. *Performance Evaluation Review, 42*(1), 407–419. doi:10.1145/2637364.2591986

Hamilton, M. C. (2012). Recent advances in energy harvesting technology and techniques. *IECON 2012 - 38th Annual Conference on IEEE Industrial Electronics Society*. doi:10.1109/IECON.2012.6389019

Harrop, P., Hayward, J., Das, R., & Holland, G. (2015). *Wearable Technology 2015-2025: Technologies, Markets, Forecasts*. Retrieved June 11, 2016, from http://www.idtechex.com/research/reports/wearable-technology-2015-2025-technologies-markets-forecasts-000427.asp

He, W., Guo, Y., Gao, C., & Li, X. (2012). Recognition of human activities with wearable sensors. EURASIP Journal on Advances in Signal Processing. doi:10.1186/1687-6180-2012-108

Khalifa, S., Hassan, M., & Seneviratne, A. (2013). *Adaptive pedestrian activity classification for indoor dead reckoning systems*. In *International Conference on Indoor Positioning and Indoor Navigation(IPIN13)*, Montbeliard- Belfort, France. doi:10.1109/IPIN.2013.6817868

Khalifa, S., Hassan, M., & Seneviratne, A. (2015a). Pervasive self-powered human activity recognition without the accelerometer. *Proceedings of the International Conference on Pervasive Computing and Communication (PerCom)*. doi:10.1109/PERCOM.2015.7146512

Khalifa, S., Hassan, M., & Seneviratne, A. (2015c). Step detection from power generation pattern in energy-harvesting wearable devices. *Proceedings of the 8th IEEE International Conference on Internet of Things (iThings 2015)*. doi:10.1109/DSDIS.2015.102

Khalifa, S., Hassan, M., & Seneviratne, A. (2016b). *Feasibility and Accuracy of Hotword Detection using Vibration Energy Harvester*. In the 17th International Symposium on A World Of Wireless, Mobile And Multimedia Networks (WoWMoM), Coimbra, Portugal. doi:10.1109/WoWMoM.2016.7523555

Khalifa, S., Hassan, M., Seneviratne, A., & Das, S. K. (2015b). Energy harvesting wearables for activity-aware services. *IEEE Internet Computing*, *19*(5), 8–16. doi:10.1109/MIC.2015.115

Khalifa, S., Lan, G., Hassan, M., & Hu, W. (2016a). *A Bayesian framework for energy-neutral activity monitoring with self-powered wearable sensors*, In the 12th IEEE PerCom Workshop on Context and Activity Modeling and Recognition, Sydney, Australia. doi:10.1109/PERCOMW.2016.7457112

Khan, A. M., Lee, Y., & Kim, T.-S. (2008). Accelerometer signal-based human activity recognition using augmented autoregressive model coefficients and artificial neural nets. *Proceedings 30th annual International Conference of the IEEE Engineering in Medicine and Biology Society*. doi:10.1109/IEMBS.2008.4650379

Kim, H., Kim, J., & Kim, J. (2011). A review of piezoelectric energy harvesting based on vibration. International Journal of Precision Engineering and Manufacturing, 12(6), 1129–1141. doi:10.1007/s12541-011-0151-3

Kwapisz J. R., Weiss G. M., & Moore S. A. (2011). Activity recognition using cell phone accelerometers. *ACM SigKDD Explorations Newsletter, 12*(2).

Lan, G., Khalifa, S., Hassan, M., & Hu, W. (2015). Estimating calorie expenditure from output voltage of piezoelectric energy harvester - an experimental feasibility study. *Proceedings of the 10th EAI International Conference on Body Area Networks (BodyNets)*. doi:10.4108/eai.28-9-2015.2261453

Lan, G., Xu, W., Khalifa, S., Hassan, M., & Hu, W. (2016). *Transportation Mode Detection Using Kinetic Energy Harvesting Wearables*. In *WiP of the International Conference on Pervasive Computing and Communication (PerCom)*, Sydney, Australia. doi:10.1109/PERCOMW.2016.7457048

Lee, D., Dulai, G., & Karanassios, V. (2013). Survey of energy harvesting and energy scavenging approaches for on-site powering of wireless sensor and microinstrument-networks. In *Proceedings SPIE 8728, Energy Harvesting and Stor age*. Materials, Devices, and Applications.

Lefeuvre, E., Badel, A., Richard, C., Petit, L., & Guyomar, D. (2006). A comparison between several vibration-powered piezoelectric generators for standalone systems. *Sensors and Actuators, 126*(2), 405–416. doi:10.1016/j.sna.2005.10.043

Mitcheson, P. D., Yeatman, E. M., Rao, G. K., Holmes, A. S., & Green, T. C. (2008, September). Energy harvesting from human and machine motion for wireless electronic devices. *Proceedings of the IEEE, 96*(9), 1457–1486. doi:10.1109/JPROC.2008.927494

Nintanavongsa, P., Muncuk, U., Lewis, D. R., & Chowdhury, K. R. (2012). Design optimization and implementation for rf energy harvesting circuits. IEEE Journal on Emerging and Selected Topics in Circuits and Systems, 2(1), 24–33. doi:10.1109/JETCAS.2012.2187106

Olivares, A., Grriz, J. M., & Olivares, G., (2010). A study of vibration-based energy harvesting in activities of daily living. *Proc. 4th International Conference on Pervasive Computing Technologies for Healthcare (PervasiveHealth).*

Orsi, G., & Tanca, L. (2010). Context modelling and context-aware querying: can datalog be of help? *Proceedings of the Datalog 2.0 Workshop.*

Osmani, V., Balasubramaniam, S., & Botvich, D. (2008). Human activity recognition in pervasive health-care: Supporting efficient remote collaboration. *Journal of Network and Computer Applications*, *31*(4), 628–655. doi:10.1016/j.jnca.2007.11.002

Poppe, R. (2010). A survey on vision-based human action recognition. Image and Vision Computing Journal, 28(6). doi:10.1016/j.imavis.2009.11.014

Qi, X., Keally, M., Zhou, G., Li, Y., & Ren, Z. (2013). Adasense: Adapting sampling rates for activity recognition in body sensor networks. *Proceedings of IEEE 19th Real-Time and Embedded Technology and Applications Symposium (RTAS).*

Raju, M. (2008). *Energy Harvesting ULP Meets Esnergy Harvesting: A Game-Changing Combination for Design Engineers*. Texas Instrument. Retrieved June 11, 2016, from http://www.ti.com/corp/docs/landing/cc430/graphics/slyy018_20081031.pdf

Rao, Y., Cheng, S., & Arnold, D. P. (2013). An energy harvesting system for passively generating power from human activities. *Journal of Micromechanics and Microengineering*, *23*(11), 114012. doi:10.1088/0960-1317/23/11/114012

Ravi, N., Dandekar, N., Mysore, P., & Littman, M. L. (2005). Activity recognition from accelerometer data. *IAAI'05 Proceedings of the 17th conference on Innovative applications of artificial intelligence.*

Singla, G., Cook, D. J., & Schmitter-Edgecombe, M. (2010). Recognizing independent and joint activities among multiple residents in smart environments. Journal of Ambient Intelligence and Humanized Computing, 1(1). doi:10.1007/s12652-009-0007-1

Sodano, H. A., Inman, D. J., & Park, G. (2005). Comparison of piezoelectric energy harvesting devices for recharging batteries. *Journal of Intelligent Material Systems and Structures*, *16*(10), 799–807. doi:10.1177/1045389X05056681

Vocca, H., & Cottone, F. (2014). Kinetic energy harvesting. In *ICT* (pp. 25–48). Energy - Concepts Towards Zero - Power Information and Communication Technology.

Wang, S., Yang, J., Chen, N., Chen, X., & Zhang, Q. (2005). *Human activity recognition with user-free accelerometers in the sensor networks*. In Neural Networks and Brain, In International Conference on Neural Networks and Brain.

Weinberg, H. (2002). Minimizing power consumption of imems accelerometers. In *Applications AN-601*. Analog Devices.

Wongpatikaseree, K., Ikeda, M., Buranarach, M., Supnithi, T., Lim, A. O., & Tan, Y. (2012). *Activity recognition using context-aware infrastructure ontology in smart home domain*. In *Seventh International Conference on Knowledge, Information and Creativity Support Systems (KICSS)*. doi:10.1109/KICSS.2012.26

Xu, G., Yang, Y., Zhou, Y., & Liu, J. (2013). Wearable thermal energy harvester powered by human foot. *Frontiers in Energy*, *7*(1), 26–38. doi:10.1007/s11708-012-0215-9

Yan, Z., Subbaraju, V., Chakraborty, D., Misra, A., & Aberer, K. (2012). Energy-efficient continuous activity recognition on mobile phones: An activity-adaptive approach. *Proceedings of the 16th Annual International Symposium on Wearable Computers (ISWC)*. doi:10.1109/ISWC.2012.23

Yun, J., Patel, S., Reynolds, M., & Abowd, G. (2008). A quantitative investigation of inertial power harvesting for human-powered devices. *Proceedings of the 10th International Conference on Ubiquitous Computing, UbiComp '08*. doi:10.1145/1409635.1409646

Zappi, P., Lombriser, C., Stiefmeier, T., Farella, E., Roggen, D., Benini, L., & Trster, G. (2008). *Activity recognition from on-body sensors: Accuracy-power trade-off by dynamic sensor selection*. In *European Conference on Wireless Sensor Networks (EWSN)*, Bologna, Italy. doi:10.1007/978-3-540-77690-1_2

Zungeru, A. M., Ang, L. M., Prabaharan, S., & Seng, K. P. (2012). Radio frequency energy harvesting and management for wireless sensor networks. In *Green Mobile Devices and Networks* (pp. 341–368). CRC Press. doi:10.1201/b10081-16

Chapter 6
Wearable Health Care Ubiquitous System for Stroke Monitoring and Alert

Allan de Barcelos Silva
University of Vale do Rio dos Sinos (UNISINOS), Brazil

Sandro José Rigo
University of Vale do Rio dos Sinos (UNISINOS), Brazil

Jorge Luis Victoria Barbosa
University of Vale do Rio dos Sinos (UNISINOS), Brazil

ABSTRACT

Research regarding stroke indicates that ensuring a short elapsed time between accident and treatment can be fundamental to allow saving patient's life and avoid future sequels. This paper describes a model for monitoring and rescuing victims in situations of possible stroke occurrence. It uses stroke symptoms that can be monitored by mobile equipment, ambient intelligence, and artificial neural networks. The model is independent of human operation and applications or third party devices, therefore adding facilities to increase the quality of life for people with stroke sequel, due to constant monitoring and follow-up provided, allowing the stroke patient to consider a recovery period with greater autonomy. A prototype based on free software platforms was developed, to assess the accuracy and the time elapsed between the prototype to detect and to send an alert. The results indicate a positive outcome for the work continuity.

DOI: 10.4018/978-1-5225-3290-3.ch006

1. INTRODUCTION

The health is one of the areas that can benefit using information for decision making. The growth of health informatics is associated in part due to advances in computing and communication technologies. Also is related to the belief that medical knowledge and patient information are not manageable by traditional methods based on paper and also because of the certainty that access to knowledge processes and decision making play a central role in modern medicine (SBIS, 2014)

Health informatics area is growing on many fronts, and one of the motivations to this advances is the information need for decision making, that is also motivation to an increase in the number of health informatics research projects. Nevertheless, only 27% of research in health informatics area is aimed at the patient monitoring for disease prevention. When the subject of monitoring is Stroke, this number is even lower (IWAYA et al., 2013).

The distribution of deaths caused by cardiovascular diseases in Brazil has shown increasing importance among people over 20 years old, also reaching the level of the first cause of mortality in the range of 40 years old and predominantly in the following age groups. Among these, is situated the cerebrovascular diseases and especially the stroke (FALCÃO et al., 2004). According to Almeida (2012), worldwide Stroke is the second leading cause of death. Also, it represents the third cause of death in industrialized countries and the major cause of disability among adults. Besides that, in the group of cerebrovascular diseases, considering this age group, acute stroke corresponds to over 80% of admissions by the Brazilian Unified Health System (FALCÃO et al., 2004).

Over the last decades, Brazil has been changing his morbidity and mortality profile, with chronic diseases the leading causes of death. Among the most significant chronic illnesses is the stroke, which is one of the main causes of hospitalization and mortality. The stroke causes, in most patients, some form of disability, either partial or complete (ALMEIDA, 2012). Regardless of type, the stroke is the cause of dissatisfaction with life and various functional limitations because of loss of autonomy resulted from disabilities resulting from the incident (FALCÃO et al., 2004).

Information maintained by the World Health Federation indicates that every year, near 15 million people worldwide suffer a stroke. From this context, approximately six million dies and five million are left permanently disabled. According to (Rosamond et al., 2008) and (Thrift et al., 2014), stroke is the second leading cause of disability, after dementia. Regarding global health, the stroke is indicated as the second most frequent cause of death for elder people, above the age of 60 years.

According to Donnan et al. (2008), stroke is the cerebrovascular disease with the highest incidence and is more morbidity, with mortality rates around 25% within one month, 33% at six months and reaching 50% within one year. In cases diagnosed with intracerebral hemorrhage, the prognosis is even worse, because only 50% of patients survive after the first month. The authors Almeida (2012), Chaves (2000), Falcon et al. (2004), Cumbler et al. (2010) claim that stroke is a medical emergency. Perlini and Mancussi and Faro (2005) point out that the doctor should work quickly in these cases because it has a limited time window to perform the necessary interventions.

Once the accident happens, after the initial treatment, the most common source of support and care is the family. Only when exhausted the resources in the family as in the support for the more specific tasks needed, is that the family seek the path of institutionalization (PERLINI; Mancussi and Faro, 2005). The treatment after occurred the stroke is crucial, but the need for prevention and rescue of the victims is one of the determining factors for survival in these cases. Throughout of this work, several studies will be observed relating the rescue time the victims as one of the deciding factors for the treatment of disease.

The adoption of Intelligent Environments technology can assist in the care, monitoring and on the comfort of these patients. The Intelligent Environments make use of embedded computing devices to integrate in an intelligent way to the environment, using models of automated decision that applies information from sensors (AARTS; WICHERT, 2009). Intelligent Environments is considered the technology that enables a new generation of systems that provide services in a flexible, transparent, sensitive a responsive. Also, these services can be relied upon daily and can anticipate events. Another aspect of this technology is that it requires minimal skills to human-computer interaction (Becker et al., 2006) (Kleinberger Thomas et al., 2007). Related to these technologies, the domain of Assisted Environments was recognized as a promising approach, thus making it currently one of the most important areas of research and development, where accessibility, usability, and learning play a major role (Kleinberger Thomas et al., 2007).

After a study in the health informatics area, it was observed that most related work required some infrastructure by the hospitals or, in some cases, was focused on the care after the victim is affected by stroke. Another line of known works didn't produce alerts in the event of detection of an impending stroke. Therefore, the development of initiatives such as the one described in this work can assist in helping people before they become victims of stroke, anticipating the emergency call and reducing the time window between the accident and the treatment. Thus the objective of this research is to evaluate the potential use of a model for monitoring and help stroke victims using mobile computing, sensors, and technologies of intelligent environments and artificial intelligence.

Some specific objectives have been identified, and they are preconditions for achieving the overall goal. First, rules for monitoring patients have been determined through interviews with medical professionals and theoretical frameworks. After using the Intelligent Environment Technologies and Artificial Intelligence together with the Arduino prototyping platform, we developed a model of monitor and rescue stroke victims, and also a partial ubiquitous prototype. The results provided through the prototype were evaluated, indicating a positive perspective of continuity of the work.

This chapter is organized into six sections, starting with the introduction and ending by the conclusion. Section two presents the theoretical foundations that describe the issues that served as the basis for the development of this research. In section three, will be addressed related studies, which will contribute to the theme of this research, as well as similar studies in the literature. During the four-section, we will describe the materials and methods used during the research, as well as the architecture of the model object of the chapter and its implementation. Finally, Section five describes and analyzes the results of the conducted tests with the prototype.

2. BACKGROUND

2.1 Stroke

The stroke is a cerebrovascular disease that presents a high incidence and morbidity. In general, it has a death rate of about 25% in a month, 33% in 6 months and 50% within one year. In cases when is confirmed the diagnosis of intracerebral hemorrhage and subarachnoid, the prognosis is that only 50% of patients survive after the first month (DONNAN et al., 2008). These numbers become even more alarming when observed the fact that most survivors display neurological deficiencies and significant disabilities, which makes the stroke the first causes of functional disability in the western world (PERLINI; Mancussi and Faro, 2005).

The authors Warlow et al. (2003) and Donnan et al. (2008) suggest that stroke is a big burden for global public health and prevention programs are essential to reduce the incidence. They say that the emergence of an epidemic in less developed countries is almost inevitable since they account for about 66% of the nearly five million deaths from the disease each year. In Brazil, the distribution of deaths by diseases of the circulatory system has been growing in importance among young adults. It assumes the rank of leading causes of death for people in the range of 40 years old and continues to be the dominant cause in the subsequent age groups. It can reach a little more of 80% of hospitalization registered by the Brazilian Unified Health System (SUS) among adults aged between 20 and 59 years old (Falcão et al.,

2004). About 7% of the 15% of Strokes occur in patients who have been hospitalized (CUMBLER et al., 2010).

According to the authors Almeida (2012), Chaves (2000), Falcão et al. (2004), Cumbler et al. (2010), Perlini and Mancussi and Faro (2005) the Stroke is a medical emergency, therefore the doctors must work quickly, because it has a limited window of time to perform a patient rescue. After the patient rescue is necessary the physicians to answer various questions such as: If the start of the symptoms was sudden and if the symptoms can be attributed to brain damage.

A result of this is the change of Brazilian morbidity and mortality profile, making the non-communicable chronic diseases group the leader among t causes of death. Also making the stroke the most important chronic illnesses as one of the leading causes of hospitalization and mortality, causing the vast majority of patients some disability, whether full or partial (ALMEIDA, 2012). In the research of Falcão et al. (2004) is assert that all cases interviewed in the study showed the change of functional capacity as the leading cause of dissatisfaction with life, due to interference in personal life, in the reduction of autonomy and self-esteem caused by stroke. This statement is also seen in the research of Ostwald (2008) which append the information that the cognitive deficits resulting from stroke may compromise a person's ability to concentrate, decrease the level of attention and causes memory loss of short or extended periods of time.

The creation and adoption of a comprehensive action to increase the satisfaction of people with disabilities caused by Stroke is suggested by Ostwald (2008), Perlini e Mancussi e Faro (2005) e Falcão et al. (2004). According to them, the permanence of disabling sequelae by imposing limitations on mobility, sensitive, sensory, understanding and expression of thoughts for patients can change the dynamics of the lives of these people. This happens not only by physical sequels that restrict the activities of daily life and make them often dependent on others to move and act with bigger or lower independence but also by compromising their ability to manage personal and family life.

Studies indicate that family is the most common source of support and care, both for formal support and more specific tasks, only when exhaust the resources it's when family seeks the path of institutionalization. The experience of taking care of someone affected by Stroke at home has become increasingly common in the daily life of families. In hospitals, the incentive policy for patient discharge as soon as possible creates a constant challenge for nurses: families and patients preparation to reorganize their lives in their homes, to enable them to take care of relatives or themselves in a few days by detecting, preventing and controlling situations that may occur. After all, the final stage of recovery will happen at home (PERLINI; Mancussi and Faro, 2005).

We can split Strokes into two broad groups: ischemic or hemorrhagic, the distinction between clinical subtypes is one of the most important and urgent steps (WARLOW et al., 2003). Is observed for Donnan et al. (2008) that the risk of recurrence of the disease is still substantially larger than previously thought, reaching 30 percent in the first month in some subgroups.

In recent decades, more accurate studies have identified modifiable and not modifiable risks for ischemic stroke and hemorrhagic (CHAVES, 2000). It is believed that approximately 80% of all strokes are ischemic (DONNAN et al., 2008).

Some risk factors can be modified, and others are not, the modifiable factors (e.g. hypertension, arterial fibrillation, diabetes and smoking) are common and affect health in many ways, have treatment and offer opportunities to modify the risk of a significant number of people. On the other hand, fixed factors such as age and heredity are less prevalent and cannot be treated. The risk factors that have been identified only explain about 60% of the causes and risk studies are needed to determine the risk factors that represent the 40% remaining, some of which may be genetic (DONNAN et al., 2008).

Identification and control of risk factors aim the primary prevention of disease in the population, but nowadays, distinguish determinants of recurrence and mortality after ischemic event became the basis of secondary prevention strategies because of the recurrence of stroke still the greatest threat to any survivor (CHAVES, 2000).

It's unanimous among the authors Chaves (2000), Donnan et al. (2008), Warlow et al. (2003), Eluf Neto, Lotufo e Lólio (1990) the notion that hypertension is the major risk factor and being associated with the disease of large and small arteries, besides the fact that 75% of patients with ischemic stroke are high blood pressure diagnosis, of which about 50% have a history of hypertension. Additionally, research such as Leonardi-Bee et al. (2002) and Goldstein et al. (2006) reinforces the relationship between higher levels of blood pressure after the accident with the increased risk of recurrence of the stroke. On the other hand, Warlow et al. (2003) and Goldstein et al. (2006) point out that atrial fibrillation is the most important factor because it's so common, carries a high risk of stroke, and is causative in many cases.

Therefore, the control of hypertension in patients with atrial fibrillation is crucial, reducing the risk of ischemic stroke and the risk of intracerebral hemorrhage complicating therapy (GOLDSTEIN et al., 2006). The research developed by Donnan et al. (2008), suggests that the leading cause of early mortality is the neurological deterioration with contributions from other causes, such as secondary infections, but later deaths are commonly caused by heart disease or complications of the stroke. After a less intensity disease episode, the risk of recurrence is still substantially larger than previously thought, reaching 30% in the first month in some subgroups. Although the epidemiological data shows a decline in mortality, it is expected that the incidence of the disease prevalence of disabilities in reverse and related mental episodes (PERLINI; Mancussi and Faro, 2005).

The challenge to the model developed here proposed is, therefore, possible to analyze how the monitoring features to be composed of smart environments, sensors, and artificial intelligence can support detection and alerting activities.

2.2 Inteligent Environments

Smart environments make use of embedded computing devices to intelligently integrate the environment using automated decision-making models, from the reading of sensor information (AARTS; WICHERT, 2009). Is considered to be the technology that enables a new generation of systems, which provide their services in a flexible, daily, transparent, responsive, receptive, and anticipation way and that requires minimum skills for human-computer interaction (BECKER et al., 2006) (Kleinberger Thomas et al., 2007). This paradigm of smart environments is used as the basis for new models of technological innovation, whose large-scale integration of electronic components in the environment allows the user to interact with the space around you in a natural way (AARTS; WICHERT, 2009).

The intelligent environments were soon recognized as a promising approach to address the problems in the field of Assisted Living or Assisted Environments. The accessibility, usability, and learning play a significant role and where future interfaces are a concern to be taken into account (Kleinberger Thomas et al., 2007).

Assisted environments apply the technology of intelligent environments to enable people with special needs such as disabled, elderly, victims of stroke have an increase in quality of life and autonomy (Kleinberger Thomas et al., 2007). It is understandable that people often want to remain in their familiar living environment, especially when they get older, or who prefer to recover at home and not in a hospital. This improves their quality of life and gives them the confidence that is some of the best things that can be achieved with modern computer technology (BECKER et al., 2006). In these cases, one of the main types of intelligent environments assisted is the Home Care Systens (HCS)(Kleinberger Thomas et al., 2007).

The goal of HCS is to allow people to live longer in their ideal environment, retaining their independence, even when they have disabilities or medical illnesses. Therefore, through early detection of future or existing problems, these systems can help people proactively in preventing problems, for example, suggesting actions or help them in critical situations, such as neurological, mental problems, loss of consciousness, or physical disability. Additionally, these systems can provide support for other systems, which also reduces the likelihood of false alarms either positive or negative. For example, the first system provides new proactive assistance, providing useful instructions. Then it integrates friends, family members or caregivers to the solution of the problem (Kleinberger Thomas et al., 2007).

Kleinberger Thomas et al. (2007) and Becker et al. (2006) suggest that as shown in Figure 1 the field of HSCs can be divided into:

- **Emergency Assistance Services:** Are intended for beginning, reliably detects, alerts and recovery spreading critical conditions that may result in emergency;
- **Improvement of Services Autonomy:** Aim to make it possible to abandon the previous manual given by medical personnel and social assistance or family and replace it with a system with adequate support;
- **Comfort Services:** Cover all areas that do not fall into the previous categories.

Of course, the comfort services do not have the same importance and social impact as the other two categories, although they may increase acceptance of HCS in the mass market, especially for people who currently have reduced autonomy and require emergency treatment (BECKER et al., 2006).

All three service categories are desirable. In general, once that make life easier for the target group and their respective environment, emergency treatment can be considered as the core of any system of domiciliary care (BECKER et al., 2006).

According to Becker et al. (2006), the HCS shall contain the following:

1. **Precision:** Must provide their emergency assistance with high precision;
2. **Adaptability:** Must be adaptable to the changing capabilities of assisted person (e.g. by adding or replacing devices) and be able to respond to all kinds of exceptional situations flexibly, abnormal without interruption of your service;
3. **Availability:** Their operation tends to be uninterrupted;

Figure 1. Home Care Systems types (KLEINBERGER THOMAS et al., 2007)
Source: KLEINBERGER THOMAS et al., 2007

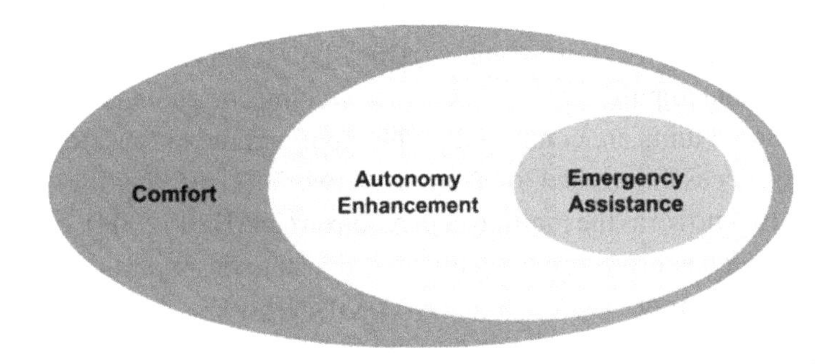

4. **Reliability:** Unreliable systems are not an acceptable alternative to human care, especially for systems for the treatment of emergencies;
5. **Adequation:** It must meet the current demand of the assisted person.

2.3 Neural Networks

According to Faceli et al. (2011) in recent decades, with the growing complexity of the problems to be treated computationally and data volume generated by different sectors, it became evident the need of more sophisticated computational tools, which were more autonomous, reducing the need for human intervention and addiction specialists.

In his book, Haykin (1999) explains that the topic artificial neural network, commonly referred to as "neural networks" has been driven from its inception by the recognition that the human brain computes in a way entirely different from a conventional computer. The brain is highly complex, nonlinear and parallel operating in processing information.

Is highlighted by Rosa (2011) and Bishop (2006) that neural network-based computing is a mathematical method that is used in the field of artificial intelligence to simulate the neural mechanism of the human brain. Bishop (2006) states that the neural networks have its widespread use in the field of intelligent computing, due to give better performance on data classification with high fault tolerance and robust besides learning capacity. As well as having the ability to organize its structural constituents, to perform certain computations (e.g., pattern recognition, perception, and motor control) (HAYKIN, 1999).

In its most general form, a neural network is a machine that is designed to model the way in which the brain performs a task or function of particular interest (HAYKIN, 1999). According to Rose (2011) neural networks obey the laws or rules, that Haykin (1999) says he is typically implemented using electronic or components is simulated in software. And to achieve a good performance. Still Haykin (1999), employ an interconnection between fundamental elements for the functioning of a neural network, the large information-processing units, also known as neurons.

The standard model of neural network consists of a set of neurons and connections. Each neuron has its activation, which is typically a binary number (1 means the presence of the input impulse and 0, absence). Each connection contains a real number, your weight. Some units are connected to the input and the output (ROSA, 2011).

The most successful model of Neural Network is the feedforward pattern recognition, also known as the multilayer perceptron (BISHOP, 2006). In this model, the ability of hidden neurons to extract higher order statistics is particularly valuable when the size of the input layer is high (HAYKIN, 1999).

Second Bishop (2006) the hidden layer can include multiple layers, and each layer can have one or more neurons, where the existence of intermediate layer allows the network develop internal representations, where the behavior of these units is automatically learned, is not pre-programmed (ROSA, 2011). The largest property of Connectionist systems is that the neural network learns not only to sort the inputs in which she is trained but also to generalize and be able to sort entries never seen (ROSA, 2011).

In his book, Haykin (1999) notes the existence of three models of neural networks, single-layer feedforward networks, multilayer feedforward networks and recurrent networks. In the first, we have a layer of source nodes entry that juts over a layer of output neurons, but not vice versa, this network is strictly a feedforward acyclic type. The second is distinguished by the presence of one or more hidden layers, and the compute nodes are correspondingly called hidden neurons or hidden units. The function of hidden neurons is intervening between the external input and the output of the network in any useful way, by adding one or more hidden layers, the network can extract higher-order statistics. And the last stands of a neural network on the input power that is, at least, have a closed circuit feedback (HAYKIN, 1999).

3. RELATED WORKS

According to Iwaya et al. (2013) most projects in Brazil are belong the categories of research and health surveillance, with a focus on primary health care, both in urban areas and in the most remote part of the country. Many of these projects are relatively general, regarding the target population. However, some of them were developed for specific groups, such as children, or for specific health conditions, such as oral health habits, dengue fever or heart disease.

In their study, Iwaya et al. (2013), suggests that we can classify the projects for mobile devices in Health Informatics as:

- Research and health surveillance;
- Patient records;
- Patient monitoring;
- Decision support systems;
- Monitoring of treatment;
- Awareness;

As seen in Figure 2, most applications in Brazil belong to the category of research and surveillance with the focus on primary care. The most patient registry projects simply provide physicians access to the history of their patients through mobile

Figure 2. Projects classification in Brazil research (IWAYA et al., 2013)
Source: IWAYA et al., 2013

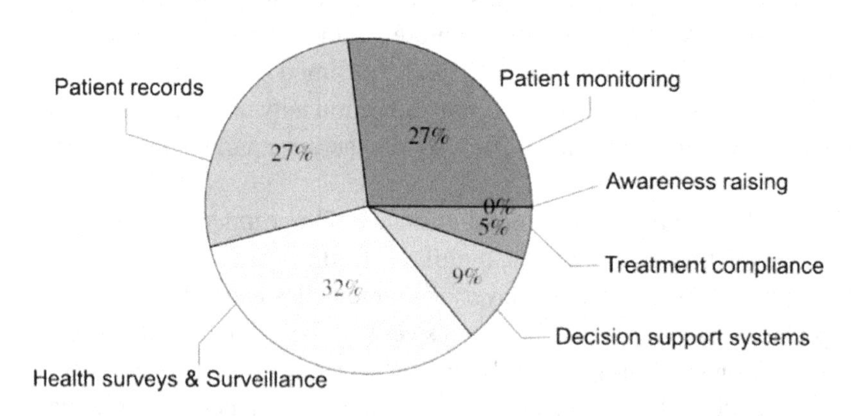

devices. Despite this, unlike the previous categories, the projects belonging to the category of monitoring of patients are focused on a particular type of disease, such as heart disease, mental disorders, and physical activity analysis (IWAYA et al., 2013).

Seen the numbers displayed in Figure 2 and Figure 3, we conclude that most of the projects are in the category of research and surveillance in health and about 42% of the projects are developed for smartphones.

Some of the projects cited by Iwaya et al. (2013) in his research and that are related to this work were the HandMed, MonGluco, TeleCardio, SIAF and AtoMS, and they will be described below.

Figure 3. Project target devices in Brazil research (IWAYA et al., 2013)
Source: IWAYA et al., 2013

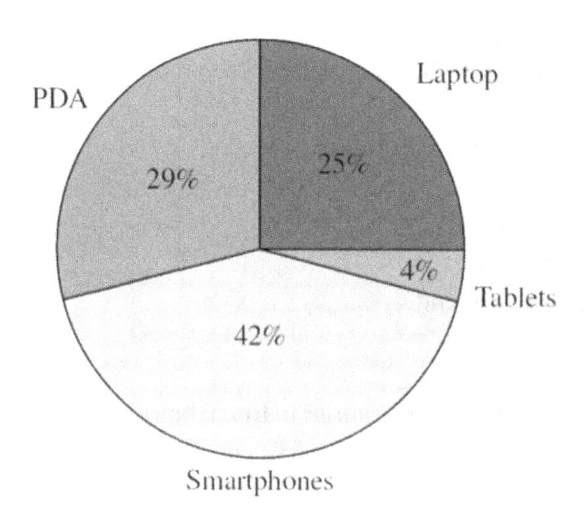

The HandMed is designed for PDA and based on forms, this application automatically identifies the patient's symptoms, to detect health problems in advance. For this, are made available to patients, automatic interrogation according to a proposed algorithm for systematic review of symptoms. It was developed by the University of Brasilia and tested with patients at the University Hospital of Brasília. The solution is part of a larger project characterized by the creation of a patient medical information manager system (GIMPA Project), in which integrates sensors on the human body that enable continuous monitoring of the patient, digital libraries and the Electronic Patient Record (CASTRO et al., 2004). It was developed by the Federal University of Minas Gerais for the continuous monitoring of patients using sensors. Apparently, never became fully operational.

The MonGluco was also developed for PDA. This application is used to monitor the health of cardiac patients in Intensive Care Units (ICU). The data was transmitted over a Wi-Fi network using the communication protocol of Health Level Seven (HL7). The information collected was stored in an electronic patient record and showed in a bedside monitor in the ICU for the medical staff can easily access the glucose readings (MURAKAMI et al., 2006). The system was developed by the Heart Institute (InCor) in São Paulo and deployed within the same institution.

TeleCardio is a system for remote monitoring of cardiac patients, which uses a platform to support mobile applications, context-sensitive analysis of and automatic generation of electrocardiographic signals alerts, enabling the expansion of access to emergency care for chronic patients (ANDREÃO; PEREIRA; CALVI, 2006). The project was developed by the Federal University of Espírito Santo (UFEES).

The MedKart is a mobile workstation that was designed as part of complete Hospital Information System (HIS) and Picture Archiving and Communication System (PACS). It allows patient admission, discharge, and transfer (ADT), the registration of medical activities, registration of diagnostics and therapy, to order entry and access to all patient data, including vital signs, images and lab tests (GUTIERREZ et al., 2008). The survey was conducted by the Heart Institute (InCor) in partnership with the University of São Paulo Medical School. Currently, the project is in InCor dependencies.

The SIAF operates in physical activity monitoring (e.g. heart rate, blood pressure, glucose) for the generation of health and performance indicators. It will support the planning of public policies to promote physical activity for the population, using sensors on human body, wireless networks and mobile devices to monitor the physical condition of the participants of these events. The data is sent to health managers, where they will evaluate the effectiveness of the program (PORTOCARRERO et al., 2010). This research project was developed by the Federal University of São Carlos, in the State of São Paulo in Brazil.

The AtoMS is a mobile system for remote monitoring and support for patients with heart disease. It provides mechanisms that allow the paramedics to interact remotely with a cardiologist, facilitating the detection of acute myocardial infarction in early stages (ZIVIANI et al., 2011). The project was integrated into a set of emergency care units in the city of Rio de Janeiro. It was developed by the National Laboratory for Scientific Computing (LNCC).

The research developed by Lucas and Aguiar (2013) proposes the use of depth sensors to improve the quality of life for patients and their families, from the creation of a monitoring and recognition system that will be applied to hemiplégicos patients.

Machado et al. (2008) proposed an architecture of a system for remote monitoring of patients through the application of mobile devices and Web Services. It consists of a modular architecture that allows you to integrate a variety of development platforms based on free software and free of charge, which can be easily extended or even appropriate to existing systems. One of the advantages of the adoption of this system is the facility for the exchange of information between the patient and the doctor without the need of physical contact between they. This allows the release of hospital beds allowing patients may remain in their households counting on quick assistance in case of emergency (MACHADO et al., 2008).

As showed in Figure 4, the major Brazilian projects has generic focus, which is described by Doukas, Pliakas and Maglogiannis (2010) as a common situation to new areas of research such as mHealth. It is expected that in the future with the

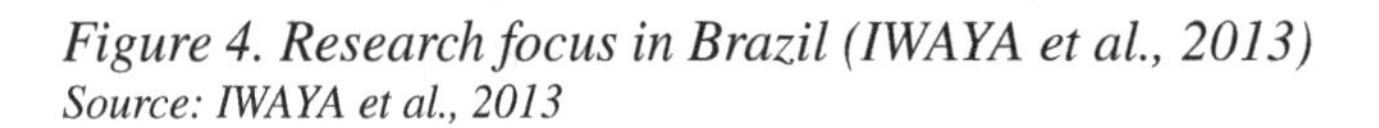

Figure 4. Research focus in Brazil (IWAYA et al., 2013)
Source: IWAYA et al., 2013

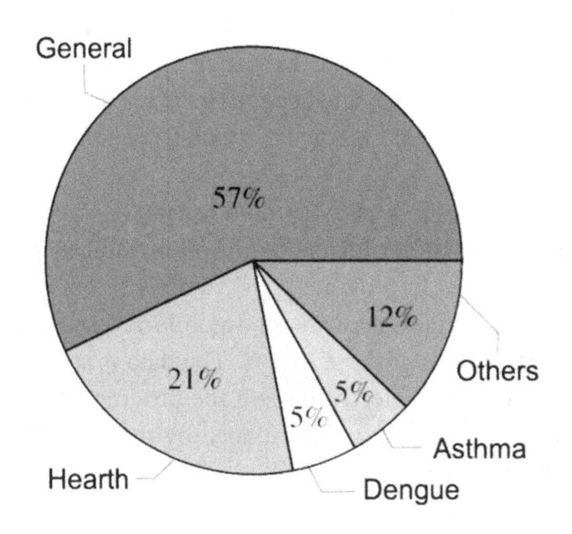

General-purpose applications becoming more established, and start the projects tailored to the specific needs of each health problem.

Most similar projects described above use remote monitoring systems as a focus and have a little or no mobility, although exceptions occur such as in cases of HandMed, MonGluco and research of Machado et al. (2008). Mostly these projects are not for use by the patient, but by doctors, which restricts its use into the hospital environment and or with the need of high level of infrastructure to support their systems.

As can be seen in Table 1, most projects need workstations and not have their device, so they require institutions support for can implement them, it can complicate the patient access to this technology. It can be observed that the projects studied have used more general focus, usually involving remote monitoring of vital signs and need to consult an expert to identify symptoms. Unlike the others, the proposed model makes use of neural networks and differs from the models studied in the literature, both in architectural aspects as also the project focus. And the use of free software and neural networks shows up as a front differential to the other works studied.

Table 1. Compative between systems

System	Focus	Device	Direct Intervention	Open software
The present study	Stroke	It has its own device	No	Yes
Machado et al. (2008)	Remote monitoring	Phones	Não	Yes
HandMed	Identification of mental disturbances	PDA	Yes	Yes
MonGluco	Heart patients in intensive care units	PDA	Yes	No
TeleCardio	Heart disease and analysis of ECG signals	Workstation	Yes	No
MedKart	HIS and PACs	Workstation	Yes	No
SIAF	Physical activity Monitoring	Sensors and mobile devices	Yes	No
AtoMS	Cardiac diseases	ERP and mobile devices	Yes	No
SRA	Hemiplegic gait identification	Kinnect and work stations	Yes	No

Figure 5. Mobility in patient monitoring through Web Services (MACHADO et al., 2008)
Source: MACHADO et al., 2008

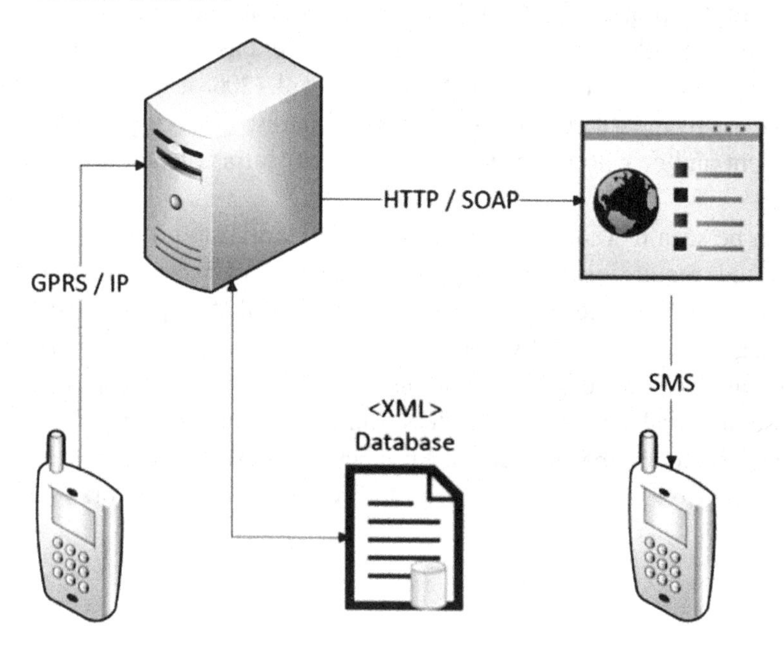

Figure 5 shows the model proposed by Machado et al. (2008), which appear the client application running on the mobile phone, by sending the data through a General Packet Radio System (GPRS) connection for a WS running in Java J2EE context GlassFish V2. The server module analyzes the information received, persist it in the database and also checks if the clinical picture is changed and if necessary notify the doctor with information obtained by sending an email. Additionally, it makes a connection to a WS outsourced that sends a Short Message Service (SMS) to the cell phone of the doctor in charge (MACHADO et al., 2008).

The same general findings showed in this section can also be observed when other related works are studied. For example, Tamura (Tamura et al., 2015) demonstrated a new model for monitoring physiological parameters for rehabilitation training system to improve team-based healthcare. The system collect data from blood pressure (estimated with a cuffless monitor), electrocardiogram, Kinect-based rehabilitation training system, and web-based care system. In this work the ECG signs are obtained through sensors mounted in the patient's clothes, to ensure precision in the data gathering about the heart rates. According to the author, the training system was tested with four hemiplegic patients and garnered a high acceptance rate.

Rofouei (Rofouei et al., 2011) developed a non-invasive wearable neck-cuff system for real-time sleep monitoring and visualization fo physiological signals. The signals were generated from sensors in a soft neck-worn collar and sent via Bluetooth to a cell phone for data storage. According to the author, the data was processed and reported to the user or uploaded to the cloud or a local PC.

Saponara (Saphonara et al., 2012) described an ICT solution to improve the provisioning of healthcare service for Chronic Heart Failure (CHF) patients able to connect in-hospital care of acute phase with out-of-hospital follow-up. The author used a set of relevant, vital parameters consists of ECG, Sp02, blood pressure and weight. Saponara uses a monitoring system formed by a bunch of three Bluetooth and battery-powered sensing devices, the commercial UA-767BT arm cuff device for blood pressure readings and the UA-321PBT digital scale by A&D Medical and the ECG-SpO2 module.

Woznowski (Woznowski et al., 2015) proposed a multi-model architecture for Ambient Assisted Living (AAL) remote healthcare monitoring in the home, gathering information from, multiple sensors. The main research goal it has to be not only capable of assisting people with diverse medical conditions, but can also recognize the habits the house members.

Sengupta (Sengupta et al., 2016) designed a new multi-class classification model to predict stroke occurrence. The model proposed by the author addresses the challenges of class inbalance.

Mcheick (Mcheick et al., 2016) proposed a Stroke Prediction System based on ontology and Bayesian Belief Networks (BBN). The system collect and analyzes data of a patient then use a wearable sensor and the mobile applications to interact with the patient and staffs.

4. THE PROPOSED MODEL

The present study used the bibliographic and research methods of prototyping, where the works studied provided the expertise to propose the model and develop the prototype, which, in turn, sought to gauge the effectiveness of the model. The methodology used was of quantitative character, where was partially developed a prototype to monitor and help victims of stroke. For this study was carried out a partial simulation of use of prototype, in order to assess several aspects. These aspects are the accuracy and the time taken by the neural network to identify the stroke, the time to capture GPS positioning, the time required to send the alert, in addition to the time between the identification of an accident until sending the alert.

The architecture developed for the model was developed on the premise that the device will constantly check the heart rate of the patient. This allows to the search of abnormalities that may trigger the stroke, avoiding that there is interruption in your services, ensuring user safety and allowing the patient to resume its autonomy

Figure 6. Workflow of the system

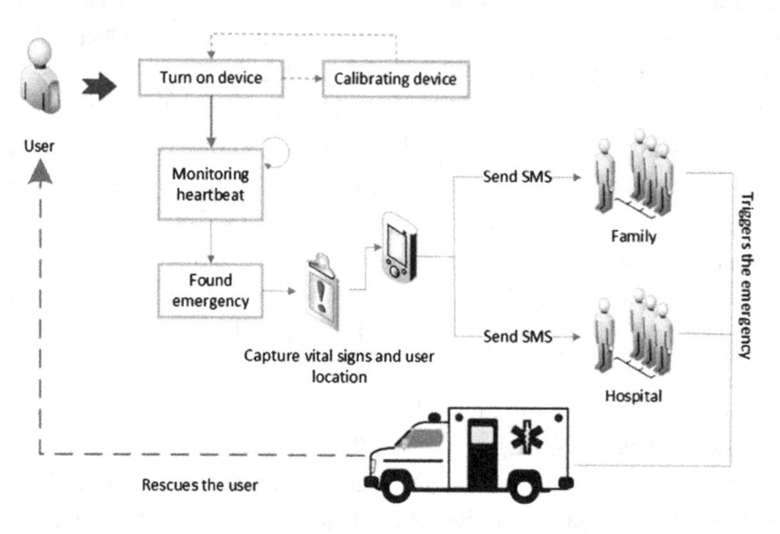

due to the ubiquitous aspect of the model. Once found any anomaly, the device will check blood pressure, confirm the symptoms, activate the GPS system and send the vitals and global positioning via SMS to the phone numbers registered in advance. Figure 6 illustrates these model elements and their operation.

The proposed model begins its operation when placed on the patient's wrist, where it will be identified by the pattern of beats and the user's blood pressure for a few minutes, to the calibration of the device, avoiding in this way an erroneous reading of the data and sending alerts.

The number of devices that will receive the alert is varied, so the patient can choose to notify one or more relatives and hospitals. This is important because, according to Perlini and Mancussi and Faro (2005), the most common source of support and care is the family, both for formal support for the more specific tasks help. In this way, the patient creates a network of contacts in case of emergency, where the closer can unite in an attempt to help the victim while awaiting the arrival of the emergency.

In the course of the architecture development, medical experts in neurology have been consulted to assess the project and assist in the development of neural networks, whose role will be to identify possible Strokes. Furthermore, aiming at the promotion and implementation of the proposed model were chosen open source hardware components, with easy access and low cost. In this way, thus enabling its use by any interested public. In the following chapter will be presented the implementation model, where are described the hardware resources used, as well as the description of the development of a prototype with the features and functionality provided in this chapter.

As can be seen in Table 1, the work described here is the only one that requires no infrastructure, different mind from other works studied requiring servers, data connection, and other resources to perform the processing.

4.1 Prototype and Evaluation

The Arduíno prototyping platform was chosen to develop the prototype of the proposed model, because it is readily available on the market, and have a significant number of users participating in communities and forums regarding its use. In addition to that, it is an open-source platform for physical computing based on a simple microcontroller Board and a development environment for writing software for the Board. The boards can be built, purchased pre-assembled or modified. These boards are usually used together with shields (boards that extend the basic functionality of the Arduino, e.g. sensors, displays, among others). With this, it is possible to develop interactive objects, taking inputs from a variety of sensors, switches, motor control and physical outputs (D AUSILIO, 2012).

The software consists of a C programming language in an IDE based on processing, where then the code must be compiled and loaded aboard. The basic structure of Arduino programming consists of at least two parts:

- Method void Setup(); In this method is located the setting of the environment variables, sensors, and inputs used in the scope.
- Method called void Run(); In this method should be written the code for operation of the Board.

Once written the code, it must be compiled and then loaded on Arduíno board memory, where you can work without the interface with computers or external software, thus ensuring complete independence, portability, and accuracy (D AUSILIO, 2012).

The Arduino Uno is based on the ATmega328 Board and has 14 digital input/output pins operating at 5v, where each pin can provide or receive a maximum of 40mA. Besides, it has six analog inputs, a 16 MHz crystal oscillator, a USB connection, a power connector. It can be powered via USB connection or with an external source of energy (i.e., a 9 V battery). The power supply is automatically selected. The Board can operate on an external supply of 6 to 20v. In addition to the Arduino platform, will be used the DFRobot and SIM908 template Shield a blood pressure meter apparatus model in everyday use, together they will compose the prototyping resources employed in this work.

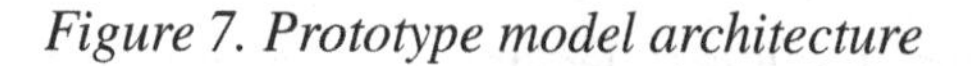
Figure 7. Prototype model architecture

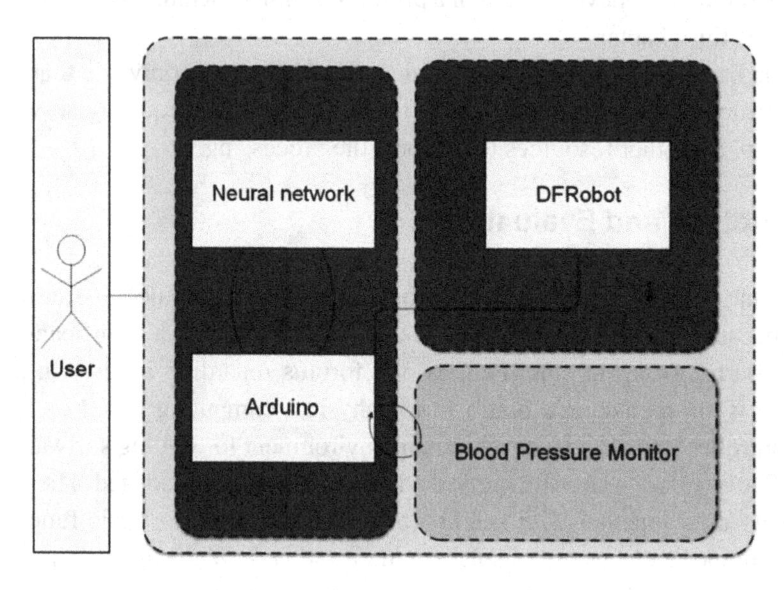

The Shield DFRobot SIM908 has coupled the components of GPS location and then Sim Card Reader, with it will be possible to obtain the patient location and send text messages. For this, the device blood pressure meter will be integrated to the Arduino Uno and DFRobot, to form the prototype hardware composition, as can be seen in Figure 7.

You can see in Figure 7 the independence of features, where Arduino receives the data of the pressure gauge and passes them to the artificial neural network to analysis the occurrence of stroke probability. The Artificial Neural Network is located in the Arduindo Board and executed in its resources. After analyzed the data, it returned to the Arduino and send to the DFRobot shield to capture GPS positioning and send SMS alerts to the families and hospitals. However, the implementation of the prototype was only partial functionality had been developed to obtain the GPS

Figure 8. Partial prototype configuration

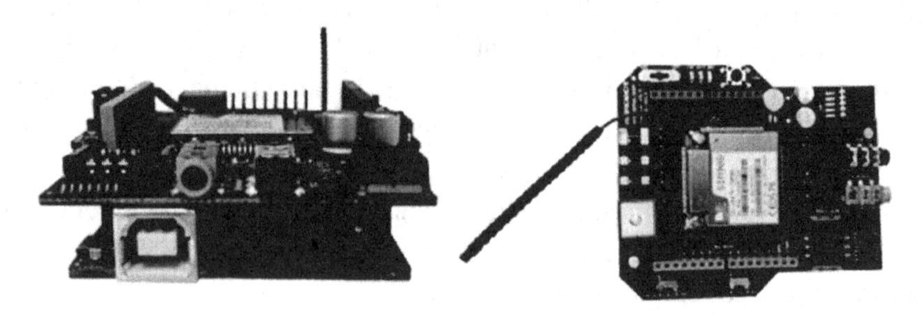

positioning, send alert via SMS and the neural network. You can see the current appearance of the prototype in Figure 8.

The neural net used was the feedforward multi-layer, cited by Haykin (1999) in his book. It was developed through the study of researchers in teh area of neural networks and interviews with experts in neurology from the medical field.

During the interviews with the specialists in neurology initially were performed the following questions: "which are the determining factors for the occurrence of stroke?", "what could be an indicator for stroke?". After answered the previous questions, the expert was asked about: "what is the approximate value of systolic and diastolic pressure would be considered at risk for the occurrence of stroke?", "how much the heartbeat may influence the occurrence of stroke?" and "What the approximate value of beats that could be considered at risk for occurrence of stroke?".

Experts said that blood pressure could be considered as a determining factor for the occurrence of stroke, but stressed the existence of other factors that also influence on the occurrence of the accident. It was said that blood pressure above 16.1 could pose a threat of stroke. Also the experts said that the heartbeat has little influence on the occurrence of the accident, but stressed that it is preceded by values above 70 beats per minute in most cases.

After consultations with experts has been set as inputs of neural network the systolic and diastolic blood pressure of the patient, both would be picked up by the unit pressure gauge attached on the prototype. Once inserted the blood pressure, the neurons in the first layer identify the possibility of stroke and generate an output [0.1].

When the neural network generates the output of the first layer with, the same value will be used as input to the second layer, with the patient's heart rate. The output layer, will determine the actual existence of the possibility of stroke and

Figure 9. Artificial Neural network model

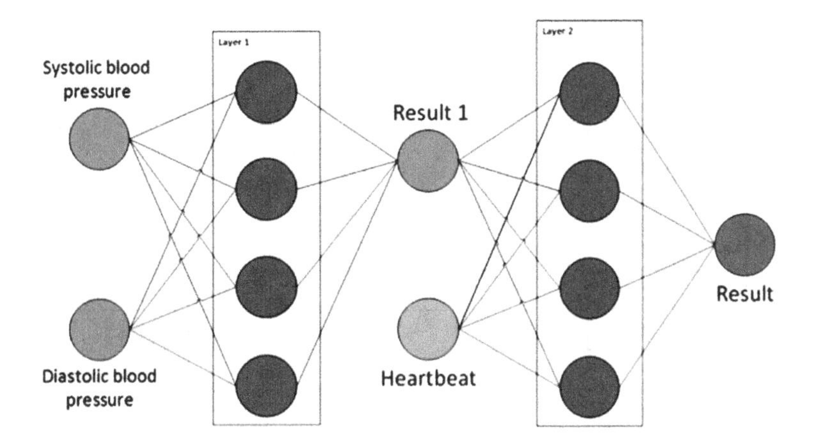

generate a second output [0.1], as seen in Figure 9. If the result obtained for the second is layer 1, then fires the other system modules for initiating the procedure of generating alerts.

The use of neural networks for the processing of patient vital signs makes it unnecessary to have recourse to external services using a data connection to identify a possible stroke. The neural network has become the key differentiator of this work. However, the identification of an accident by the neural network until the alert is considered critical to attest to the operation of the model and prototype.

The first step in the development of the prototype, model validation was the construction of a neural network created by studies in the area and with the help of three medical professionals that are specialists in neurology. The initial idea was to engage the automatic blood pressure monitor with IntelliSense to prototype, providing thus the systolic and the diastolic pressure, both the first neural network inputs. However, the scope of work proved to be extensive because the time available for the coupling between components, and therefore, to the neural network training were inserted eight values, 0 to 19, 9 simulating the systolic and diastolic pressure. As well as the simulated values of heart rate in the range of 60 to 120 beats.

The results of the neural network had adjustments to translate your learning rate to 0.025. After, 210 were inserted input values for the training of the network during 300 times. The learning process took about 4 hours and 30 minutes, with the error rate at 0.02.

The performance of the neural network can be viewed in Table 2. As can be seen, the network discards the remaining steps of the process the values obtained at exit one that is less than 0.5, since they do not represent a threat of stroke. In the table, there is the accuracy of the network and the average time of 1.657 seconds to identify possible strokes. Therefore, the next step if recognized the possibility of stroke is the patient location by GPS module installed on the prototype.

Table 2. Results of the neural network

Blood pressure	*Expected value*	*Output 1*	*Heart beat*	*Expected value*	*Result*	*Elapsed time*
180,10	1	0.98	80	1	0.98	0.332
130,80	0	0.0	-	-	-	0.164
150,60	0	0.0	-	-	-	0.166
100,90	0	0.0	-	-	-	0.164
210,90	1	0.99	120	1	0.98	0.384
160,90	0	0.72	70	1	0.98	0.330
20,80	0	0.0	-	-	-	0.164

The accuracy of the DFRobot Shield was described in the documentation from the manufacturer and suggests that the accuracy of this module is no more that five meters from the place where it was fired.

After the capture of the location via the GPS module, the next step is to send the coordinates in conjunction with blood pressure and heart rate through the SIM Card module during the execution of the tests. It was noted that the SIM Card module takes about 9.25 seconds for processing and sending the alert, but have 100% success rate when sending the message. It is necessary to emphasize that service availability varies according to the sign of the carrier, as well as the difference between the time of sending and receipt of the alert.

Table 3 displays the time required for each step in the process of monitoring and alerting of stroke, where you can see the almost immediate action of the prototype to send alerts to hospitals and or registered family members at about 27.4 seconds.

The results of the tests described in this section operate in order to contribute to the short window of time to perform the necessary interventions in cases of stroke by authors Almeida (2012), keys (2000), Hawk et al. (2004), Cumbler et al. (2010), Perlini and Mancussi and Faro (2005).

Table 3. Operation time evaluation

Activity	*Time Required(s)*
Neural network	0.384
GPS	17.83
Sim Card	9.25
Total	27.4

Figure 10. Prototype in an evaluation session

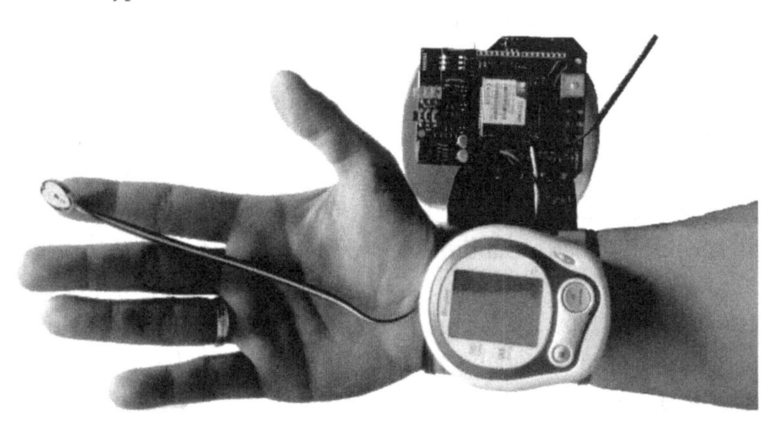

The Figure 10 ilustrates the possibilities of this prototype, as it is shonw in the case of an user session in an evaluation experiment.

In addition to the alert in the event of stroke was found the possibility of improvement in patient's autonomy, since it will not be necessary even to be in a fixed location, and may move from home to work, to the market, physical activities, among others. In this way will make the patient more independent and also increase your quality of life, as Hawk et al. (2004) the stroke is cause of dissatisfaction with life and of various functional limitations, because of loss of autonomy due to the disability.

5. FUTURE RESEARCH DIRECTIONS

Technology projects related to the health area has been growing in recent years. It is common to observe business StartUps developing new applications and devices for monitoring and health care. High expectations in health technologies applied area are due to the arrival of the internet of things where the integration and communication of several wearable sensors and environmental data collection are possible. Through a greater amount of data, you can promote the automation of some stages of the health system such as current history, screening, treatment, monitoring and diagnosis of patients. However, it is strictly necessary that bodies such as the Food and Drug Administration (FDA) to make the regulation and testing of such technologies, as the population may use such guidelines to identify applications with credibility and acting actually.

After the popularization of prototyping and more affordable hardware resources platforms, mobile applications for health initiated the development of joint solutions between applications for smartphone and gadgets, the latter consisting of a sensor group. Among these technologies, one of the most striking is the wearable computing, which can include a set of sensors to capture the most different user data. Such information contributes directly to the discovery of knowledge through data analysis and enables the capture of these in real time.

The main objective of this integration between sensors is the collection of data and transfer this information to interpretation by health professionals, to promote closer doctor-patient relationship and support the diagnosis. However, the need for data analysis by medical professionals must be questioned in future work, as the use of artificial intelligence capabilities to mobile devices facilitate identification of simpler disease without the need for an expert consultation. With this, the number of patients in health clinics and hospitals would be reduced to more complex cases.

This paper presented a model containing the concepts described and what the authors believed the trend in the development of applications for health, in addition to the partial construction of a prototype to validate the model. However, some points still need improvement. The first point is to build a model of the hardware described, not necessarily following the prototype, but always seeking the application of new Technologies. Secondly, for enhancement of the neural network used, to use other parameters that could be evaluated for the network entry, so as to increase the accuracy of the result. As a third point, the authors highlight the inclusion of communication with smartphones or other resources that you can store user data. The fourth and final point are to apply machine learning techniques to discover patterns and knowledge of the user. The raised patterns may act directly on the adaptation of the predictive model to the user.

6. CONCLUSION

This study aimed to present a model for monitoring and help for stroke victims, making use of neural networks and intelligent environments technologies. It is known that most of the victim's stroke survivors are left with sequels, and many of them need care, generally provided by members of the family group who have no instruction necessary to carry them. The situation of the victim could worsen in the event of negligence or lack of preparation by the caregiver who can not understand the symptoms of a new Stroke and consequently fail to provide adequate relief, may cause the patient's death.

One of the main motivations to monitor the Stroke victims is due to the fact that there is ahigh probability of the person become a victim again, case in such the short time to deliver the necessary attention is critical to ensure better conditions to recover (Thrift et al., 2014).

From the analysis of results obtained from tests performed on a prototype, it is observed that the creation of the model-based system is possible. The results of the neural network tests were positive and in conjunction with location tests show the possibility of using the model in the development of prototypes and equipment to be used on a daily basis, aiming to help people in stroke risk group. In this way, these people would have part of their autonomy re-established due to the fact of constant monitoring by the system.

Future possibilities to improve the presented model includes the integrations of the prototype with smartphones or mobile phones, in a manner that this equipment could be used as a gateway of the system. Therefore, besides other advantages, this strategy could provide the necessary functionalities to record the ongoing information for the users and to apply analytics processes into it, ensuring that a set of users could be monitored and their signals integrated to improve the detection process.

REFERENCES

Almeida, S. (2012, January). Análise epidemiológica do Acidente Vascular Cerebral no Brasil. *Revista Neurociências*, *20*, 481–482. doi:10.4181/RNC.2012.20.483ed.2p

Andreão, R., Pereira, J. G., & Calvi, C. Z. (2006). TeleCardio: Telecardiologia a Serviço de Pacientes Hospitalizados em Domicílio. In Congresso da Sociedade Brasileira de Informática em Saúde (vol. 1, pp. 1267-1272). Florianópolis, SC.

Becker, M., Werkman, E., Anastasopoulos, M., & Kleinberger, T. (2006). Approaching Ambient Intelligent Home Care Systems. *Pervasive Health Conference and Workshops*, 1–10.

Bishop, C. M. C. C. M. (2006). *Pattern recognition and machine learning* (Vol. 4). Springer.

Castro, L., Branisso, H., Figueiredo, E., Nascimento, F., Rocha, A., & Carvalho, H. (2004). HandMed: an integrated mobile system for automatic capture of symptoms. In *IX Congresso Brasileiro de Informatica em Saúde* (*vol. 9*, pp. 1-6). Ribeirão Preto, SP: UNIFESP.

Chaves, M. L. F. (2000). Acidente vascular encefálico: conceituação e fatores de risco. *Revista Brasileira de Hipertensão*, 372–382.

Cumbler, E., Anderson, T., Neumann, R., Jones, W. J., & Brega, K. (2010). Stroke alert program improves recognition and evaluation time of in-hospital ischemic stroke. *Journal of Stroke and Cerebrovascular Diseases*, *19*(6), 494–496. doi:10.1016/j.jstrokecerebrovasdis.2009.09.007 PMID:20538480

DAusilio, A. (2012). Arduino: A low-cost multipurpose lab equipment. *Behavior Research Methods*, *44*(2), 305–313. doi:10.3758/s13428-011-0163-z PMID:22037977

Donnan, G. A., Fisher, M., Macleod, M., & Davis, S. M. (2008). Stroke. *Lancet*, *371*(9624), 1612–1623. doi:10.1016/S0140-6736(08)60694-7 PMID:18468545

Eluf Neto, J., Lotufo, P. A., & de Lólio, C. A. (1990). Tratamento da hipertensão e declínio da mortalidade por acidentes vasculares cerebrais. *Revista de Saude Publica*, *24*(4), 332–336. doi:10.1590/S0034-89101990000400013 PMID:2103653

Faceli, K., Lorena, A. C., Gama, J. A., & Carvalho, A. C. P. L. F. (2011). *Inteligência artificial: uma abordagem de aprendizado de máquina* (1st ed.). LTC.

Falcão, I. V., de Carvalho, E. M. F., Barreto, K. M. L., Lessa, F. J. D., & Leite, V. M. M. (2004, March). Acidente vascular cerebral precoce: Implicações para adultos em idade produtiva atendidos pelo Sistema Único de Saúde. *Revista Brasileira de Saú de Materno Infantil*, *4*(1), 95–101. doi:10.1590/S1519-38292004000100009

Goldstein, L. B., Adams, R., Alberts, M. J., Appel, L. J., Brass, L. M., Bushnell, C. D., & Sacco, R. L. et al. (2006, June). Primary prevention of ischemic stroke: a guideline from the American Heart Association / American Stroke Association Stroke Council: cosponsored by the Atherosclerotic Peripheral Vascular Disease Interdisciplinary Working Group; Cardiovascular Nursing Counc. *Stroke*, *113*(24).

Haykin, S. (1999). *Neural Networks - A Comprehensive Foundation* (2nd ed.). Prentice Hall.

Iwaya, L. H., Gomes, M. A. L., Simplício, M. A., Carvalho, T. C. M. B., Dominicini, C. K., Sakuragui, R. R. M., & Håkansson, P. et al. (2013). Mobile health in emerging countries: A survey of research initiatives in Brazil. *International Journal of Medical Informatics*, *82*(5), 283–298. doi:10.1016/j.ijmedinf.2013.01.003 PMID:23410658

Kleinberger, T., Becker, M., Ras, E., Holzinger, A., & Müller, P. (2007). Ambient Intelligence in Assisted Living: Enable Elderly People to Handle Future Interfaces. In Universal access in human-computer interaction. ambient interaction (pp. 103–112). Academic Press.

Leonardi-Bee, J., Bath, P. M. W., Phillips, S. J., & Sandercock, P. A. G. (2002). Blood pressure and clinical outcomes in the International Stroke Trial. *Stroke*, *33*(5), 1315–1320. doi:10.1161/01.STR.0000014509.11540.66 PMID:11988609

Lucas, I. R. B., & Aguiar, C. S. R. (2013). *Sistema de Reconhecimento de Ações - Uso para Monitoramento de Pacientes Hemiplégicos Vítimas de Acidente Vascular Cerebral (Monograph)*. Universidade de Brasília.

Machado, A., Padoin, E. L., Salvadori, F., Righi, L., Campos, M., Sausen, P. S., & Dill, S. (2008). Mobilidade no monitoramento de pacientes através de Serviços Web. In VII Simpósio de Informática da Região Centro do RS, 8.

Mcheick, H., Nasser, H., Dbouk, M., & Nasser, A. (2016). Stroke Prediction Context-Aware Health Care System. *2016 IEEE First International Conference on Connected Health: Applications, Systems and Engineering Technologies (CHASE)*, 30–35.

Perlini, N. M. O. G., & Mancussi e Faro, A. C. (2005). Taking care of persons handicapped by cerebral vascular accident at home: The familial caregiver activity. *Revista da Escola de Enfermagem da U S P.*, *39*, 154–163. doi:10.1590/S0080-62342005000200005 PMID:16060302

Rofouei, M., Sinclair, M., Bittner, R., Blank, T., Saw, N., DeJean, G., & Heffron, J. (2011). A Non-invasive Wearable Neck-Cuff System for Real-Time Sleep Monitoring. *2011 International Conference on Body Sensor Networks*, 156–161. doi:10.1109/BSN.2011.38

Rosamond, W., Flegal, K., Furie, K., Go, A., & Greenlund, K. (2008). Heart Disease and Stroke Statistics—2008 Update. *Circulation*, *2008*(117), e25–e146. doi:10.1161/CIRCULATIONAHA.107.187998 PMID:18086926

Saponara, S., Donati, M., Bacchillone, T., Sanchez-Tato, I., Carmona, C., Fanucci, L., & Barba, P. (2012). Remote monitoring of vital signs in patients with Chronic Heart Failure: Sensor devices and data analysis perspective. *Sensors Applications Symposium (SAS)*, 1–6. doi:10.1109/SAS.2012.6166310

Sengupta, A., Rajan, V., Bhattacharya, S., & Sarma, G. R. K. (2016). A statistical model for stroke outcome prediction and treatment planning. *38th Annual International Conference of the IEEE Engineering in Medicine and Biology Society (EMBC)*, 2516–2519. doi:10.1109/EMBC.2016.7591242

Tamura, T., Sekine, M., Tang, Z., Yoshida, M., Takeuchi, Y., & Imai, M. (2015). Preliminary study of a new home healthcare monitoring to prevent the recurrence of stroke. *37th Annual International Conference of the IEEE Engineering in Medicine and Biology Society (EMBC)*, 5489–5492. doi:10.1109/EMBC.2015.7319634

Thrift, A. G., Cadilhac, D. A., Thayabaranathan, T., Howard, G., Howard, V. J., Rothwell, P. M., & Donnan, G. A. (2014). Global stroke statistics. *International Journal of Stroke*, *9*(1), 6–18. doi:10.1111/ijs.12245 PMID:24350870

Warlow, C., Sudlow, C., Dennis, M., Wardlaw, J., & Sandercock, P. (2003). Stroke. *Lancet*, *362*(9391), 1211–1224. doi:10.1016/S0140-6736(03)14544-8 PMID:14568745

Woznowski, P., Fafoutis, X., Song, T., Hannuna, S., Camplani, M., Tao, L., & Craddock, I. et al. (2015). A multi-modal sensor infrastructure for healthcare in a residential environment. *2015 IEEE International Conference on Communication Workshop (ICCW)*, 271–277. doi:10.1109/ICCW.2015.7247190

Ziviani, A., Correa, B., Gonçalves, B., Teixeira, I., & Gomes, A. (2011). AToMS: A ubiquitous teleconsultation system for supporting ami patients with prehospital thrombolysis. *International Journal of Telemedicine and Applications*, 12.

Chapter 7
Wearable Antennas:
Breast Cancer Detection

Amal Afyf
University of Maine, France

Etienne Gaviot
University of Maine, France

Larbi Bellarbi
University of Maine, France

Lionel Camberlein
University of Maine, France

Mohamed Latrach
University of Maine, France

Mohamed Adel Sennouni
Hassan I University, Morocco

Nourdin Yaakoui
University of Maine, France

ABSTRACT

Having the merits of being light-weight, energy efficient, in addition to low manufacturing cost, reduced fabrication complexity, and the availability of inexpensive flexible substrates, flexible and wearable technology is being established as an appealing alternative to the conventional electronics technologies which are based on rigid substrates. This chapter is organized as follow into three major sections. In the first part, a detailed review of wearable antennas including applications and antenna families is presented. The second part of this project deals with the flexible antennas materials and fabrication methods. A wearable antenna prototype for medical applications, more accurately, early breast cancer detection, is discussed in the last section of this chapter.

DOI: 10.4018/978-1-5225-3290-3.ch007

INTRODUCTION

Flexible and wearable electronics, which are lightweight, bendable, rollable, portable, reconfigurable and potentially foldable, would substantially expand the applications of modern electronic devices, are beginning to enjoy tremendous popularity thanks to the great advances in materials science and electronics manufacturing and packaging. This technology is recognized as one of the fastest growing technologies in today's world. According to a recent market analysis, the global revenue of this technology is estimated to be 45 billion USD in 2016 and over 300 billion USD in 2028 (Afyf et al., 2015). Moreover, recent developments in miniaturized and printable energy storage, flexible photovoltaic, and green (self-powered) electronic components have paved the road for the success of this technology. Consistently, wearable and flexible devices would often require the integration of antennas operating in specific frequency bands to provide wireless connectivity which is greatly demanded by modern information-oriented consumers. The aim of this chapter is to provide a comprehensive guide to various technologies and methods applied in the realization of flexible and wearable technologies along with state of the art antenna designs and implementations. Moreover, this document serves as an extensive reference in wearable topics. An example of a wearable flexible antenna for early breast cancer detection is designed.

WEARABLE ANTENNAS BACKGROUND AND OVERVIEW

Wearable antennas have been a topic of interest for more than the past decade, and hundreds of scientific papers can be found on the subject. This large number of publications asks for some classification in order to get an overview of the trends and challenges. To this aim, an overview of wearable antennas according to the applications, antenna families, materials, and technology, is proposed.

1. Applications of wearable antennas:
2. Security (Military and Police) and Rescue Service Applications

Our main focus will be orientated towards wearable robust antennas intended to operate in various harsh environments. These antennas are mainly used for security and defense applications (Psychoudakis, Lee, Chen, & Volakis, 2010) and within different rescue services like firefighters (Hertleer et al., 2009), mountain and water rescue workers, (Figure 1), (Serra, Nepa, & Manara, 2011), (Corner, 2013).

Frequency bands intended for security and rescue applications are regulated by special government regulatory offices. In our case, we will mainly consider

Figure 1. Illustrative examples of wearable antennas applications: (a) Army forces and (b) Firefighters

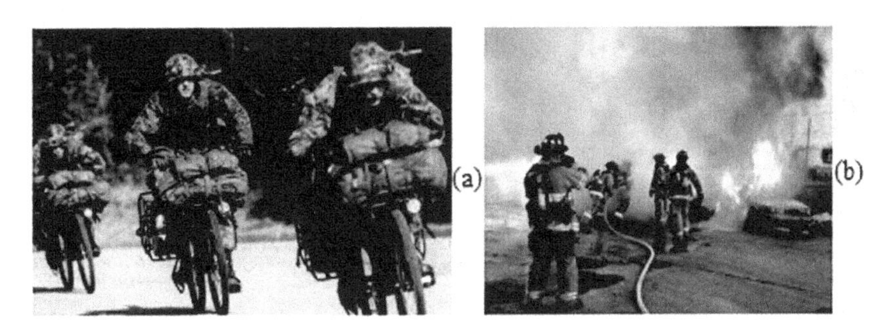

the regulations and frequency allocations of the Swiss authorities (National Swiss Plan, 2014). A band of interest is the Tetrapol communication band, 380-430 MHz, widely used by different security and emergency services within the Private Mobile Radio (PMR) applications (Tetrapol, 2015). There are two common requirements for most of the wearable antennas used in security and rescue service applications: the miniaturization size and the resilience of the antenna to the environment. The wavelength of the antennas operating inside the Tetrapol band, if observed in a free space, is rather large. Antennas proposed for these applications should be small, non-protrusive, low profile and conformable so that can be easily mounted inside the uniform and placed around the wearer's body. Apart from the antenna size, another important aspect is the environment and weather conditions where wearable antennas operate. As most of the wearers from the indicated groups operate outdoor, robustness against water, rain, moisture, mud, etc., is among the antenna requirements. Moreover, taking into account that the uniforms need to be washed, an integrated antenna should withstand this process. Several antenna prototypes exist in literature and on the market (Psychoudakis et al., 2010) (Hertleer et al., 2009). All the reported antennas have a size suitable for placement on the wearer. Some of the antennas are waterproof and most of them are too specific and customized and cannot be widely used for different situations. For example, the antenna presented in (Psychoudakis et al., 2010) is exclusively designed for armor vests used by the military and law enforcement agencies. It is an asymmetric dipole antenna that can be integrated only in the specified vests. Another dipole wearable antenna type is presented by the same authors in (Psychoudakis & Volakis, 2009), a conformal Asymmetric Meandered Flare (AMF) dipole antenna. The antenna's conductive surface is fabricated by using precise embroidering techniques in order to produce fully flexible antenna conformal to the wearer and his clothes, see (Figure 2). More wearable antenna prototypes operating between 0.1-1 GHz, mainly for military applications, are presented in (Matthews & Pettitt, 2009), exploiting different

Figure 2. Asymmetric Meandered Flare (AMF) wearable dipole antenna (Psychoudakis & Volakis, 2009)

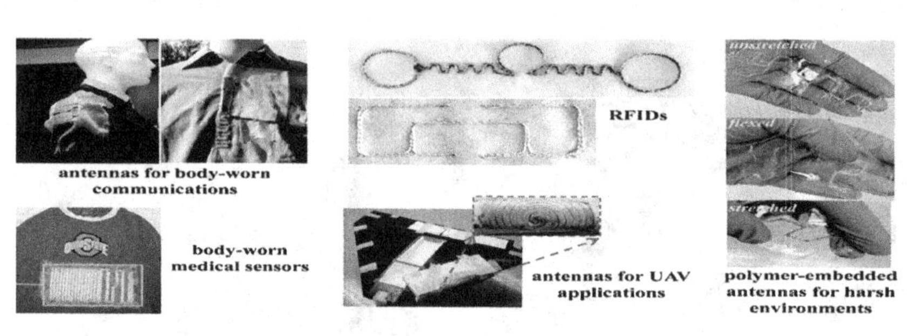

conductive and textile materials. A set of textile wearable antennas intended for firefighter services have been introduced by Hertler et al. A textile GPS wearable antenna and the antenna intended for voice communication between the firefighters are presented in (Vallozzi et al., 2010).

The importance of the wearable devices in the everyday operations of modern security and rescue services is reported in (Curone et al., 2007). Different parameters can be monitored with the units integrated into their garments like psychological parameters, actual position or the conditions on the field where they operate. In summary, all the concerned applications mentioned above, such as military, firefighters or rescue services require an involvement of different electronic devices and equipment in their daily activities. Depending on the type of application and activities they perform, different wearable antennas are needed. Therefore, wearable antennas play an important role in ensuring robust and reliable communication links and thus performance of these services.

MEDICAL APPLICATIONS

These applications are mainly used for health monitoring of different categories of patients and elderly people. They can monitor 24 hours a day different parameters and detect changes, for instance, blood pressure, heart rate or body temperature (Popovic, Momenroodaki, & Scheeler, 2014). With the more emphasized use of smartphones, collected data from the medical applications can be directly sent to the hospital or responsible medical doctors. An illustrative example of wearable medical applications (Ullah & Kwak, 2011) is presented in (Figure 3). Before showing some of the typical wearable antenna prototypes found among the medical applications, a short overview of their dedicated standards and frequency bands will be presented. A large number of bands indicate how widely these applications are

Figure 3. Wireless Body Area Network for Ubiquitous Health Monitoring (Ullah & Kwak, 2011)

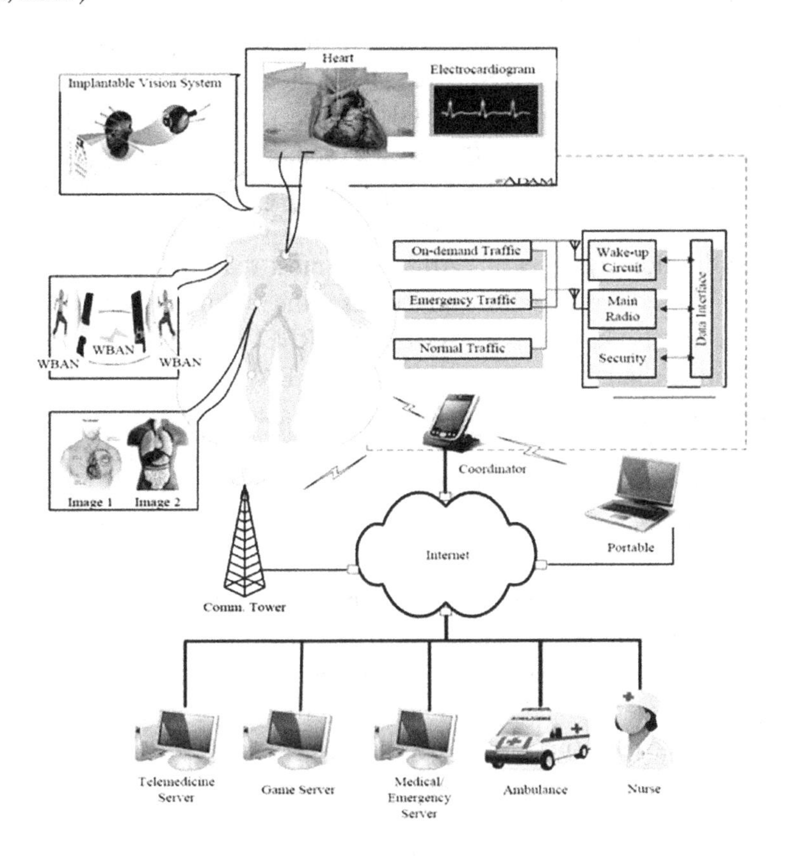

spread, emphasizing their importance. Depending on the type of devices (implanted or on the body) and the required communication. In-, On- or Off-body) different standards have been defined. The Federal Communications Commission (FCC) has established a Medical Implant Communication Service (MICS) like the band for communication with medical implants ("Federal Communications Commission (FCC)," 2014), while in Europe the same standard is regulated by the European Telecommunications Standards Institute (ETSI). In 2009, the Medical Device Radio Communications Service (Med-Radio) range, 401-406 MHz, has been created by the FCC for diagnostic and therapeutic purposes in implanted medical devices and devices worn on the body. A large number of wearable applications operate inside one of the FCC proposed Industrial Scientific Medical (ISM) bands, 902-9281MHz, 2.4-2.4835 GHz and 5.725-5.875 GHz. Some of the bands are used on a secondary basis because they are already utilized in other applications. These

ISM bands are submitted to certain restrictions, like short distance operation or low emitted power, in order to allow their co-existence in the spectrum with minimal interference (“Federal Communications Commission (FCC),” 2014), (“Medical Device Radiocommunications Service (MedRadio),” 2009).

In 2002, a spectrum from 3.1-10.6 GHz has been allocated by the FCC for Ultra Wideband (UWB) applications (“Federal Communications Commission (FCC),” 2014). Many researches and different types of wearable medical applications can be found in this band, making it one of the most exploited frequency ranges for WBAN applications (Hao et al., 2012). The most recent frequency band with a potential to be used for wearable applications is the millimeter wave range between 57-64 GHz, and also known as a 60 GHz band (REED, 2007). Some research about the propagation characteristics at 60 GHz for WBAN has been done in ((Alipour et al. 2010), (Constantinou et al., 2012)) while antenna designs are proposed in ((Pellegrini et al., 2013), (Wu et al., 2010)) giving promising results. The large choice of frequency bands used for medical WBAN applications leads to various wearable antenna families. Several implantable and wearable antennas are presented within the following subsection. Implantable antennas are a specific group challenging to design, fabricate and package and implant. The antenna and the accompanying electronic components need to be implanted into the human or animal body, leading to a complex procedure. The antenna needs to be small, compact, built out of appropriate biocompatible materials and when implanted providing a reliable communication link towards the external units and devices (Skrivervik & Merli, 2011). An overview and analysis on the development and implantation of these antennas are presented in (Merli, 2008). Several other implantable antenna designs are shown in ((Kim & Samii, 2004), (Xia, Saito, Takahashi, & Ito, 2009), (Kiourti et al., 2012), (Curto et al., 2009)). Wearable antennas on the other side, offer more freedom in terms of size, shape and communication link because they are placed on the wearer’s body. Different antenna types used for medical purposes are found in the literature, e.g. patch antenna (Chandran, Conway, & Scanlon, 2008), monopole type antenna (Ma, Edwards, & Bashir, 2008), PIFA (Lin et al., 2012), Planar Inverted Conical Antenna (PICA) (Alomainy et al., 2009), printed strip antenna (Alomainy, Hao, & Pasveer, 2007), classical dual-band patch antenna (Samii, 2007), etc.

Medical applications will remain one of the main consumers of various wearable devices. Constant progress of the wearable allows monitoring of various health activities and parameters. The fact that nowadays smartphones and other wireless devices (e.g. tablets, laptops, etc.) can act as reliable hubs and transceivers; make the wearable applications even more favorable. Wearable antennas will continue to be the crucial connection between the on-body devices and the external units, providing reliable and on time communication.

SPORT, ENTERTAINMENT AND FASHION APPLICATIONS

Another attractive market for wearables is sport, entertainment and fashion applications (Figure 4). Nowadays consumers enjoy keeping track of performances in any kind of sports activity they do. Different wearable applications are able to track and record speed, elevation, distance and other similar parameters. Most of the tracking can be performed with the currently existing smartphones, but some separate wearable devices also exist. Several wristwatches with built-in GPS antennas, that can measure time, distance, position of the wearer or even the heart rate, can be found on the market ((“Running GPS Watches,” 2014),(“Micoach,” 2014)). Wearable GPS antennas ((Chen, Kuster & Chavannes, 2012), (Zürcher, Staub, & Skrivervik, 2000), (Skrivervik & Zürcher, 2007)) play an important role in the performance of such systems. Furthermore, for the purposes of professional sportsmen, wearable devices are used in a more extensive manner. All the collected data from different wearable sensors are analyzed and used for further improvement in the sportspeople performances (Lapinski, Feldmeier, & Paradiso, 2011). Interesting examples of wearable antennas intended for integration into the shoes and thus potential use for sports applications are presented in ((Gaetano et al., 2013).

Entertainment applications are also a market for the wearable technologies. They offer a lot of possibilities to this industry. Some examples of the entertainment wearable’s present on the market include a bluetooth jacket (“Zegna Sport Bluetooth iJacket incorporates smart fabric,” 2007), intelligent shoes (“Verb for shoe - very intelligent shoes,” 2004) or a ski monitoring system, flaik (“Snow flaik tracks skiers,” 2011). The list of wearable applications used in sports, entertainment and fashion is growing rapidly, occupying a significant part of the market (Ranck, 2012.).

Figure 4. Illustrative examples of wearable antennas applications in Sports, Entertainment and Fashion: (a) Skiing (“Snow flaik tracks skiers,” 2011), (b) Swimming and (c) Intelligent shoes (“Verb for shoe - very intelligent shoes,” 2004)

(a)

(b)
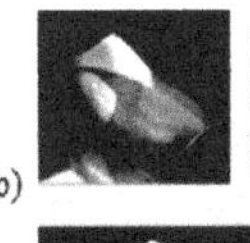

(c)

ANTENNA FAMILIES

Different types of antennas are used in different situations. Wearable antennas follow similar trends and thus a wide number of various antenna types can be found in the literature. The following four antenna families are considered for the purposes of this thesis:

- Dipole antennas.
- Monopole antennas.
- Patch antennas.
- Planar Inverted-F Antennas (PIFAs).

The antennas' size plays an important role for their placement on the wearer. The gain and effective area are directly proportional to the antenna's size measured in wavelengths (Balanis, 2005). Usually, antennas that are smaller than the quarter free space wavelength are considered to be electrically small or miniature antennas (Staub et al., 1999). Before considering each of the indicated antenna family, a short overview of the miniaturization techniques will be presented.

Introduction to Miniaturization Techniques

In order to be practical, wearable antennas should often be small in size, which can be problematic for the designs especially in the UHF band. Therefore, a short overview of miniaturization techniques will be presented. Antenna miniaturization is always a matter of compromise between size and radiation characteristics (Staub et al., 1999). The fundamental physical limits on antenna EM performance related to size are well known since the early days of wireless communication ((Wheeler, 1975), (Chu, 1948), (Harrington, 1960)). The study of these limits regained a huge interest in the nineties with the boom of mobile phones (see for instance (McLean, 1996)). Lately, these limits have been refined taking into account antennas shape (Gustafsson, Sohl, & Kristensson, 2007). Apart from restrictions in bandwidth, gain, efficiency and polarization purity (Fujimoto, 2008), the process of miniaturization brings to the surface problem of feeding a small antenna efficiently. In general, antennas are either balanced or unbalanced, and their feeding should be in accordance to this. The majority of small antennas are of an unbalanced character and the reduction of their size is always followed by the reduction of their ground (often a plane). A reduced ground cannot absorb the charge flow as the proper ground would lead to spurious ground currents altering the radiation characteristics of the antenna (Staub, Zürcher, & Skrivervik, 1998). Electrically small antennas require thus specific measurement techniques.

The main techniques available for antenna miniaturization are presented in (Staub et al., 1999):

- Material or lumped element loading, usually at the expense of the bandwidth of the antenna.
- Geometrical loading, by using bands and slots.
- Using grounds and short circuits.
- Using the antenna's environment to enhance the radiation.

Examples of miniaturization techniques (Staub et al., 1999) are illustrated in (Figure 5 and 6), ((Bokhari et al., 1996), (Zurcher, Skrivewik, & Staub, 2000)), where an environment (human body) is used for the miniaturization.

The Dipole and Related Antennas

Let us first consider the dipole as a candidate for a body-worn antenna. The dipole is the most basic design of an antenna; it is a symmetric structure, consisting of a straight thin wire conductor of a certain length and a center feeding, see for instance

Figure 5. Miniaturization techniques (Staub et al., 1999): (a) Geometrical transformation steps (bending) of a monopole into Inverted-F Antenna (IFA) and (b) A capacitor–plate antenna (loaded monopole)

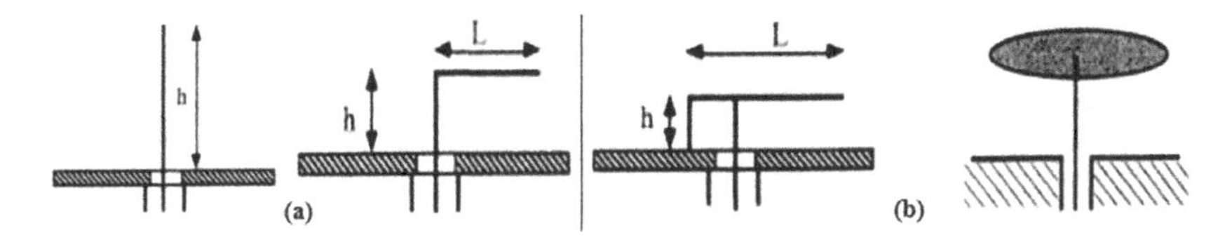

Figure 6. (a) Printed patch antenna for GPS for watch application (Bokhari et al., 1996) and (b) SMILA (Smart Mono-bloc-Integrated L-Antenna) (Zurcher et al., 2000)

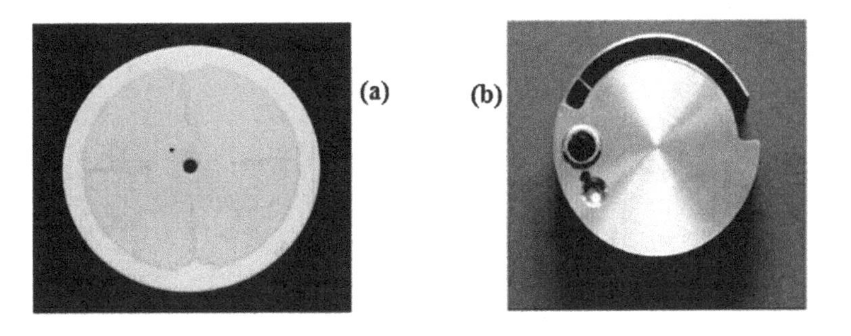

(Figure 7). The maximal radiation power is achieved in the plane perpendicular to the dipole. The typical radiation pattern is therefore in the shape of a torus. The dipole is a typical resonant antenna, having well-defined resonances precisely located in frequency.

Dipole antennas are widely used for wearable applications and several antenna designs are presented in the literature. In (D. Psychoudakis et al., 2010), a wearable asymmetric dipole antenna was introduced for military applications. Compared to the conventional center-fed dipole, asymmetry provides a broader frequency band and a slightly smaller size. The model is designed to be placed on the armor vest, and miniaturization techniques are applied (bending) to reduce the size. The main limitation of this antenna is that it has to be placed on flat surfaces, see (Figure 8). Similar wearable antennas, exclusively dedicated for armor vests, are presented in (Lee et al., 2011).

Volakis et al., have investigated more conventional models of dipole antennas that can be used in a larger field of applications as they are conformable. Their mounting around the wearer's body is more convenient compared to the previous

Figure 7. Dipole antenna

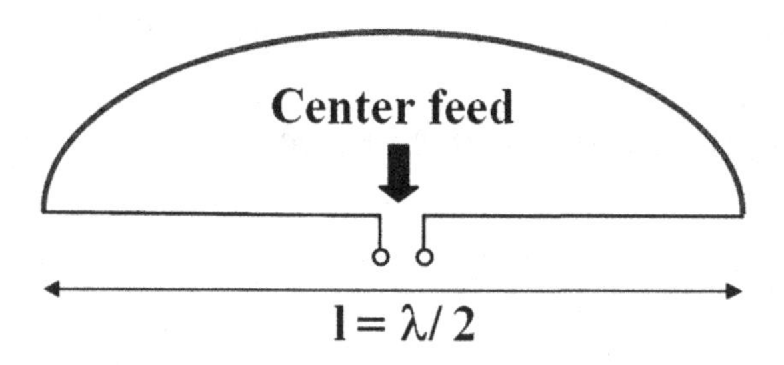

Figure 8. Asymmetric wearable dipole antenna for armor vests (D. Psychoudakis et al., 2010)

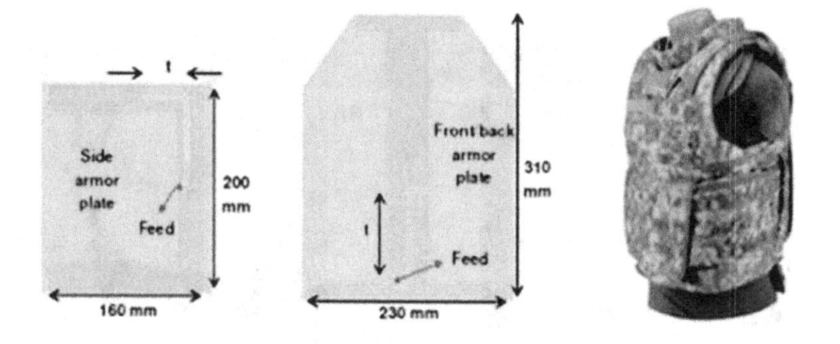

examples. The presented antennas are based on a printed wide dipole, with slight differences between them. The first is the so-called bow-tie dipole, the second a flare dipole, while the last one is an Asymmetric Meandered Flare (AMF) and they are all shown in (Figure 9). Apart from being conformal, the dimensions of those antennas, 305 x 38 mm^2, allow placing several of them on the same wearer. Their overall size makes them suitable for placing at different positions of the human body like chest, back, shoulders, etc.

Table 1 gives an overview of the advantages and disadvantages of dipole antennas while placed on the wearer's body.

Figure 9. Wide printed dipole antennas for wearable applications, from left to right: Bow-Tie, Flare and Asymmetric Meandered Flare (AMF) dipole (Psychoudakis & Volakis, 2009).

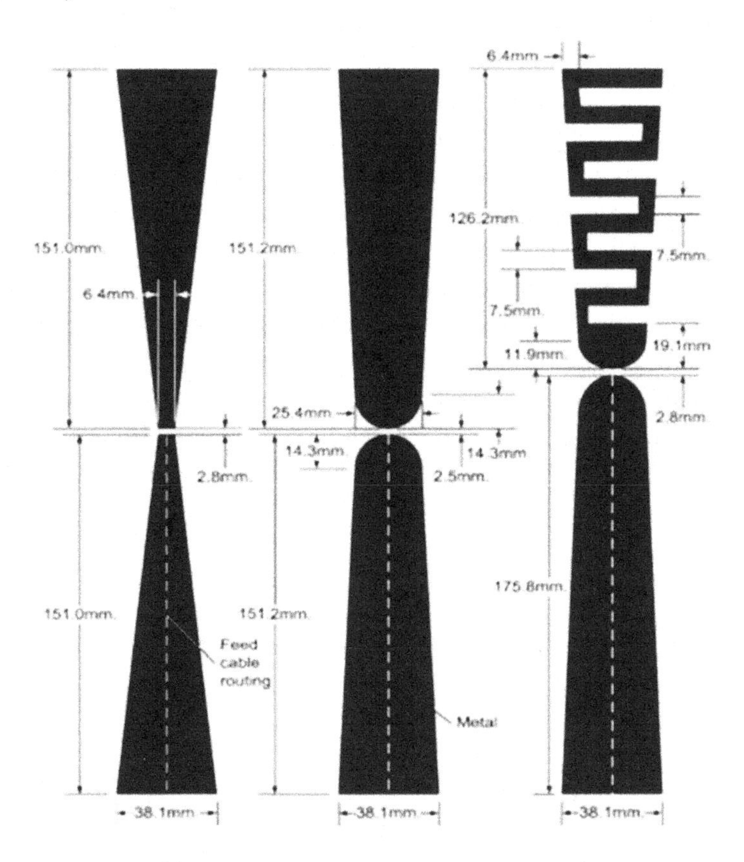

Table 1. A summary of the positive and negative aspects for dipole antennas used for wearable applications (Trajkovikj, Fuchs, & Skrivervik, 2011)

Advantages ☺	Disadvantages ☹
Single and sharp resonant behavior, good for filtering input signal	Due to the lossy body nature the antenna requires wider operating frequency bandwidth
Radiation pattern - "Torus shape" perpendicular to the wearer's body	Radiation pattern - "Torus shape" part of the energy still penetrates into the body
Most of the models are flexible and conformal for mounting	Narrow fractional bandwidth, up to 10%
Miniaturization - bending or meandering the antenna structure	Balanced feeding, need for baluns in some cases
	The lack of the ground plane and polarization parallel to the wearer increases the penetration of the radiation inside the body

Monopole Antennas

The monopole antenna can be near-resonant (length is approximately a quarter-wavelength) or electrically short (length is much shorter than a quarter-wavelength). Compared to the dipole, the monopole requires a ground plane in its structure. The dimensions of the ground plane may vary from a fraction of a wavelength to many wavelengths (Weiner, 2003).

In the case of wearable applications, the presence of the ground is an advantage, since it allows some shielding of the body. The protrusion of the antenna orthogonally to the body, however, is a serious disadvantage and miniaturization is required to make the monopole low profile. Most of the monopole antennas found in the literature have planar characteristics because it is much easier to integrate planar structure into the garment or any kind of uniform ((Ma et al., 2008), (Lin et al., 2012)).

Figure 10. Monopole antenna: (a) Quarter-wavelength monopole and (b) Electrically short monopole

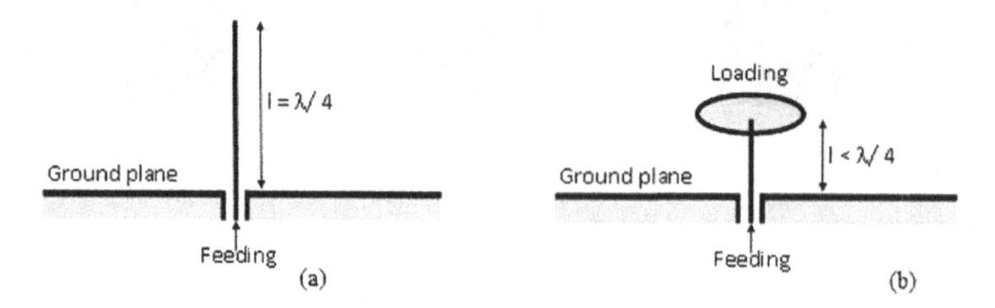

Button Antenna as a Special Monopole Antenna Case

An especially interesting sub-group of monopole antennas is the button antenna. It is designed by loading the monopole antenna with a certain load, composed of metal in most of the cases, leading to a decrease of its height. This loading method is successfully applied to many button antennas, as it will be seen from the examples shown in this section. The existing literature gives some examples of button antennas integrated into wearable applications. For example in (Izquierdo et al., 2010), a dual-band metallic button antenna is shown for body area network applications, for covering both Off- and On-body communications. The operating frequencies of this antenna are 2.45 GHz and 5.5 GHz. For the specified frequencies the dimensions of the antenna are the following: ground plane size of 50 x 50 mm^2, top disc diameter 16 mm, and a total height of 8.3 mm. The 3D model of the antenna is presented in (Figure 11 a).

More complex loading shapes are also present in literature. An antenna loaded with a G-shaped metallic structure has been reported in (Salman & Talbi, 2010). This antenna operates at ISM bands and it is a dual band antenna with an overall size of 40 x 40 mm^2 of the ground plane and 10 mm in the height. The geometrical shape of the loading can be adapted to the desired feature of the antenna. Another monopole loaded antenna has been introduced by Koohestani et al., where a cross (Swiss flag symbol) is used as a shape for loading a vertically polarized UWB antenna (Koohestani et al., 2014), see (Figure 11 b).

Most of these antennas operate inside the ISM or UWB bands. Communication distance is aimed for a short range, from a few centimeters up to 10 meters (Alomainy et al., 2009), where the majority of applications using button antennas are for indoor communication. Button antennas remain good candidates for applications where a compact antenna integrated into different parts of the clothes is needed for short range communication.

Figure 11. (a) Monopole button antenna loaded with a disc (Izquierdo et al., 2010) and (b) Monopole antenna loaded with a cross shape structure (Swiss flag cross)

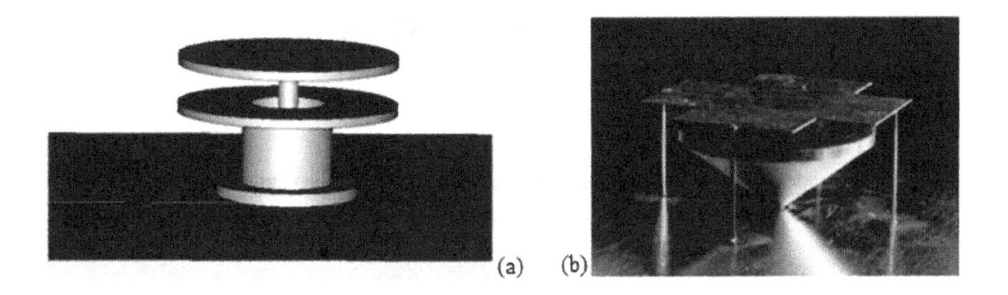

Table 2. A summary of the positive and negative aspects for the monopole and button antennas when used for wearable applications (Trajkovikj et al., 2011)

Advantages ☺	Disadvantages ☹
Presence of a ground plane	Protrusive structure
Polarization perpendicular to the wearer	
Printed monopole, planar structure	Polarization perpendicular to the wearer
Button compact antenna → easy integration	Mostly for ISM & UWB bands short distance communication

Patch Antennas

The microstrip patch is a planar antenna, consisting of a radiating patch on one side of a dielectric substrate, and the ground plane on the other. The radiating elements are made of conducting materials (e.g. copper, usually photo-etched). Patches can be found in any geometrical shape, but the most common have some regular shapes as square, rectangular, circular, triangular, elliptical, etc. They radiate primarily because of the fringing fields between the patch and the ground plane, see for instance (Figure 12). The radiation pattern of these antennas is perpendicular to the plane of the patch (Balanis, 2005).

Different structures can be used for feeding the patch antennas, such as microstrip line, coaxial probe, aperture coupling or proximity coupling (Balanis, 2005). The variety of the feeding techniques provides some degrees of freedom in the design phase, as well as in later stages while mounting them on the wearer's body. The microstrip patch is a candidate for any narrowband wearable application, as it has a low profile and can be made conformal for integration into the clothing. The presence of the ground plane is an additional advantage when these antennas are used in wearable applications, preventing the radiation into the wearer's body. Patch antennas are mainly used in GPS and ISM bands. Hertler et al. have introduced

Figure 12. Cross section of a microstrip patch antenna

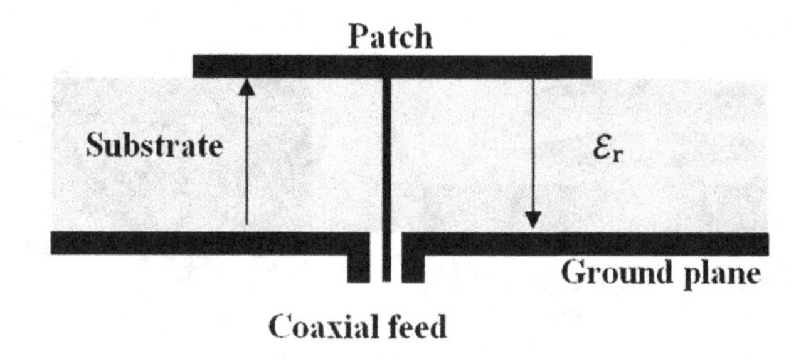

several patch based antennas operating inside the ISM and GPS bands, intended for wearable applications (Vallozzi, Van Torre, et al., 2010). All the proposed antennas are built out of specific textile materials, conductive and dielectric, and are mainly used for firefighter service applications.

An alternative shape to the conventional rectangular patch antenna can be a circular disk microstrip antenna. The main advantage of circular configuration, compared to the rectangular geometry, is that the circular disk occupies the less physical area, leading to an easier integration in arrays for instance. In (Sankaralingam & Gupta, 2009), circular disk microstrip Wireless Local Area Network (WLAN) wearable antenna is presented, operating at 2.45 GHz. Using a copper as a conducting layer, this antenna achieves a gain of 5.7 dBi, slightly lower than its rectangular counterpart which has a gain of 5.9 dBi. A complete textile patch antenna is shown in (Carter et al., 2010), where both the substrate and the patch are textile materials with good conductive and dielectric characteristics respectively. Zelt3 material is used for building the patch and the ground plane while Felt 4 is used as a substrate material. Both materials will be considered in more details in the section considering the materials. Figure 13 depicts the top view of the discussed antenna. The dimensions of this patch are 13.5 x 13.3 cm^2, i.e. half wavelength at 920 MHz. A summary about the suitability of patch antennas, when used in wearable applications, is shown as an overview of advantages and disadvantages (Table 3).

Figure 13. Patch antenna using Zelt fabric as conductor and Felt as a substrate (Carter et al., 2010)

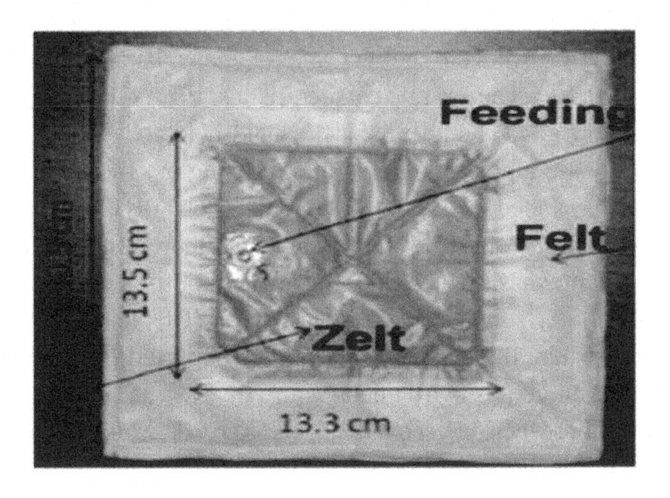

Table 3. A summary of the positive and negative aspects for patch antennas when used for wearable applications (Trajkovikj et al., 2011)

Advantages ☺	Disadvantages ☹
Light weight and low volume	Narrow bandwidth
Conformability	Low efficiency
Low cost of fabrication	Low gain
Supports linear and circular polarization	Extraneous radiation from feeds and junctions
High level of integrability	Low power handling capacity
Mechanically robust	Surface wave excitation
Capable of multiband operations	

Planar Inverted-F Antennas (PIFAs)

A Planar Inverted-F Antenna (PIFA) is the last family of antennas enclosed in this overview which is a subject of consideration. Basically, PIFA is a kind of inverted-F antenna (IFA) with the wire radiator element replaced by a plate to expand the bandwidth. On one hand, the PIFA is derived from the monopole antenna, where by bending and shortening, its characteristic shape is achieved (Salonen et al., 1999), while on the other hand, it can also be considered as a short-circuited patch antenna (Fujimoto, 2008). The addition of the shorting strip allows a good impedance match to be achieved with a top plate, which is typically less than λ/4 long. The PIFA consists of a ground plane, a top plate element, a feed wire attached between the ground plane and the top plate, and a shorting wire or strip that is connected to the ground plane and the top plate. Figure 14 illustrates a typical PIFA configuration.

PIFA is a widely deployed antenna among many Personal Communication System (PCS) devices because of its characteristics. PIFAs have many advantages that make them very good candidates for the hand-held and mobile devices (Z. N. Chen, 2007). They are highly integrable into a casing or other components of the environment and have a reduced backward radiation towards the users' body. In this way, the SAR is significantly reduced, which is an important aspect when it comes to the body-worn antennas (Wong, 2002). They exhibit moderate to medium gain in both vertical and horizontal states of polarization. This feature is very useful in certain situations where the antenna does not have a fixed position and different reflections are present in its environment.

For the applications where larger bandwidth is required, several different techniques can be used for its enhancement (Firoozy, 2011):

Figure 14. Configuration of PIFA

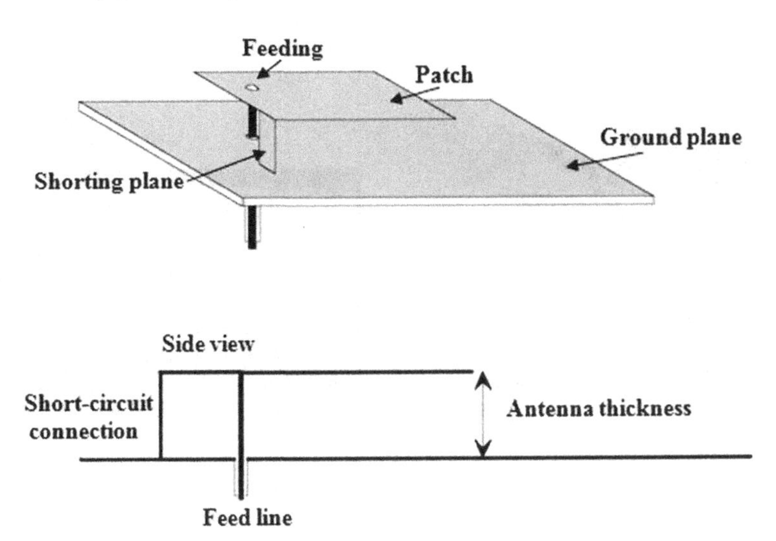

- Varying the size of the ground plane.
- Use of a thick air substrate.
- Use of parasitic resonators with resonant lengths close to main resonant frequency.
- Adjusting the location and spacing between the shorting posts.
- Use of stacked elements.
- PIFAs have two main advantages making them suitable for wearable applications:
- The presence of the ground plane in its structure which insulates the antenna from the wearer, thus reducing the coupling (favorable for off-body communications).
- Antennas low profile allows easy integration within the wearer's garments.

As an interesting candidate for UHF wearable applications, an overview of PIFAs used in WBANs is presented. A simple PIFA prototype, built out of textile materials, intended for wearable applications is presented in (Soh et al., 2010). The antenna operates at 2.45 GHz, with the overall dimensions 50 x 24 x 10 mm^3 (Figure 15). The presented antenna is characterized in free space and when placed in the vicinity of the human body.

Simulated and experimental results indicated that the textile PIFAs are able to operate with a reflection coefficient below -10 dB, and a fractional bandwidth of

Figure 15. PIFA wearable antenna (Soh et al., 2010): (a) A 3-D model and (b) Fabricated PIFAs using ShieldIt conductive textiles

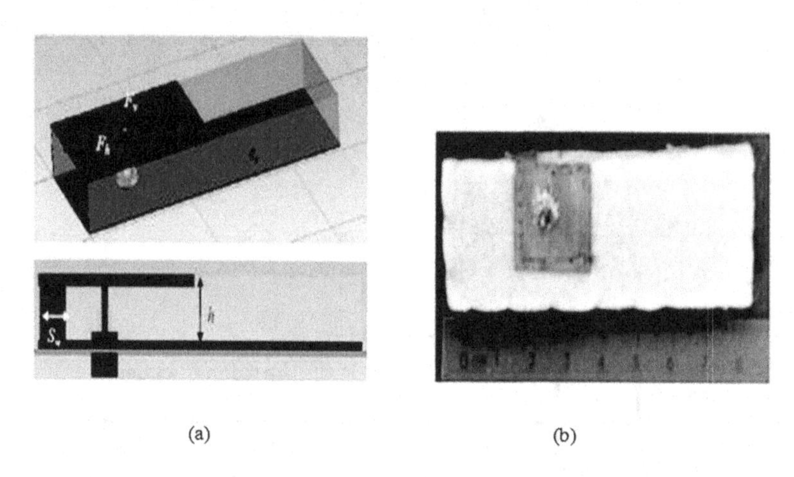

around 27%. On average the measured gain is 1.6 dBi while the efficiency ranges between 77- 85% (measurements obtained in free space, not on body or phantom). Two more antennas based on the PIFA concept, operating at 2.45 GHz and 5.2 GHz, are presented in the literature, the dual-band Sierpinski fractal textile PIFA and a broadband all-textile slotted PIFA (Soh et al., 2012). A modified E-shaped PIFA (Cibin et al., 2004), has been one of the pioneering antennas in the frame of this thesis project (Figure 16). It has been intended to operate inside the Tetrapol communication band, 380-400 MHz. The antenna exhibits -12 dBi gains while mounted on the body, which is better than the conventional half-wavelength dipoles. The proposed antenna has overall dimensions of 30 x 30 x 2 cm^3, which makes it suitable for mounting, but the area occupied is still fairly large. However, this antenna can be seen as a good starting example for the development of UHF wearable antennas. Table 4 summarizes the most relevant advantages and disadvantages of the PIFAs when used for wearable applications.

Materials and Fabrication Methods

Materials play an important role in the design and fabrication of wearable antennas. The integration of the antennas into the wearers' clothes indirectly imposes the potential choice of conductive and substrate materials. A wide range of materials can be found in the literature: textiles specially treated yarns, printed conductors, polymers, etc. (Salonens et al., 2004).

Figure 16. Modified E-PIFA structure (Cibin et al., 2004)

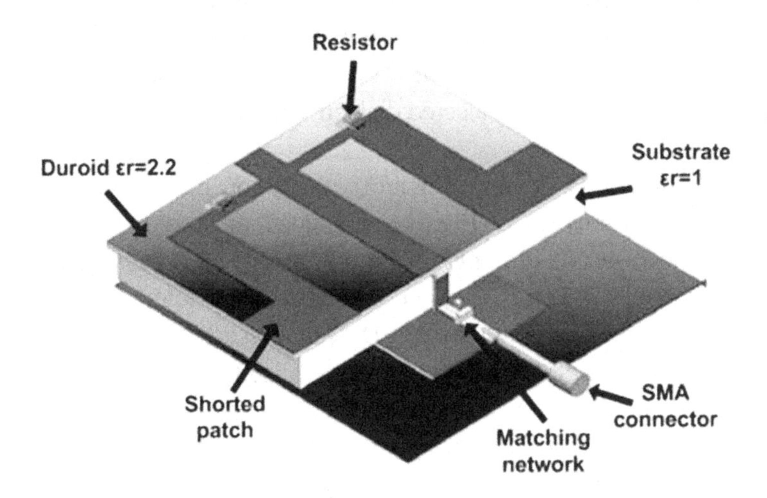

Table 4. A summary of the positive and negative aspects for the PIFAs when used for wearable applications (Trajkovikj et al., 2011)

Advantages ☺	**Disadvantages ☹**
PIFA structure: good compromise fc vs. size (low profile) vs. on-body placement	PIFA structure: at UHF frequencies the size still fairly large
Moderate to medium gain	Moderate to medium efficiency
Reduced coupling to the wearer	

Selected materials need to be combined with the set of technological and fabrication processes in order to build the final antenna structures. A set of embroidering techniques, inkjet printing, Substrate Integrated Waveguide (SIW) technology and a combination of polymers and different conductive materials are presented ((Santas, Alomainy, & Hao, 2007), (Rais et al., 2009)).

The operating environment also influences the performance of the antenna. Apart from the body's influence on the antennas, they can be also exposed to the different harsh environment, manifested through different weather conditions like rain, humidity, snow, mud, ice, etc. This is an additional reason why a careful selection of the materials should be considered (Kaija, Lilja, & Salonen, 2010).

1. Material Selection
2. Dielectric Materials

Different types of dielectrics can be used as antenna substrates. From the EM point of view, two important characteristics matter: low permittivity and low losses improve the antenna's radiation characteristics, i.e. bandwidth and radiation efficiency (Balanis, 2005). Therefore, the efforts will be concentrated on lowering the permittivity and losses of the selected dielectric materials.

1. For soft PCB processes, flexible films are the major core materials for supporting overlays, such as polyimide (PI) films ((Khaleel et al., 2012), (Virtanen et al., 2010)), polyester (commonly refers to Polyethylene TerePhthalate (PET) films ((Koo et al., 2011), (Casula et al., 2013)), and liquid crystal polymer films (LCP) ((Nikolaou et al., 2006), (DeJean et al., 2005)). These materials have the merits of high flexibility, low loss tangent, and availability of low thicknesses. Table 5 summarizes the advantages and disadvantages of each material (Numakura, 2007). The Kapton, as a high-performance PI film, shows good soldering tolerance for flexible antenna fabrication and withstands high temperature, which is required in thermal annealing of inkjet-printed antennas (Y. Kim, Kim, & Yoo, 2010). These soft films can hold deposited pure metal materials and conductive inks ((Virtanen et al., 2010), (Casula et al., 2013)).
2. Textile in clothing (nonconductive fabric) can be utilized as a platform for antennas, especially when combined with metal-plated textile conductors. Various types of textiles are employed, such as cotton, silk, wool, viscose, and felt ((Y. Kim et al., 2010), (Monti, Corchia, & Tarricone, 2013)). The relative permittivity and loss tangent of such materials are highly dependent on construction (knit or woven), constituent materials, and thickness (Ouyang & Chappell, 2008). Some textiles have anisotropic qualities, such as Cordura and Ballistic fabrics (Kaivanto et al., 2011). Therefore, parameters characterization is quite essential for the selected clothing textile before the antenna design and simulation step. Nonconductive textiles can sometimes be designed with

Table 5. Properties comparison of popular film materials

Property	PI Films	LCP Films	PET Films
Thermal rating	200°C	70°C	90°C
Soldering	Applicable	Difficult	Possible
Wire bonding	Possible	No	Difficult
Moisture absorption	High	Low	Low
Dimensional stability	High	High	High
Cost	Moderate	Low	Low

metallic pieces such as zippers and metallic buttons as the radiating parts (Mantash et al., 2011).

3. Other flexible substrates are also employed for special purposes. For example, paper substrates can be utilized in screen-printing and inject printing processes based on conductive inks. The paper is widely deployed since it is a low cost, environmental friendly material and can be modified to have hydrophobic and fire retardant properties. However, it is also loss and frequency-dependent. A dispersion model is required for accurate RF simulation and analysis (Cook & Shamim, 2013).

Conductive Materials

Similar to conventional antennas, a typical flexible/wearable antenna consists of two parts: conductive (ground plane and radiating element) and dielectric (substrate, which acts as a platform for the radiating element).

For conductive materials, the following requirements need to be fulfilled:

- Low resistivity/ high conductivity.
- Deformability, flexibility: such as the capability for bending, crumpling and stretching.
- Weatherproof: resistant to material degradation due to environmental factors such as oxidization and corrosion.
- Tensile strength: the material must be able to withstand repeated pressure, deformation, etc.
- Integration with textiles: the ability of the material to be sewed or embroidered.
    - Pure metallic materials are widely adopted in wearable textile-based antennas, such as silver paste (Tesla, 1892), copper gauze, and copper foils (Brown, 1984). The advantages of using such materials include high conductivity, cost effectiveness, and fabrication simplicity. It is worth noting that when the above-mentioned materials are integrated with clothing textiles, adhesive laminates or supporting foams are usually utilized instead of sewing and embroidering.
    - Metal-plated textile is another widely used conductive material in the fabrication of wearable and flexible antennas; it is often termed "electro-textile" and "E-textile". Metal-plated textiles possess the property of high ductility and can be sewed directly into clothing using textile yarns. Soft materials such as Kevlar, Nylon, and Vectran are coated with metals. The effective electrical conductivity of such textiles can

reach up to 1E+6S/m; the conductive thread is the basic component of E-textiles. Different kinds/brands of E-textiles are exploited in recent published literature, such as (Jabbar, Song, & Jeong, 2010), less EMF (Rishani et al., 2012), and Zelt (Haskou et al., 2012).

- Conductive ink, made of carbon or metal particles, is a promising material for flexible antenna design. Conductive inks have the merits of fabrication simplicity, compatibility with standard screen printing and inkjet printing process, and low cost. The effective conductivity is dependent on the material's intrinsic property, added solvent impurities, and the thermal annealing process (Declercq et al., 2010).

Material Characterization

The propagation and loss properties at the desired frequency band(s) need to be known for the candidate material prior to antenna design and fabrication. For conductive materials, conductivity and surface resistance have to be characterized, while permittivity and loss tangent have to be characterized for substrate materials. For clothing textile materials with different constructions and thicknesses, most of the parameters are unknown and need to be measured.

1. Conductive materials can be characterized using waveguide cavity method and microstrip resonator method (Reina., Roa, & Prado, 2006). In waveguide cavity method, Figure 17 a, the quality factors (Q) and transmission coefficient S21 with and without the conductive textile could be measured. Then, the conductive loss is calculated, and thus, the conductivity and surface resistance of the conductive textile can be extracted. In the second method, the microstrip resonator is used instead of the waveguide cavity, as shown in Figure 17 b and the measurement steps are similar to that of the waveguide method. However, the dielectric loss of the substrate should be obtained first. Transmission line

Figure 17. (a) Waveguide cavity method and (b) microstrip resonator method (Wu et al., 2010)

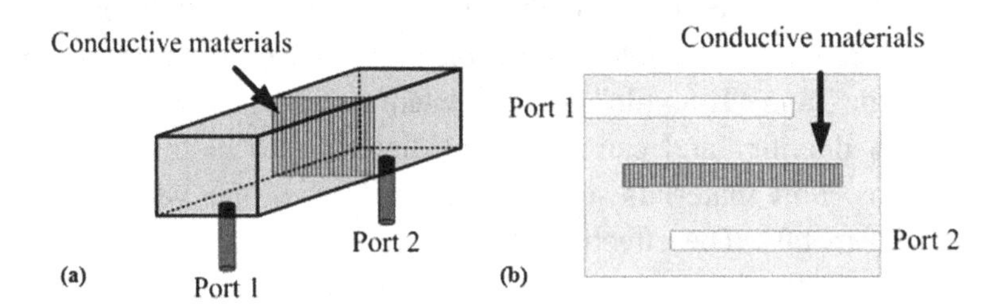

method is feasible for the same purpose (Mandal, Turicchia, & Sarpeshkar, 2010).

2. For dielectric textiles, the permittivity and loss tangent properties are mainly of interest. The most popular characterization method is the resonator method. A T-resonator microstrip line method is proposed in (Wang et al., 2009) and shown in Figure 18 a. The scattering transfer cascade matrix is used to present an additional parameter S_L, as shown in Figure 18 b. The propagation factor $e\gamma^{(l_1 - l_2)}$ (where l_1 and l_2 are the lengths of the two lines) is determined by the eigenvalue of scattering matrix of S_L. Effective permittivity and loss tangent can then be extracted.

1. ANTENNA FABRICATION

Based on the description of conductive and substrate materials in the previous section, there are several widely adopted fabrication processes of flexible and wearable antennas. This section reviews the commercial methods in addition to techniques used by the research and development sector. An overview of each technique, in addition to their advantages and drawbacks, is discussed:

Line Patterning

Line patterning is one of the simplest and most inexpensive solutions for fabricating RFIDs and flexible electronics (Carmo et al., 2006). The design of a negative image of the desired pattern is first developed using a computer-aided design program, followed by depositing a conductive polymer on the substrate. The last step involves taking out the printed mask by soni-cating the substrate (by applying an ultrasonic energy) in a toluene solution for about 10 seconds. Flexible field effect transistors, filters, resistors, and RFIDs are amongst the components produced using this method.

Figure 18. (a) T-resonator method (Skrivervik & Merli, 2011) and (b) matric-pencil two-line method (Kim & Samii, 2004)

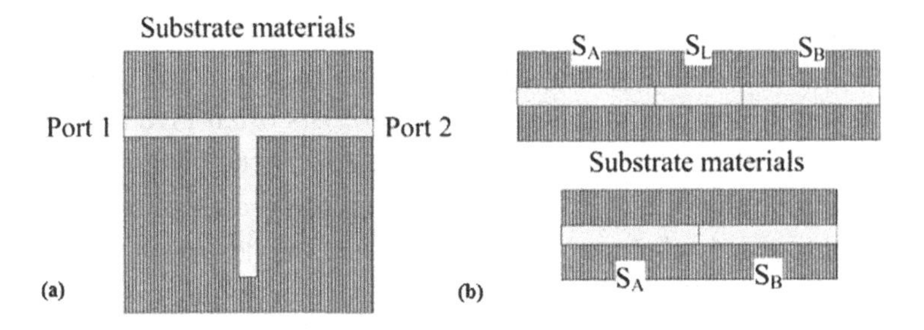

Flexography

In flexography, a print-making process of an image is involved, which is performed by inking a protuberant surface of the printing plate matrix while the recessed (suspended) areas are left free of ink (Lilja & Salonen, 2009). Flexography gained a significant interest by RFID antenna manufacturers due to its relatively high resolution, cost-effectiveness, and roll-to-roll production capability. Furthermore, this technique requires a lower viscosity ink than the inks used in the screen-printing method, which yields dry patterned films of a thickness of <2.5 μm. In contrast to the inks used in screen-printing, inks used in flexography must have higher conductivity to compensate for the difference in sheet resistance since the efficiency of the fabricated antennas relies directly on the electrical conductivity of the radiating element.

Screen-Printing

Screen-printing is another cost-effective technique used by flexible electronics manufacturers. This technique is characterized by its simplicity and being an additive process, which makes it environmentally friendly. A mask with the desired pattern is developed first and then applied directly on a flexible substrate/film where the conductive ink is administered and thermally treated to evaporate the excess solvent. Flexible transparent antennas and RFIDs have been successfully prototyped using this technique (Vallozzi et al., 2010). It is worth mentioning that there are some drawbacks associated with this technique, which includes the limited control over the thickness of the deposited ink, number of layers, and resolution of the deposited patterns.

Photolithography

Photolithography-based manufacturing has emerged in the 1960s targeting the PCB industry. It involves using a photoresist and chemical agents to etch away the unwanted area corrosively to produce the desired metallic patterns. This technique had gained notable popularity due to its capability of accurately producing patterns with fine details (Cibin et al., 2004). Currently, the fabrication of antennas and RF circuits based on photolithography is preferred to be conducted utilizing positive photoresists due to the fact that negative photoresists often give rise to the edge-swelling phenomenon, which compromises the consistency and resolution of the resulting pattern. When photolithography is used to produce flexible electronics, single or double-sided substrates are utilized where the desired pattern is obtained by etching regions of either or both sides. It is worth mentioning that stacking multi-flexible patterned layers are also possible using this technique. The major drawbacks of photolithography are low throughput, the involvement of hazardous chemicals, and production of by products; hence, it is not suitable for commercial production.

Thermal Evaporation

Thermal evaporation is a physical vapor deposition process, which is recognized amongst the most widely used thin-film deposition techniques. This method involves a vacuum process where a pure material coating is administered over the film surface. This process is conducted by heating a solid material inside a vacuum chamber where vapor pressure is created. Consequently, the evaporated material is deposited on the substrate. In antenna and RFID applications, the coating material is usually a pure atomic metal. It should be noted that this process is usually accompanied by a photolithographic process (Izquierdo, Huang, & Batchelor, 2006).

Sewing and Embroidering

Using sewing or embroidering machine is mainly employed in textile-based antennas, which is preferred over direct adhesion of E-textile over the fabric since no adhesive materials are introduced which may affect the electrical properties of the material (Izquierdo & Batchelor, 2008). Wrinkling and crumpling should be minimized to maintain the material qualities. For wearable antennas on clothing application, this is the most preferred fabrication process.

Inkjet Printing

Inkjet printing of antennas and RF circuits using highly conductive inks based on nanostructural materials are gaining extreme popularity nowadays. Inkjet material printers operate by releasing pico-litre sized ink droplets, which give rise to high-resolution patterns and compact designs (Izquierdo & Batchelor, 2008).

ANTENNA PROTOTYPES

Wearable Antenna Parameters

Conventional antenna parameters include impedance bandwidth, radiation pattern, directivity, efficiency and gain which are usually applied to fully characterize an antenna. These parameters are usually presented within the classical situation of an antenna placed in free space. However, when the antenna is in or close to a lossy medium, such as human tissue, the performance changes significantly and the parameters defining the antenna need to be revisited and redefined. In a medium with complex permittivity and non-zero conductivity, the effective permittivity $\mathcal{E}_{eff}$ and conductivity σ_{eff} are usually expressed as:

$$\varepsilon_{eff} = \varepsilon^{'} - \frac{\sigma^{''}}{\omega^{'}} \tag{1}$$

$$\sigma_{eff} = \sigma^{'} - \omega\varepsilon^{''} \tag{2}$$

where the permittivity and conductivity are composed of real and imaginary parts,

$$\varepsilon = \varepsilon^{'} - j\varepsilon^{''} \tag{3}$$

$$\sigma = \sigma^{'} - j\sigma^{''} \tag{4}$$

The permittivity of a medium is usually scaled to that of the vacuum for simplicity, $\mathcal{E}_0$ is given as 8854×10−12 F/m.

$$\varepsilon_r = \frac{\varepsilon_{eff}}{\varepsilon_0} \tag{5}$$

The equations above indicate the differences between free space and lossy material, hence, the imaginary part of the permittivity includes the conductivity of the material which defines the loss that is usually expressed as dissipation or loss tangent,

$$\tan\delta = \frac{\sigma_{eff}}{\omega\varepsilon_{eff}} \tag{6}$$

The biological system of the human body is an irregularly shaped dielectric medium with frequency dependent permittivity and conductivity. The distribution of the internal electromagnetic field and the scattered energy depends largely on the body's physiological parameters, geometry as well as the frequency and the polarization of the incident wave. Figure 19 shows measured permittivity and conductivity for a number of human tissues in the band 1–11 GHz. The results were obtained from a compilation study presented in (Harrison & Williams, 1965), which covers a wide range of different body tissues. Therefore, one major difference that can be identified directly when placing an antenna on a lossy medium, in this case, the human body, is the deviation in wavelength value from the free space one. The effective wavelength λ_{eff} at the specified frequency will become shorter since the wave travels more slowly in a lossy medium

$$\lambda_{eff} = \frac{\lambda_0}{\mathrm{Re}\left[\sqrt{\varepsilon_r - j\sigma_e / \omega\varepsilon_0}\right]} \tag{7}$$

where λ_0 is the wavelength in free space. However, the effective permittivity depends on the distance between the antenna and the body and also on the location of the human electric properties which are different for various tissue types and thicknesses. The general rule of thumb is that the further the antenna is from the body the closer its performance to that in free space. This also depends on the antenna type, its structure and the matching circuit. Wire antennas operating in standalone modes and planar antennas directly printed on the substrate will experience changes in wavelength and hence deviation in resonance frequency, depending on the distance from the body. On the other hand, antennas with ground planes or reflectors incorporated in their design will experience less effect when placed on the body from operating frequency and impedance matching factors independent of distance from the body.

Figure 19. Human tissue (a) permittivity and (b) conductivity for various organs as measured (Harrison & Williams, 1965)

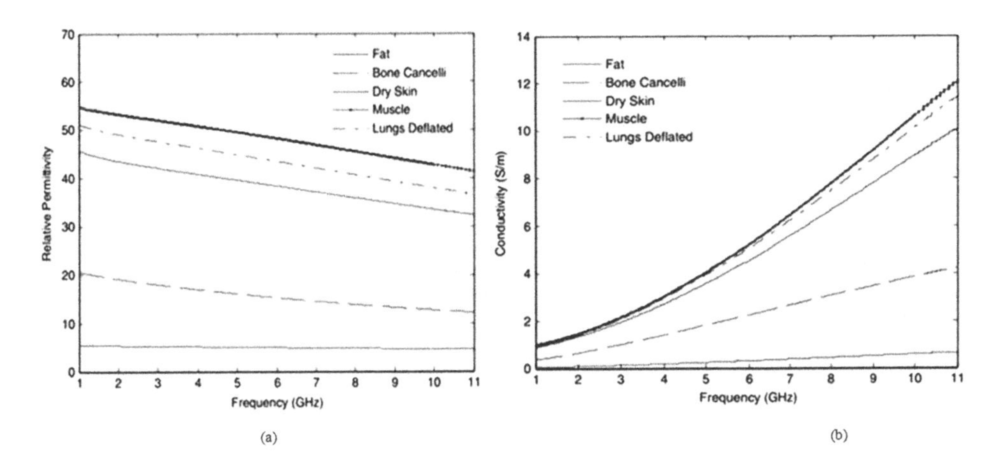

An important factor in characterizing antennas is the radiation pattern and hence, gain and efficiency of the antenna. The antenna patterns and efficiency definitions are not obvious and cannot be directly derived from conventional pattern descriptors when the antenna is placed in or on a lossy medium. This is due to losses in the medium that cause waves in the far-field to attenuate more quickly and finally to zero.

Antenna efficiency is proportional to antenna gain (Hertel & Smith, 2003),

$$G\left(\theta,\phi\right) = \eta.D\left(\theta,\phi\right) \tag{8}$$

Where η is the efficiency factor and D (θ,ϕ) is the antenna directivity which is obtained from the antenna normalized power pattern P_n that is related to the far-field amplitude F,

$$D\left(\theta,\phi\right) = \frac{P_n\left(\theta,\phi\right)}{P_n\left(\theta,\phi\right)_{average}} = \frac{\left|\vec{F}\left(\theta,\phi\right)\right|^2}{\left|\vec{F}\left(\theta,\phi\right)\right|^2_{average}} \tag{9}$$

The wearable antenna efficiency is different from that in free space, due to changes in antenna far-field patterns and also in the electric field distribution at varying distances from the body. However, the radiation efficiency of an antenna in either lossless or loss medium can be generalized as:

$$Efficiency_{radiaation} = \frac{RadiatedPower}{DeliveredPower} \tag{10}$$

An important quantity (which is in direct relation with antenna patterns and of great interest in wearable antenna designs) is the front–back ratio. This ratio defines the difference in power radiated in two opposite directions wherever the antenna is placed. The ratio varies depending on antenna location on the body and also on antenna structure.

EXAMPLE OF FLEXIBLE ANTENNA FOR EARLY BREAST CANCER DETECTION

Breast cancer is the second leading cause of women's mortality. Early detection increases the chances of successful treatment. The capabilities of current medical devices to detect breast cancer are agreeably insufficient for society's needs (Amal, Larbi, & Anouar, 2016). Noninvasive sensors have a huge potential for disease prevention in medicine and for diagnosis. Integration of such sensors into clothing can, therefore, enhance home healthcare and disease prevention.

This section presents a miniaturized flexible antenna for early breast cancer detection using a non-invasive technique based on the self thermal radiation generated by the breast; this radiation can indicate the existence of a cancerous tumor.

Antenna Design

The main goal is to develop a receiving antenna with good performances working in S-band precisely at 3 GHz which can provides a good penetration into the breast tissue according to the equation below (11). Then to achieve this goal, several optimization processes were applied by using an optimization solver in CST Microwave Studio. Table 6 presents the various optimized parameters of the proposed elementary antenna (Figure 20). Kapton Polyimide film was chosen as the antenna substrate due to its good balance of physical, chemical, and electrical properties with a low loss factor over a wide frequency range (tan δ= 0.002). Furthermore, Kapton Polyimide offers a very low profile (50.8 μm), very robust with a temperature rating of -65 to 200°C.

$$\delta_m = \frac{1}{\sqrt{\pi f \mu \sigma_m}} \quad (11)$$

where δm the skin depth and f the frequency.

In order to increase the antenna gain and directivity by preserving the operating frequency in S-band at 3GHz, authors have started from the elementary antenna optimized before and authors have modeled an array of two-Element printed on the top layer of the same substrate used for the elementary antenna presented above, with

Table 6. Physicals dimensions of the elementary antenna

Parameters	Values (mm)	Parameters	Values (mm)
W	15	L_S	9
L	20	W_S	0.5
W_A	14	H_S	0.125
L_A	12	T_M	0.035
L_G	5	G	0.25
W_F	3.122	L_F	7

Figure 20. Elementary flexible CPW antenna, (a) Front view, (b) Bottom view, and (c) Fabricated elementary antenna

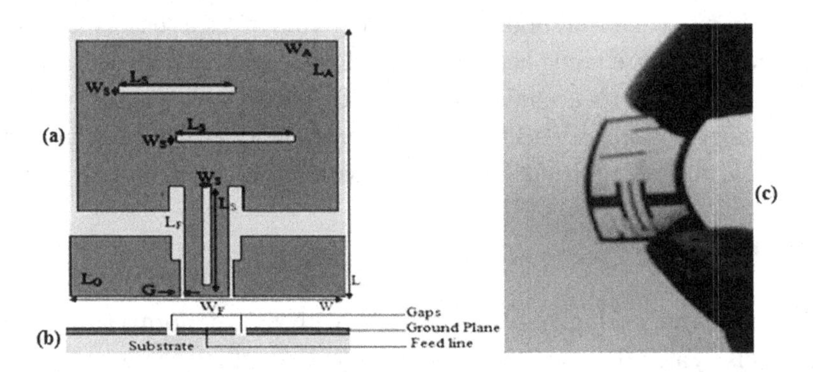

the arrangement shown in Figure 21. The authors have chosen this arrangement to minimize the effects of coupling and the generation of higher modes, also to reduce the array occupation area. A T-power junction has been used to transmit power to the array elements, in addition the CPW feed line with a characteristic impedance of 50 Ω is also used to excite the global antenna array. Further a tuning stub with an optimized dimensions and position is added to the CPW feed to improve the antenna array performances. The proposed antenna array has an overall size of about 5.3cm x5cm with the geometry along with the parameters of the antenna array which are shown in Figure 23 c.

Simulation and Measurement Results

As shown in Figure 22.a, the return loss of the elementary antenna is about -22.58dB with a large bandwidth of 480 MHz at center frequency of 3.09 GHz, which means a

Figure 21. Proposed antenna array (a) with simple ground, (b) with shaped ground, (c) with a tuning stub. (G= 0.15mm, WA= 53mm, LA= 50mm, R=10mm, WF×LF=4.8mm×12.5mm, A×B=2mm×4mm, and (d) Fabricated antenna array

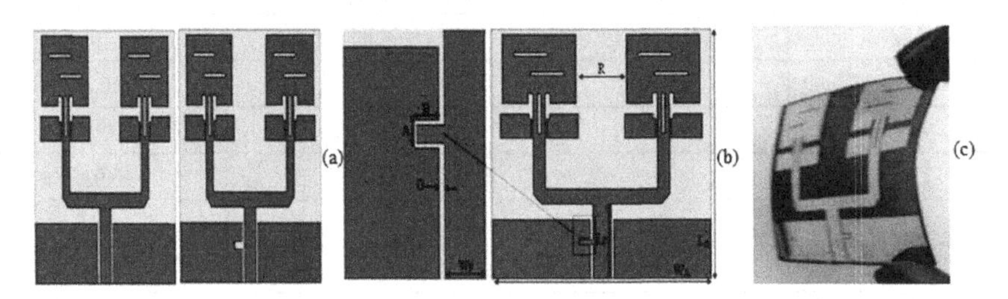

Figure 22. (a) Return Loss of the elementary antenna, and (b) 3D. Radiation pattern @3GHz

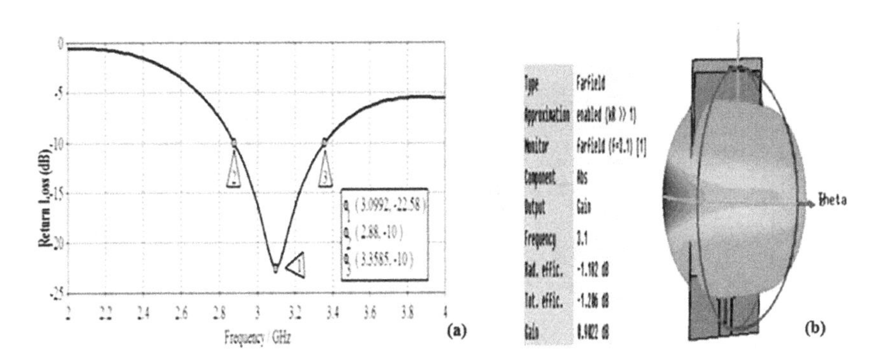

good matching input impedance is achieved at the operating frequency, also Figure 22 b presents the 3D-radiation pattern with the structure below at 3.1GHz. The gain is about 1dB around 3 GHz which is not sufficient for microwave thermography.

To show the effect of the tuning stub, a comparative study has been applied (Figure 23 a. b). It is seen that the tuning stub can shift the operating frequency from 3.4 GHz to 3 GHz providing a good impedance matching for the antenna array with a return loss of about −35 dB at the operating frequency of 3GHz and an important bandwidth of 480MHz which is from 2.72 GHz to 3.2 GHz.

The radiation pattern taken for the far-field at 3GHz is indicated in Figure 23 b, Results indicates that the antenna array provides a directional behavior in E-plan (for PHY=0°) and omnidirectional behavior in H-plan (for THETA=0°), also it design provides an important gain of about 6 dB in the frequency range below 3 GHz.

Furthermore, after fabrication and measurements shown in Figure 24, results show that the proposed structure is well matched at 3GHz (S11< -30 dB).

Figure 23. (a) Return Loss of the antenna array, and (b) 3D. Radiation pattern @3GHz

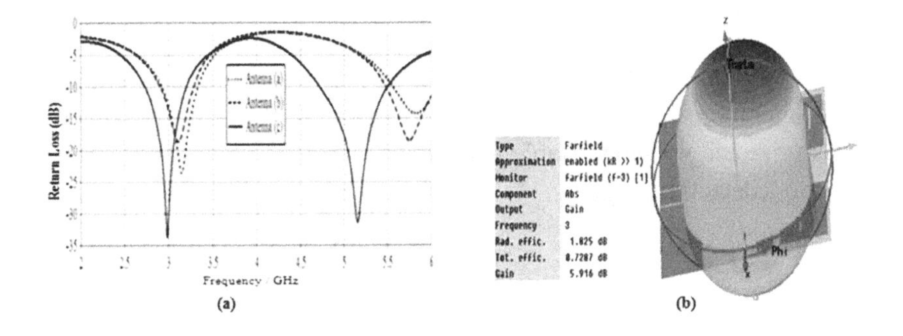

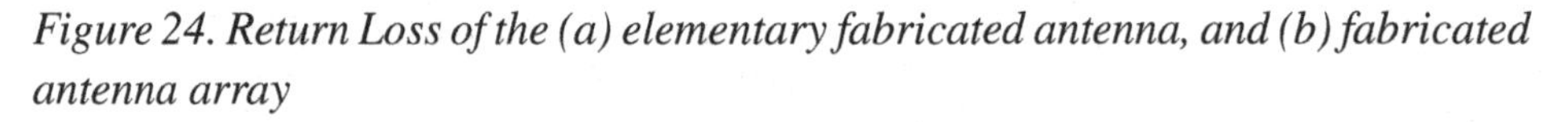

Figure 24. Return Loss of the (a) elementary fabricated antenna, and (b) fabricated antenna array

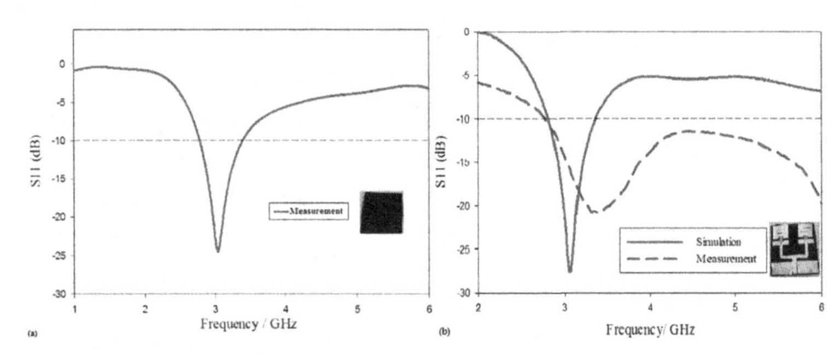

The performance criteria extracted from the software and measurement includes return loss and radiation pattern provide a clear indication that the proposed design, has the required performances to be investigated in a microwave radiometry system as well as for wearable applications, due to its miniature size (5.3cm x5 cm), low profile and weight and very thin substrate. Also the important gain (6dBi) and the large bandwidth (480MHz around the center frequency of 3 GHz), provided by the developed antenna; are good features to improve the radiometer sensitivity at very low power densities transmitted by the self-radiation of abnormal breast tissue.

CONCLUSION

This chapter provided a review of the most recent wearable and flexible antennas found in the literature. Applications such as medical, sport and Security are presented. Many antennas family included: Dipole, Monopole, Patch, and Planar Inverted-F Antennas (PIFAs) antenna, are also discussed. An overview of the electrical characterization of flexible materials used in the fabrication of the wearable antennas was given; as well, all fabrication techniques. Finally, taking into account advantages and disadvantages of, using materials and fabrication methods, authors built an example of a simple patch antenna prototype for biomedical applications, especially, for early breast cancer detection. The flexible antenna array is designed using CST Microwave Studio and fabricated in Kapton polyamide substrate. Due to it performances, the proposed antenna design is a suitable candidate which can investigated to construct a low cost, size and weight portable microwave thermography radiometer for early breast tumors detection.

REFERENCES

Afyf, A., Bellarbi, L., Riouch, F., Achour, A., Errachid, A., & Sennouni, M. A. (2015). Antenna for Wireless Body Area Network (WBAN) Applications Flexible Miniaturized UWB CPW II- shaped Slot. In *IEEE Third International Workshop on RFID And Adaptive Wireless Sensor Networks* (pp. 52–56).

Alipour, S., Parvaresh, F., Ghajari, H., & Kimball, D. F. (2010). Propagation Characteristics for a 60 GHz Wireless Body Area Network (WBAN). In *The 2010 Military Communications Conference - Unclassified Program - Waveforms and Signal Processing Track* (pp. 719–723).

Alomainy, A., Hao, Y., & Pasveer, F. (2007). Numerical and experimental evaluation of a compact sensor antenna for healthcare devices. *IEEE Transactions on Biomedical Circuits and Systems*, *1*(4), 242–249. doi:10.1109/TBCAS.2007.913127 PMID:23852005

Alomainy, A., Sani, A., Rahman, A., Santas, J. G., & Hao, Y. (2009). Transient characteristics of wearable antennas and radio propagation channels for ultrawideband body-centric wireless communications. *IEEE Transactions on Antennas and Propagation*, *57*(4 part 1), 875–884. doi:10.1109/TAP.2009.2014588

Amal, A., Larbi, B., & Anouar, A. (2016). Miniaturized Wideband Flexible CPW Antenna with Hexagonal Ring Slots for Early Breast Cancer Detection. In S. S. M. Singapore (Ed.), *Ubiquitous Networking* (pp. 211–222). Springer. doi:10.1007/978-981-287-990-5_17

B., I., & Batchelor, J. C. (2008). A dual band belt antenna. In *International Workshop on Antenna Technology: Small Antennas and Novel Metamaterial* (pp. 374– 377).

Balanis, C. A. (2005). *Antenna Theory Analysis and Design* (3rd ed.). John Wiley Sons, Inc.

Bokhari, S. A., Zürcher, J. F., Mosig, J. R., & Gardiol, F. E. (1996). A small microstrip patch antenna with a convenient tuning option. *IEEE Transactions on Antennas and Propagation*, *44*(11), 1521–1528. doi:10.1109/8.542077

Brown, W. C. (1984). The history of power transmission by radio waves. *IEEE Transactions on Microwave Theory and Techniques*, *32*(9), 1230–1242. doi:10.1109/TMTT.1984.1132833

Carmo, J. P., Dias, N., Mendes, P. M., Couto, C., & Correia, J. H. (2006). Low-power 2.4-GHz RF transceiver for wireless EEG module plug-and-play. In *13th IEEE International Conference on Electronics, Circuits and Systems* (pp. 1144–1147). doi:10.1109/ICECS.2006.379642

Carter, J., Saberin, J., Shah, T., Sai Ananthanarayanan, P. R., & Furse, C. (2010). Inexpensive fabric antenna for off-body wireless sensor communication. In *IEEE International Symposium on Antennas and Propagation* (pp. 9–12). http://doi.org/ doi:10.1109/APS.2010.5561753

Casula, G., Montisci, G., & Mazzarella, G. (2013). A wideband PET inkjet-printed antenna for UHF RFID. *Antennas and Wireless Propagation Letters*, *12*, 1400–1403. doi:10.1109/LAWP.2013.2287307

Chandran, A. R., Conway, G. A., & Scanlon, W. G. (2008). Compact Low Profile Patch Antenna For Medical Body Area Netowrks at 868 MHz. In *IEEE Antennas and Propagation Society International Symposium* (pp. 8–11).

Chen, X. L., Kuster, N., Tan, Y. C., & Chavannes, N. (2012). Body effects on the GPS antenna of a wearable tracking device. In *6th European Conference on Antennas and Propagation, EuCAP* (pp. 3313–3316). http://doi.org/ doi:10.1109/ EuCAP.2012.6205889

Chen, Z. N. (2007). *Antennas for Portable Devices* (1st ed.). Wiley-VCH. doi:10.1002/9780470319642

Chu, L. J. (1948). Small antennas. *Journal of Applied Physics*, *19*(12), 1163–1175. doi:10.1063/1.1715038

Cibin, C., Leuchtmann, P., Gimersky, M., & Vahldieck, R. (2004). Modified E-Shaped Pifa Antenna for Wearable Systems. In *URSI International Symposium on Electromagnetic Theory (URSI EMTS)* (pp. 873–875).

Cibin, C., Leuchtmann, P., Gimersky, M., Vahldieck, R., & Moscibroda, S. (2004). A flexible wearable antenna. In IEEE Antennas and Propagation Society (pp. 3589–3592). doi:10.1109/APS.2004.1330122

Constantinou, C., Nechayev, Y., Wu, X., & Hall, P. (2012). Body-area Propagation at 60 GHz. In Loughborough Antennas & Propagation Conference (pp. 1–4).

Cook, B., & Shamim, A. (2013). Utilizing wideband AMC structures for high-gain inkjet-printed antennas on lossy paper substrate. *IEEE Antennas and Wireless Propagation Letters*, *12*, 76–79. doi:10.1109/LAWP.2013.2240251

Corner, A. A. (2013). Body-Worn Antennas Making a Splash : Lifejacket-Integrated Antennas for Global Search and Rescue Satellite System. *IEEE Antennas and Propagation Magazine, 55*(2).

Curone, D., Dudnik, G., Loriga, G., Luprano, J., Magenes, G., Paradiso, R., ... Bonfiglio, A. (2007). Smart Garments for Safety Improvement of Emergency/Disaster Operators. In *29th Annual International Conference of the IEEE Engineering in Medicine and Biology Society (EMBS)* (pp. 3962–3965). doi:10.1109/IEMBS.2007.4353201

Curto, S., McEvoy, P., Bao, X., & Ammann, M. J. (2009). Compact patch antenna for electromagnetic interaction with human tissue at 434 MHz. *IEEE Transactions on Antennas and Propagation, 57*(9), 2564–2571. doi:10.1109/TAP.2009.2027040

Declercq, F., Couckuyt, I., Rogier, H., & Dhaene, T. (2010). Complex permittivity characterization of textile materials by means of surrogate modeling. In *IEEE Antennas and Propagation Society International Symposium* (pp. 1–4).

DeJean, G., Bairavasubramanian, R., Thompson, D., Ponchak, G. M. T., & Papapolymerou, J. (2005). Liquid crystal polymer (LCP): A new organic material for the development of multilayer dual-frequency dual-polarization flexible antenna arrays. *IEEE Antennas and Wireless Propagation Letters, 4*(1), 22–26. doi:10.1109/LAWP.2004.841626

Federal Communications Commission (FCC). (2014). Retrieved from http: //transition.fcc.gov/bureaus/engineering_technology/orders/2002/fcc02048.pdf

Firoozy, N., & Shirazi, M. (2011). Planar Inverted-F Antenna (PIFA) Design Dissection for Cellular Communication Application. *Journal of Electromagnetic Analysis and Applications, 03*(10), 406–411. doi:10.4236/jemaa.2011.310064

Fujimoto, K. (2008). Mobile Antenna Systems Handbook. Artech House.

Gaetano, D., McEvoy, P., Ammann, M. J., Browne, J. E., Keating, L., & Horgan, F. (2013). Footwear antennas for body area telemetry. *IEEE Transactions on Antennas and Propagation, 61*(10), 4908–4916. doi:10.1109/TAP.2013.2272451

Gustafsson, M., Sohl, C., & Kristensson, G. (2007). Physical limitations on antennas of arbitrary shape. In Royal Society a-Mathematical Physical and Engineering Sciences (Vol. 463, pp. 2589–2607). doi:10.1098/rspa.2007.1893

Hao, Y., Alomainy, A., Hall, P. S., Nechayev, Y. I., Parini, C. G., & Constantinou, C. C. (2012). *Antennas and Propagation for Body Centric Wireless Communications* (2nd ed.; pp. 38–41). Norwood, MA: Artech House, Inc.

Harrington, R. F. (1960). Effect of antenna size on gain, bandwidth, and efficiency. *Journal of Research of the National Bureau of Standards, Section D. Radio Propagation*, *64D*(1), 1–12. doi:10.6028/jres.064D.003

Haskou, A., Ramadan, A., Al-Husseini, M., Kasem, F., Kabalan, K. Y., & ElHajj, A. (2012). A simple estimation and verification technique for electrical characterization of textiles. In *Middle East Conference on Antennas and Propagation* (pp. 1–4). doi:10.1109/MECAP.2012.6618190

Hertel, T. W., & Smith, G. S. (2003). On the dispersive properties of the conical spiral antenna and its use for pulsed radiation. *IEEE Transactions on Antennas and Propagation*, *51*(7), 1426–1433. doi:10.1109/TAP.2003.813602

Hertleer, C., Rogier, H., Member, S., Vallozzi, L., & Van Langenhove, L. (2009). A Textile Antenna for Off-Body Communication Integrated Into Protective Clothing for Firefighters. *IEEE Transactions on Antennas and Propagation*, *57*(4), 919–925. doi:10.1109/TAP.2009.2014574

J., R.-T., Roa, L. M., & Prado, M. (2006). Design of antennas for a wearable sensor for homecare movement monitoring. *IEEE Engineering in Medicine and Biology Society*, 5972–5976.

Jabbar, H., Song, Y. S., & Jeong, T. T. (2010). RF energy harvesting system and circuits for charging of mobile devices. *IEEE Transactions on Consumer Electronics*, *56*(1), 247–253. doi:10.1109/TCE.2010.5439152

Jr, H., & Williams, C. S. (1965). Transients in wide-angle conical antennas. *IEEE Transactions on Antennas and Propagation*, *13*(2), 236–246. doi:10.1109/TAP.1965.1138399

Kaija, T., Lilja, J., & Salonen, P. (2010). Exposing textile antennas for harsh environment. In *Military Communications Conference (MILCOM)* (pp. 737–742).

Kaivanto, E., Berg, M., Salonen, E., & Maagt, P. (2011). Wearable circularly polarized antenna for personal satellite communication and navigation. *IEEE Transactions on Antennas and Propagation*, *59*(12), 4490–4496. doi:10.1109/TAP.2011.2165513

Khaleel, H., Al-Rizzo, H. M., Rucker, D. G., & Mohan, S. (2012). A compact polyimide based UWB antenna for flexible electronics. *Antennas and Wireless Propagation Letters*, *11*, 564–567. doi:10.1109/LAWP.2012.2199956

Kim, J., & Rahmat-Samii, Y. (2004). Implanted antennas inside a human body: Simulations, designs, and characterizations. *IEEE Transactions on Microwave Theory and Techniques*, *52*(8), 1934–1943. doi:10.1109/TMTT.2004.832018

Kim, Y., Kim, H., & Yoo, H. (2010). Electrical characterization of screen-printed circuits on the fabric. *IEEE Transactions on Advanced Packaging, 33*(1), 196–205. doi:10.1109/TADVP.2009.2034536

Kiourti, A., Member, S., Costa, J. R., Member, S., Fernandes, C. A., Member, S., & Member, S. et al. (2012). Miniature Implantable Antennas for Biomedical Telemetry : From Simulation to Realization. *IEEE Transactions on Bio-Medical Engineering, 59*(11), 3140–3147. doi:10.1109/TBME.2012.2202659 PMID:22692865

Koo, T., Kim, D., Ryu, J., Seo, H., Yook, J., & Kim, J. (2011). Design of a label typed UHF RFID tag antenna for metallic objects. *Antennas and Wireless Propagation Letters, 10*, 1010–1014. doi:10.1109/LAWP.2011.2166370

Koohestani, M., Zurcher, J.-F., Moreira, A. A., & Skrivervik, A. K. (2014). A Novel, Low-Profile, Vertically-Polarized UWB Antenna for WBAN. *IEEE Transactions on Antennas and Propagation, 62*(4), 1888–1894. doi:10.1109/TAP.2014.2298886

Lapinski, M., Feldmeier, M., & Paradiso, J. A. (2011). Wearable wireless sensing for sports and ubiquitous interactivity. *IEEE Sensors, 1425–1428.* doi:10.1109/ICSENS.2011.6126902

Lee, G. Y., Psychoudakis, D., Chen, C. C., & Volakis, J. L. (2011). Omnidirectional vest-mounted body-worn antenna system for UHF operation. *IEEE Antennas and Wireless Propagation Letters, 10*, 581–583. doi:10.1109/LAWP.2011.2158381

Lilja, J., & Salonen, P. (2009). Textile material characterization for software antennas. In *IEEE Military Communications Conference* (pp. 1–7).

Lin, C. H., Li, Z., Ito, K., Takahashi, M., & Saito, K. (2012). A small tunable and wearable planar inverted-F antenna (PIFA). In *6th European Conference on Antennas and Propagation, EuCAP* (pp. 742–745). http://doi.org/ doi:10.1109/EuCAP.2012.6206554

Ma, L., Edwards, R., & Bashir, S. (2008). A wearable monopole antenna for ultra wideband with notching function. In *IET Seminar on Wideband and Ultrawideband Systems and Technologies: Evaluating current Research and Development* (pp. 1–5). doi:10.1049/ic.2008.0695

Mandal, S., Turicchia, L., & Sarpeshkar, R. (2010). A low-power, battery-free tag for body sensor networks. *IEEE Pervasive Computing / IEEE Computer Society [and] IEEE Communications Society, 9*(1), 71–77. doi:10.1109/MPRV.2010.1

Mantash, M., Tarot, A., Collardey, S., & Mahdjoubi, K. (2011). Wearable monopole zip antenna. *Electronics Letters, 47*(23), 1266–1267. doi:10.1049/el.2011.2784

Matthews, J., & Pettitt, G. (2009). Development of flexible, wearable antennas. In *3rd European Conference on Antennas and Propagation (EuCAP)* (pp. 273–277).

McLean, J. S. (1996). A re-examination of the fundamental limits on the radiation Q of\nelectrically small antennas. *IEEE Transactions on Antennas and Propagation*, *44*(5), 672–676. doi:10.1109/8.496253

Medical Device Radiocommunications Service (MedRadio). (2009). Retrieved from www.fcc.gov

Merli, F. (2008). Implantable antennas for biomedical applications. In *Antennas and propagation society international Symposium* (Vol. 5110, pp. 1642–1649). doi:<ALIGNMENT.qj></ALIGNMENT>10.5075/epfl-thesis-5110

Micoach. (2014). Retrieved from http://micoach.adidas.com/

Monti, G., Corchia, L., & Tarricone, L. (2013). UHF wearable rectenna on textile materials. *IEEE Transactions on Antennas and Propagation*, *61*(7), 3869–3873. doi:10.1109/TAP.2013.2254693

National Swiss Plan. (2014). *Swiss National Frequency Allocation Plan and Specific Assignments*. Author.

Nikolaou, S., Ponchak, G., Papapolymerou, J., & Tentzeris, M. (2006). Conformal double exponentially tapered slot antenna (DETSA) on LCP for UWB applications. *IEEE Transactions on Antennas and Propagation*, *54*(6), 1663–1669. doi:10.1109/TAP.2006.875915

Numakura, D. (2007). Flexible Circuit Applications and Materials. In Printed Circuit Handbook (6th ed.). McGraw Hill.

Ouyang, Y., & Chappell, W. (2008). High-frequency properties of electro-textiles for wearable antenna applications. *IEEE Transactions on Antennas and Propagation*, *56*(2), 381–389. doi:10.1109/TAP.2007.915435

Pellegrini, A., Brizzi, A., Zhang, L., Ali, K., Hao, Y., Wu, X., & Sauleau, R. et al. (2013). Antennas and propagation for body-centric wireless communications at millimeter-wave frequencies: A review. *IEEE Antennas and Propagation Magazine*, *55*(4), 262–287. doi:10.1109/MAP.2013.6645205

Popovic, Z., Momenroodaki, P., & Scheeler, R. (2014). Toward wearable wireless thermometers for internal body temperature measurements. *IEEE Communications Magazine*, *52*(10), 118–125. doi:10.1109/MCOM.2014.6917412

Psychoudakis, D., Lee, G., Chen, C., & Volakis, J. (2010). Military UHF body-worn antennas for armored vests. In *6th European Conference on Antennas and Propagation (EuCAP) IEEE* (pp. 1–4).

Psychoudakis, D., & Volakis, J. L. (2009). Conformal Asymmetric Meandered Flare (AMF) Antenna for Body-Worn Applications. *IEEE Antennas and Wireless Propagation Letters*, *8*, 931–934. doi:10.1109/LAWP.2009.2028662

Rahmat-Samii, Y. (2007). Wearable and implantable antennas in body-centric communications. In *2nd European Conference on Antennas and Propagation (EuCAP)* (pp. 1–5).

Rais, N., Soh, P., Malek, F., Ahmad, S., Hashim, N., & Hall, P. (2009). A review of wearable antenna. In *Loughborough Antennas and Propagation Conference(LAPC)* (pp. 225–228).

Ranck, J. (n.d.). *The wearable computing market: A global analysis*. Retrieved from http://go.gigaom.com/rs/gigaom/images/ wearable-computing-the-next-big-thing-in-tech.pdf

Reed, F. (2007). 60 GHz WPAN Standardization within IEEE. In *International Symposium on Signals, Systems and Electronics, (ISSSE '07)* (pp. 103–105).

Rishani, N. R., AI-Husseini, A., E.-H., & Kabalan, K. Y. (2012). Design and relative permittivity determination of an EBG-based wearable antenna. In Progress in Electromagnetics and Radio Frequency (pp. pp. 96–99). Moscow, Russia: PIERS.

Running G. P. S. Watches. (2014). Retrieved from http://sports.tomtom.com/en_us/

Salman, L. K. H., & Talbi, L. (2010). G-shaped wearable cuff button antenna for 2.45 GHZ ISM band applications. In *2010 14th International Symposium on Antenna Technology and Applied Electromagnetics and the American Electromagnetics Conference, ANTEM/AMEREM 2010* (pp. 14–17). doi:<ALIGNMENT.qj></ ALIGNMENT>10.1109/ANTEM.2010.5552573

Salonen, P., Rahmat-Samii, Y., Hurme, H., & Kivikoski, M. (2004). Effect of conductive material on wearable antenna performance: A case study of WLAN antennas. *Antennas and Propagation Society*, *1*, 455–458. doi:10.1109/APS.2004.1329672

Salonen, P., Sydanheimo, L., Keskilammi, M., & Kivikoski, M. (1999). A Small Planar Inverted-F Antenna for Wearable Applications. In *IEEE Conference Publications* (pp. 95–100). doi:10.1109/ISWC.1999.806679

Sankaralingam, S., & Gupta, B. (2009). A circular disk microstrip WLAN antenna for wearable applications. In *IEEE India Council Conference* (pp. 3–6). doi:10.1109/INDCON.2009.5409355

Santas, J., Alomainy, A., & Hao, Y. (2007). Textile Antennas for On-Body Communications: Techniques and Properties. In *2nd European Conference on Antennas and Propagation (EuCAP)* (pp. 1–4.). doi:10.1049/ic.2007.1064

Sanz-Izquierdo, B., Huang, F., & Batchelor, J. C. (2006). Small size wearable button antenna. In *First European Conference on Antennas and Propagation* (pp. 1–4).

Sanz-Izquierdo, B., Miller, J. A., Batchelor, J. C., & Sobhy, M. I. (2010). Dual-Band Wearable Metallic Button Antennas and Transmission in Body Area Networks. *IET Microw. Antennas Propag.*, *4*(2), 182–190. doi:10.1049/iet-map.2009.0010

Serra, A. A., Nepa, P., & Manara, G. (2011). Cospas Sarsat rescue applications. *IEEE International Symposium on Antennas and Propagation (APSURSI)*, *3*, 1319–1322. doi:10.1109/APS.2011.5996532

Skrivervik, A. K., & Merli, F. (2011). Design strategies for implantable antennas. In *Loughborough Antennas and Propagation Conference* (pp. 1–5). doi:10.1109/LAPC.2011.6114011

Skrivervik, A. K., & Zürcher, J. F. (2007). Miniature antenna design at LEMA. In *19th International Conference on Applied Electromagnetics and Communications* (pp. 4–7). doi:<ALIGNMENT.qj></ALIGNMENT>10.1109/ICECOM.2007.4544410

Snow flaik tracks skiers. (2011). Retrieved from http://news.discovery.com/tech/snow-flaik-tracks-skiers.htm

Soh, P., Boyes, S., Vandenbosch, G., Huang, Y., & Ma, Z. (2012). Dual-band Sierpinski textile PIFA efficiency measurements. In *6th European Conference on Antennas and Propagation (EuCAP)*, (pp. 3322–3326).

Soh, P. J., Vandenbosch, G. A. E., Volski, V., & Nurul, H. M. R. (2010). Characterization of a Simple Broadband Textile Planar Inverted-F Antenna (PIFA) for on Body Communications. In *22nd International Conference on Applied Electromagnetics and Communications (ICECom)* (pp. 1–4).

Staub, O., Zürcher, J.-F., & Skrivervik, A. (1998). Some considerations on the correct measurement of the gain and bandwidth of electrically small antennas. *Microwave and Optical Technology Letters*, *17*(3), 156–160. doi:10.1002/(SICI)1098-2760(19980220)17:3<156::AID-MOP2>3.0.CO;2-I

Staub, O., Zurcher, J.-F., Skrivervik, A. K., & Mosig, J. R. (1999). PCS antenna design: the challenge of miniaturisation. *IEEE Antennas and Propagation Society International Symposium*, 1. doi:10.1109/APS.1999.789198

Tesla, N. (1892). Experiments with alternate currents of high potential and high frequency. *Journal of the Institution of Electrical Engineers*, *21*(97), 51–162. doi:10.1049/jiee-1.1892.0002

Tetrapol. (2015). *Tetrapol Factsheet Trunked radio system for emergency services*. Author.

Trajkovikj, J., Fuchs, B., & Skrivervik, A. (2011). *LEMA internal report*. Academic Press.

Ullah, S., & Kwak, K. S. (2011). *Body Area Network for Ubiquitous Healthcare Applications: Theory and Implementation*. 10.1007/s10916-011-9787-x

Vallozzi, L., Van Torre, P., Hertleer, C., Rogier, H., Moeneclaey, M., & Verhaevert, J. (2010). Wireless communication for firefighters using dual-polarized textile antennas integrated in their garment. *IEEE Transactions on Antennas and Propagation*, *58*(4), 1357–1368. doi:10.1109/TAP.2010.2041168

Vallozzi, L., Vandendriessche, W., Rogier, H., Hertleer, C., & Scarpello, M. L. (2010). Wearable textile GPS antenna for integration in protective garments. In *4th European Conference on Antennas and Propagation (EuCAP)* (pp. 1–4).

Vallozzi, P., Van Torre, P., Hertleer, C., Rogier, H., Moeneclaey, M., & Verhaevert, J. (2010). Wireless communication for firefighters using dual-polarized textile antennas integrated in their garment. *IEEE Transactions on Antennas and Propagation*, *58*(4), 1357–1368. doi:10.1109/TAP.2010.2041168

Verb for shoe - very intelligent shoes. (2004). Retrieved from http://www.gizmag.com/go/3565/picture/7504/

Virtanen, J., Björninen, T., Ukkonen, L., & Sydänheimo, L. (2010). Passive UHF inkjet-printed narrow-line RFID tags. *Antennas and Wireless Propagation Letters*, *9*, 440–443. doi:10.1109/LAWP.2010.2050050

Wang, Y., Li, L., Wang, B., & Wang, L. (2009). A body sensor network platform for in-home health monitoring application. In *4th International Conference on Ubiquitous Information Technologies Applications* (pp. 1–5). doi:10.1109/ICUT.2009.5405731

Weiner, M. M. (2003). *Monopole Antennas* (1st ed.). Taylor and Francis. doi:10.1201/9780203912676

Wheeler, H. A. (1975). Small antennas. *IEEE Transactions on Antennas and Propagation*, *23*(4), 462–469. doi:10.1109/TAP.1975.1141115

Wong, K. (2002). *Compact and Broadband Microstrip Antennas* (1st ed.). Wiley-VCH. doi:10.1002/0471221112

Wu, X. Y., Akhoondzadeh-Asl, L., Wang, Z. P., & Hall, P. S. (2010). Novel Yagi-Uda antennas for on-body communication at 60GHz. In *Loughborough Antennas* (pp. 153–156). Propagation Conference. doi:10.1109/LAPC.2010.5666188

Xia, W., Saito, K., Takahashi, M., & Ito, K. (2009). Performances of an implanted cavity slot antenna embedded in the human arm. In IEEE Transactions on Antennas and Propagation (Vol. 57, pp. 894–899). doi:10.1109/TAP.2009.2014579

Zegna Sport Bluetooth iJacket incorporates smart fabric. (2007). Retrieved from http://www.gizmag.com/go/7856/

Zurcher, J.-F., Skrivewik, A. K., & Staub, O. (2000). SMILA: a miniaturized antenna for PCS applications. *IEEE Antennas and Propagation Society International Symposium*, *3*, 1646–1649. doi:10.1109/APS.2000.874556

Zürcher, J. F., Staub, O., & Skrivervik, A. K. (2000). SMILA: A compact and efficient antenna for mobile communications. *Microwave and Optical Technology Letters*, *27*(3), 155–157. doi:10.1002/1098-2760(20001105)27:3<155::AID-MOP1>3.0.CO;2-P

Chapter 8
Wearable Technologies for Helping Human Thermophysiological Comfort

Radostina A. Angelova
Technical University of Sofia, Bulgaria

ABSTRACT

The thermophysiological comfort is one of the aspects of the human comfort. It is related to the thermoregulatory system of the body and its reactions to the temperature of the surrounding air, activity and clothing. The aim of the chapter is to present the state of the art in the wearable technologies for helping the human thermophysiological comfort. The basic processes of body's thermoregulatory system, the role of the hypothalamus, the reactions of the body in hot and cold environment, together with the related injuries, are described. In the second part of the chapter smart and intelligent clothing, textiles and accessories are presented together with wearable devices for body's heating/cooling.

INTRODUCTION

The thermophysiological comfort is one of the aspects of the human comfort both indoors and outdoors. It is related to the thermoregulatory system of the body and its reactions to the temperature of the surrounding air, body activity, and clothing.

It is assumed that the human body is in a state of *thermophysiological comfort* when the heat, generated by the body through metabolic processes at a cellular level, is equal to the heat emitted, to the surroundings (heat losses from the body). The

DOI: 10.4018/978-1-5225-3290-3.ch008

thermophysiological comfort can be also defined as a condition, in which the body heat storage is equal to zero (Havenith, 1999; Angelova, 2016).

Very often in literature the terms "thermophysiological comfort" and "thermal comfort" are used as synonyms. According to the definition of "thermal comfort", given in the international standard ISO 7730 (1995), it could be assumed that the thermal comfort is much more a personal judgment about the ambient temperature (often related to indoor environment), presence of local draft and local temperature discomfort, etc., while the thermophysiological comfort is directly related to the reactions of the human body, caused by the activity of the thermoregulatory system.

In fact, the human body is designed to function in an environment, which temperature is around 20 °C. At the same time, the human beings inhabit different climates: from very cold in the arctic regions to very hot in the deserts. Without shelter and clothing, people could not survive in these rigorous climates. This is even more valid for the modern world, as people nowadays are working in extreme temperature conditions both outdoors and indoors. Such activities could not be performed effectively or performed at all without the presence of clothing, which provides thermal protection for the individual. Today, however, new items and devices are added to the traditional protection, given by clothing and shelter: the wearable technologies that help the thermophysiological comfort of the body.

Wearable devices are among the most important hi-tech items nowadays. They increasingly represent innovative products that are used in one or another field of human life, work and security. One of these applications is the protection from adverse environmental conditions in terms of low / high temperatures, supporting the processes of thermoregulation, and helping, ultimately, the human thermophysiological comfort.

Clothing and textiles have always been an interface between the human body and the environment: they are the most frequently used commodities to protect the human from the environment, "working" as an insulation barrier between the body and the surrounding air. They have truly wearable "characteristics". High-tech clothing and textiles obtain properties as a result of complex, interdisciplinary production processes and know-how. Even traditional fashion items today can be provided with additional and unusual performance characteristics: sensors, light, heating, color variations, etc. Smart and intelligent textiles are high-tech solutions to improve the role of textiles and clothing as insulating barriers between the human body and the environment.

Figure 1 illustrates the relationship between wearable electronics and textiles. Though the Technical Report CEN / TR 16298 (2011) does not provide a distinction between smart and intelligent textiles, the works of Park and Jayaraman (2003) and Van Langenhove and Hertleer (2004) clearly explain their particular properties. The essential difference is that the smart textiles involve electronic components

Figure 1. Relationship between ICT and Textiles

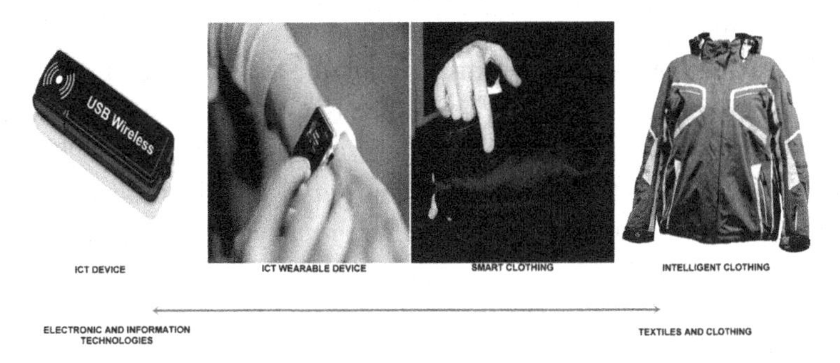

and devices in their structure and / or process information electronically. While intelligent textiles do not need electronics to have specific technological performance and function properly in the application they have been designed for. Certainly, they can be equipped with electronic components, sensors, actuators, and to transfer elements of information as any other textiles.

The aim of the chapter is to give to the reader on overview of the state of the art in the wearable technologies, used for helping the human thermophysiological comfort. There are many solutions, which can be found in the form of single wearable electronics, heating or cooling outfits in smart clothing and smart textiles, footwear or accessories, as well as intelligent clothing and intelligent textiles for support of the thermoregulatory system. Having in mind the fast moving nature of wearable technologies, the examples given here can be followed by new items shortly after the release of this book.

THE THERMOREGULATION OF THE HUMAN BODY

Heat enters the human body through three main sources:

- Solar Energy: It can be "absorbed" by the body directly from the Sun and indirectly, via diffusion radiation, conductive heat from the ground and radiative heat from the surrounding air;
- Heating sources of different type (mostly in the indoor environment);
- Metabolism of food

The food metabolism determines the human as an endothermic being, which uses the metabolism for heat generation. So, like birds and other mammals, the man keeps its body temperature in most of the cases higher than the temperature of the surrounding air, but it allows him to inhabit hazardous (in terms of temperature) environments, like deserts and polar regions.

The thermoregulatory system aims to retain the temperature of the human core body about 37 °C. The core body includes the brain and all major organs that find room in the torso. In cold or hot environment body's thermal regulation strives to ensure this temperature for all tissues, organs, and systems in the torso and head, providing the normal working temperature for the enzyme systems and vital organs. At the same time, the surface of the skin and extremities can suffer from thermal discomfort and even injuries (mainly cold-induced injuries in low temperatures environment).

There are two ways of dealing with human thermal discomfort:

- Physiological reactions and
- Behavioral reactions.

The physiological reactions are related to the response of the thermoregulatory system and they are different in the hot and cold environment.

The behavioral reactions are associated with searching for shelter, contact with a heat source, cooling with water, drinking water or liquids to hydrate the body, etc.

Animals, which belong to the group of endotherms, have a third mode to deal with the thermal discomfort: physical reactions. They have developed different physical features, which protect them in the warm or cold environment: fur, big ears, or layers of fat.

A similar feature in the world of humans is the clothing, but the selection of proper clothing items for a particular environment is considered as a behavioral reaction.

The Hypothalamus

The temperature in the human body is controlled mainly by the hypothalamus, a gland in the human brain, and partly by the spinal cord. The hypothalamus integrates into its center signals, which continuously receives from two groups of sources (Havenith, 2002; Parsons, 2003):

- Signals on the back side of the hypothalamus, which register the temperature of the blood that washes the brain area around the gland.
- Signals on the front side of the hypothalamus, which are received from the peripheral nerves, associated with the temperature receptors for the skin temperature.

The temperature receptors register the skin sensation for cold / warmth. They are very sensitive: the receptors for warmth are able to register temperature increase of 0.007 °C, and the receptors for cold can detect decrement of the temperature with 0.012 °C.

Having the integrated signal in its center, the hypothalamus sends output signals to different parts of the body, aiming to increase or decrease the core body temperature (Nielsen, 1978; Tarlochan & Ramesh, 2005):

- **Signals to the Smooth Muscles in the Walls of the Arterioles in Order to Change the Cross Section of the Blood Vessels:** The vasodilatation (expansion of the blood vessels) increases the blood flow to the peripheral parts of the skin and extremities, helping the process of heat release from the body to the surrounding air in the hot environment. The vasoconstriction (constricting of the blood vessels) decreases the blood flow to the extremities and surface parts of the skin, so as to keep the core body as warmer as possible in cold environment.
- **Signals to the Adrenal and Thyroid Glands to Increase the Metabolism:** The increased metabolism provokes heating of the body in the cold environment.
- **Signals to the Erector Pili Muscles to Cause Pricking:** These muscles, located below the surface of the skin, provoke rising of the skin hair in the cold. This way higher quantity of air is "trapped" near the skin in a form of motionless "boundary layer" between the skin surface and the air, surrounding the body.
- **Signals to the Skeletal Muscles to Start Shivering:** Via the uncontrolled contraction of the muscles the body increases its heat production in cold environment. However, this could have an effect up to two hours after the exposure to low temperatures, as the glycogen in the muscles is constantly depleting.
- **Signals to the Sweat Glands to Increase the Water Vapor Release From the Skin Surface in a Hot Environment:** The heat losses can be even increased if a sweat in a liquid form appears on the skin surface, as the liquid "layer" can increase the cooling of the body up to 5 times.

Basal Metabolic Rate

The basal metabolic rate is the metabolism of a resting human in a state of thermophysiological comfort. It reflects the energy, needed by the body during relax (no physical or mental work), at least 12 hours after the last food intake. There are several factors, which can affect the human metabolism and consequently – the reactions of the body towards protection of the thermophysiological comfort. The most frequently commented in the literature are (Harris & Benedict, 1918; Nesbitt, Lam, & Thompson, 1999; Gropper & Smith, 2012):

- **Cold / Hot Environment:** People, leaving in a cold environment have higher basal metabolic rate compared to people, living in a warmer environment.
- **Sex:** Females have lower basal metabolic rate compared to men, due to the difference in the quantity of the muscle tissues.
- **Hormones:** Some hormones provoke increment of the basal metabolic rate. Adrenaline has a short-term effect on the metabolic increment, while the thyroxine leads to long-term augmentation of the metabolism.
- **Age:** The basal metabolic rate decreases gradually with aging.
- **Physical Activity:** Frequent physical exercises lead to an increment of the basal metabolic rate due to the active involvement of muscles.
- **Dieting:** The reduction of food calories or starvation is responsible for a significant decrement of the basal metabolic rate.

The total amount of the heat, produced by the body, depends on body's volume. The rate of the heat losses is preconditioned by the skin surface (surface of the body).

The factors, which narrow the heat losses, are:

- The rate of the blood flow between the core body and the skin;
- The evaporation rate from the skin surface to the surroundings;
- The insulation ability of the clothing / textile layers.

To ensure proper thermoregulation, all endothermic creatures, including the humans, require a continuous supply of food, oxygen, and water. 80 to 90% of the metabolic energy is used to deal with the thermoregulation of the body (Díaz & Becker, 2010).

Body Reactions in Cold Environment

All activities in cold environments pose a significant risk to the human body in terms of comfort, performance, and health. When the body is exposed to low temperatures, the thermoregulatory system triggers the mechanisms for generating additional thermal energy by increasing the rate of metabolic processes in liver and muscle tissues. That's why people in colder countries have increased metabolism compared to those, living in warmer climates. At the same time, the thermoregulatory system reduces the heat losses from the lungs and skin surface via constriction of blood vessels. Redirecting the blood flow to the brain and torso, combined with restraint heat losses, results in a temporary increment in body temperature with 1-2 °C. The vasoconstriction lowers the temperature of the extremities, thus causing shivering of the muscles and new increment of the heat production, in addition to the generation of heat from metabolic processes. The behavioral reaction of the individual to add extra layers of clothing / textiles also contributes to the rise in body temperature through decrement of the heat loss from the skin to the surrounding air.

A protective barrier between the body and the environment, consistent with the low ambient temperature, must be used to overcome the influence of cold on human thermophysiological comfort, health, and productivity (Angelova, 2007; Holmér, 2009). Clothing and textiles for indoor and outdoor use play the role of such a barrier. They have to prevent the heat exchange between the body and the environment having high enough insulating capacity. Along with this, they have to avert the entry of cold air and water to the body and limit the radiation of heat from the skin surface to the surroundings (Morrissey & Rossi, 2013; Celcar, 2014).

The cold-related problems, which may occur in the human body, can be summarized as follows:

- **Thermal Discomfort and Pain Sensation (Mainly in the Extremities):** They come into sight when heat losses from parts of the body (i.e. uncovered face or hands) or the entire body appear.
- **Low Efficiency:** Fine finger movements are hampered even at a moderate level of cold stress. The efficiency and energy during activity decrease when the temperature of the torso falls below 36 °C.
- **Cardio-Respiratory Effects:** The inhalation of cold air cools the lining of the upper respiratory tract and subsequently can cause irritation, occurrence of micro-inflammatory reactions and even a bronchospasm. Bronchospasms are common reactions to cold and typical for people with asthma or sensitive airways. Cooling of the airways can also cause pain to people with cardiovascular disease or circulatory disorders.
- **Different Types of Health Disorders:** Cold injuries, aggravation of existing cardiovascular disorders, increased risk of accidents.

In the absence of protection from the low air temperature, a person, exposed to cold, may suffer from direct tissue damages. Frostbite usually affects the extremities and the fingers, in particular, as well as unprotected parts of the face (nose, ears). If the cold exposure is prolonged necrosis of the tissues can appear. The mild hypothermia is associated with muscles shivering, cold, pale and dry skin, and respiratory changes: from hyperventilation to slow and shallow breaths. In prolonged cold exposure severe hypothermia can appear, related to drop of the core body temperature and depletion of the glycogen: the shivering is stopped, and different symptoms come into sight, like loss of reflexes, loss of consciousness, slow heartbeat, muscle stiffness, possible bruises of the lips, ears, fingers, and feet.

Cold allergy is also frequently observed: red and itchy pimples with tips arise on places, exposed to cold air (and usually disappear after about half an hour in a warm environment). More severe states of cold allergy can cause seizures, fever, increment of the heartbeat, and swelling of the limbs or torso.

Cooling causes thermal discomfort and begins to affect mental state, neuromuscular function, and performance. The low temperature of the surrounding air (either natural or artificial cold) provokes a delay in sensory response and reactions as well as drowsiness of the person. Both are natural reactions of physiological processes, provoked by the decrement of the blood flow rate to the brain. A result is the large number of errors in the task being performed.

Natural and often effective strategy for coping with the cold-related stress is putting on an extra layer of clothing to regulate heat exchange with the environment (Holmer, 2005).

Body Reactions in Hot Environment

When the body is exposed to a hot environment, the thermoregulatory system actuates the mechanisms for increasing the heat losses that can decrease the temperature of the core body: vasodilatation and evaporation. The higher core body temperature leads to augmentation of the pulse rate and enhancement of the respiration rate. Thus the increased heart rate accelerates the transfer of blood to the extremities and the skin surface. The process is supported by the expansion of the cross-section of the peripheral blood vessels, which helps the cooling of the blood near the skin surface.

The blood moves away from internal organs to the periphery (skin). The blood flow towards the extremities and the skin surface stimulates the heat losses through the skin by radiation and / or convection. Perspiration and evaporation of sweat from the skin are even more effective reactions of the thermoregulation as they accelerate the heat losses from the wet skin surface, which again leads to a lowering of the temperature of the core body.

Despite its sensitive and very complex thermoregulatory system, the human body is protected from the dangerous temperature surroundings in a relatively narrow temperature range. As the decrement of the core temperature of the body, hypothermia, the reverse process, hyperthermia, have a disadvantageous influence on body's functions. The hot living or working environment provokes fatigue, which is closely related to the faster cardiac rhythm. The threshold of the sensory perceptions moves up, the time for reaction runs high, while the concentration decreases. The final result is similar to that in the cold environment: the number of errors in the human performance increases.

Many of the hot-related injuries of the human body are related exactly with the increased sweating from the skin surface. The evaporation of sweat is not effective in a hot and humid environment. The prolonged sweating process may lead to fast dehydration and coupled disorders like heat cramps and hypernatremia (excess of sodium in the blood flow). Other hot related injuries are preconditioned by the sun radiation (sunburn), which can undoubtedly increase the core body temperature as well. Heat exhaustion appears when plenty of liquid sweat appears over the whole body in combination with cooler and pale skin, headache, fast pulse, and vomiting.

When the thermoregulatory system cannot overcome the severe attacks of the hot environment and the sweating process is not possible anymore, the body temperature rises quickly, provoking heatstroke. The heatstroke can lead the person to unconsciousness.

WEARABLE DEVICES FOR HELPING THE THERMOPHYSIOLOGICAL COMFORT

EMBR Labs (http://www.embrlabs.com/)

EMBR Labs have announced the creation of the *Wristify* bracelet that can locally cool or warm the body. The developed by MIT-trained engineers technology has gained first prize in the annual competition of MIT in Materials Engineering for 2013 and can be applied in different consumer products. *Wristify* could both cool and warm the wrists, depending on the temperature of the environment, assuring thermophysiological comfort, akin to the action of heating, ventilation, and air-conditioning devices.

The new technology seems to play with the skin thermoreceptors: it gives to the wearer the local sensation of warmth or cold, similar to the effect of radiation heat or a contact with cold water. During the development of the wearable device, the team has found that the rate of temperature change of 0.1 °C a second can provoke the entire body to feel colder (or warmer) with few degrees. A calculation has also shown, that the change of the temperature of the indoor environment even with 1 °C (cooler in winter time or hotter in summertime) will decrease the heating / cooling building expenditure with 100 kWh per month (Quick, 2013).

Digitsole (http://www.digitsole.com/)

The company is already known with its smart line innovative footwear products. Its second generation series *Digitsole Warm* have been recognized as one of the top innovations at CES International Forum in Las Vegas. Apart from its heating abilities, which are very important for warming the feet, *Smartshoes* of Digitsole are shock absorbent and interactive. They can be controlled by a smartphone via Bluetooth connection and special application, designed for the whole *Warm Series*. The application collects data about the feet temperature and can control their comfort, changing the heating rate. Even more, the wearer may control the temperature of each sole separately, adjusting the temperature with a single click and sending a signal to the built-in thermostat. This function is very useful if the person is experiencing a situation of thermal asymmetry.

The *Smartshoes* can also play the role of a pedometer: their accelerometer, built-in in the sole, can measure precisely the number of steps and calculate the distance walked, as well as the calories burned.

Digitsole insole is another product from *Digitsole Warm Series*. It is lightweight (around 100 grams), with a thickness of 13 mm at the heel and 5 mm at the toe zone. Polyurethane, which is extremely vibrations absorbing, is used for molding the sole platform. The uniform spreading of the heat is guaranteed by a high-quality micro-fiber layer of polyurethane.

Venture Heat (http://www.ventureheat.com/)

Venture Heat has developed series of US patented far infrared heating products, which can heat the muscles and body tissues, helping healing and recovery processes in the body. The wearable items are designed for different parts of the body: chest, shoulder, neck, wrist line, elbow, waistline, etc. Though they are not directly related to the thermoregulatory system of the body, their performance influences the vasodilatation of the blood vessels, increases the blood flow circulation (and oxygen transfer), thus helping the thermophysiological comfort as well. The adapter, connected with the battery, stops the heating automatically after 30 minutes, but if the person likes to continue with this modern, non-invasive form of healing and rehabilitation, he can switch on the device again.

Nike's Sports Research Lab (http://nike.com)

Face and head are more sensible to temperature changes than other parts of the body. During physical activity or mental stress, the temperature of the head increases as in a very hot environment. After an event, many athletes pour cold water on their heads to take off part of the unpleasant heat.

Following its experience in the development of the *PreCool Vest* Nike's Sports Research Lab (NSRL) is developing a prototype of a cooling hood. This wearable device has a simple design: water is used to fill in the inner layer of the hood, which then cools the head and the face. The structure around head and face keeps the "cooling" mesh on its place, applying a kind pressure on the skin to assure its contact with the cold surface. The hood prototype will probably be used during the 2016 Olympic Games in Brazil.

Institute for Basic Science (IBS) in Seoul (http://www.ibs.re.kr)

The Korean scientists from the Center for Nanoparticle Research at IBS have developed a wearable device for heating and thermotherapy (Choi et al., 2015). It is actually a mesh, made of silver nanowires.

In the modern medicine, silver is considered as a natural, effective antimicrobial material. It eliminates many of the common types of bacteria and fungi that cause infections (Foulger & Gregory, 2003). As a conductive element, silver offers several advantages: the molecules of silver assist the transfer of heat from the skin to the environment or vice versa, helping the thermophysiological comfort. In combination with its antimicrobial effect, silver has particular application in medical products, as it shortens the time for healing of wounds, chronic pains etc. while providing excellent hygienic conditions and thermophysiological comfort.

The diameter of the silver nanowires, used for the mesh heater, is approx. 150 nm and their length is around 30 μm. A liquid elastic material is applied to mix the nanowires; after drying this material becomes stretchy, but soft. 2D interlock pattern, similar to that, used in the production of knitted textiles, is applied to obtain firmness to the structure and to ensure that the material will stay tight, where necessary. The heated mesh is subsequently placed between two layers of insulation. The result is breathable, lightweight heating device that stays firmly on joints (knees or wrists) and warms the body tissues. There is no effect of undesired cooling with the time flow (as in the case of hot water bottles usage), or needs for central electricity supply (as it is with the electric heating pads). The wired heater is battery powered and ensures the desired temperature during the exploitation.

The inventors claim other possible applications of the wearable device, beyond the heating and thermotherapy: for heating car seats or heating element in smart clothing.

UNIST (http://www.unist.ac.kr)

A research team from the South-Korean Ulsan National Institute of Science and Technology (UNIST) has developed a wearable transparent heater, which electrodes are visible over the skin (An et al. 2015). The heater is attached to the hands or clothing and provides additional help for the thermophysiological comfort of the body.

The transparent electrodes are made of metallic glass: a kind of alloy, which is a newly developed material. The research team has managed to produce a large scale of an electrode for the heating device. The metallic glass is elastic, resistant to oxidation and corrosion, and with good electric conductivity. A key to the developed technology is the rolling process (Roll-to-Roll) of a very thin film to make a circuit.

The inventors claim that many companies can reduce their production costs by introducing the transparent electrodes technology. The same technology has been used by the team to produce another wearable device: a gas detector, which can be worn on the finger and sense any harmful gas much better than the body's olfactory system.

Cool On The Go (http://www.coolonthego.com)

Cool On The Go® is a wearable fan, which can be used everywhere to cool the body down, if the temperature of the surrounding air is high. In summer periods it is very suitable for people, who suffer from cardiovascular diseases, and also for people under mental or physical stress.

Developed in South Florida, this wearable device provides portable personal protection from the warm environment. The cooling fan is attached to the body, using the provided clip, belt loop, lanyard, or armband. It features USB adapter and a D/C input, so it is up to the wearer and the wearing conditions which mode of operation will be used. The producer states that 4 AA alkaline batteries power *Cool On The Go®* for up to 4 hours. The velocity of the air flow can be regulated via a speed control adapter.

Black Ice (https://www.blackicecooling.com)

Founded in 2005, Black Ice is a company, focused on research, development, production and marketing of personal cooling devices and technologies. *Cool Collar CCX-S®* is a neck cooling personal system that helps the body thermoregulation in a hot environment. The place of the wearable device is very well selected, as all the blood flow to the brain passes through the neck. The cooling down of the blood, directed to the brain, helps the thermophysiological comfort of the wearer and prevents the body from hot related injuries.

The *Cool Collar CCX-S®* can assure hands-free, uninterrupted effective cooling at a temperature of 14 °C. It is based on molecular alloy technology, which does not allow the device to exchange heat with the body. The result is that the temperature of the collar remains constantly low. By contrast with other wearable cooling gadgets, it works well in high humidity environment and can provide a pleasant cool sensation of up to 90 min, followed by 20 min quick recharging in ice water.

Black Ice also works on development and production of cryotherapy items with several benefits for the human health. These products can successfully replace frozen gels and ice that are too painful to be used on the skin in case of injuries. The *CoolTherapy Packs®* ensure uniform cooling with 14 °C, which is the beneficial temperature for cold therapy and not uncomfortable for the wearer.

SMART AND INTELLIGENT CLOTHING AND ACCESOIRES

GydeSupply Co. (http://gydesupply.com/)

Microwire™ is a patented technology of Gerbing, applied to protect the human body from extreme cold conditions. Very thin conductive filaments (75% thinner than the human hair), are used to build the *Microwire*™. In addition, the filaments are covered with *Teflon®* coating, thus ensuring the waterproofness of the system. The *Microwire*™ elements are incorporated in clothing garments as heating panels: one on the back, two on the chests and one on the collar. Powered by a battery (7 V or 12 V), each heating panel may reach 57 – 63 °C. The heating temperature has 3 optimum levels. The battery is placed on a top side pocket of the garment to allow free motions during any kind of activity.

One of the advantages of the new generation smart clothing is the wireless heat control, which is also applied in *Microwire*™. Gerbing has developed *Bluetooth Thermogauge*™ adapter and a smartphone application that works with all Android or iOS devices. The adapter can control and manage several heating devices that are paired with him in one and the same time.

The *Microwire*™ technology is applied in the production of gloves. The warm hands are one of the main preconditions for avoiding cold-related injuries and sustaining as long as possible the ability of the fingers to perform tasks in the cold environment. In addition to the heating panels in each glove in the pair, powered by 7 V dual mode batteries, the gloves are designed using *PrimaLoft®* micro-fiber insulation and breathable *Aquatex*™ membrane, which ensures the waterproofness of the gloves. An index finger, compatible to touch screen technologies, is also added.

One of the Gerbing's collections (*12 Volt Collection*) is supplied with interconnecting system, which allows the smart clothing items to be plugged directly into the electrical systems of all means of transports: snowmobiles, motorcycles, airplanes, boats, etc. Thus the body can be kept warm during any activity and any temperature with practically unlimited power supply.

EXO² (http://exo2.co.uk)

The main production line of the EXO² Company consists of heated clothing and accessories for different target groups of consumers. EXO² have developed their own technology for heated clothing, called *FabRoc®*, which corresponds to the high-quality standards in the field. The core of the technology is far infrared (FIR) energy. A lightweight polymer (*FabRoc®*) is used, as it is able to produce FIR heating when a low voltage passes through it. Either rechargeable batteries or power adaptor can be applied for energy supply. The heat rate is controlled by the wearer, who can adjust the temperature of the clothing.

The *FabRoc®* technology is also used for heat therapy. The far infrared energy, produced by the heating panels, heats not only the skin surface, helping the thermophysiological comfort of the body, but also heats the muscle tissues. The wearable technology of EXO² can be used for FIR therapy of muscle spasms and pain, venous disease, arthritis, joint stiffness, etc. The radiative heat can penetrate the skin at a distance of up to 9 cm, thus increasing the healing and regeneration process of the body tissues. A person can use the *FabRoc®* wearable items as long as he wants and in any situation. The proper design can be applied in any particular case, so as the infrared heating system to fit the particular part of the human body and to deliver infrared radiation to the tender spot in a safe manner. The control system is designed to avoid any malfunctions and overheating, even in situations of overuse.

Venture Heat (http://www.ventureheat.com/)

Venture Heat produces smart clothing for low temperatures protection, aiming to cover the whole range of items, used in cold environment: heated jackets, heated hoodies, heated basic line shirts and pants, heated gloves. The clothing design involves heating panels, which are ultra thin and made of conductive micro alloy fibers. Battery power (rechargeable Lithium-ion battery) is used to feed the panels. The number of the heating panels and their position depend on the type of the clothing item. For a heated jacket, there are three heaters: one on the back and two on the chests, while the basic line heated pants have one heater on rear hips and two heaters on the knees.

To assure maximum comfort, the heated smart clothing of Venture Heat involves highly breathable textile layers from inside, which are soft and warm, and waterproof and breathable outer layers for maximal protection from the outdoor environment.

Venture Heat has also designed a heated hoodie, powered by a USB power bank with at least 2.0 A power output. A built-in controller allows the user to set three different heating temperatures from 35 °C to 52 °C. The hoodie is equipped with a power bank of 10400 mAh, so as to provide heat from 3 to 12 hours, depending on the heating level. If the person is warmed enough, the USB port may be used for power supply to mobile devices and any other wearable hi-tech items. Certainly, the heating panels inside the hoodie are safely embedded so as to be worn without any problems and even to be processed in washing and drying machines as regular textiles.

The heated gloves, produced by using the same technology, are able to protect the hands in a cold environment from 2.5 to 5 hours, depending on the outside temperature, the activity, and the heating power applied. The gloves are touch screen compatible and highly insulated, in order to protect both the hands and the batteries (hidden behind a waterproof zipper). *Dintex Membrane* is applied in the design to keep the hands dry. *3M Thinsulate*™ microfiber insulation is also used to entrap the natural heat, produced by the body, inside the gloves, while assuring the transfer of water vapor.

An interesting addition to the range of products of Venture Heat is the heated scarf, made of 100% polyester fibers. Powered by a 7 V, 1800 mAh battery, the scarf may protect the body from the wind-chill from 2 to 6 hours. The different color on the LED light of the controller shows the intensity of the heat released and the duration of the battery. The heating element of *Z-Stretch Carbon-Fiber* is placed in the centre of the neck.

Flexwarm (http://www.flexwarm.com)

Flexwarm is a young company, which has started its research on smart clothing for cold protection in 2013. In 2016 they have received the 2016 IF Design Award.

The smart jacket, designed by Flexwarm, uses wearable technology that can help the thermoregulatory system, taking the control on the thermophysiological comfort of the wearer to the next level. The main goal of the clothing items, produced by Flexwarm, is they to control the body temperature, adjusting the heating level respectively. That means that the sensors control the temperature of the environment, together with sensors that measure the temperature of the body.

The thermal control is split between the different heating panels in the smart jacket, so as the person can regulate separately the heating of the chest and the back, for example. The built-in Bluetooth connection allows the regulation to be done via an application for iOS and Android devices: tablets, smart phones, and even smart watches.

The technology for production of the *Flexwarm®* heating layer is based on the technology for thick film production, which is typical for chip resistors, sensors for the opening of airbags, etc. The *Flexwarm®* patented substrate is placed between two thermal resistant layers, and then it is sintered and sealed. The result is very thin, flexible, durable, and absolutely water resistant heating layer. The thickness of the heating panels is around 0.5 mm, which is similar to the thickness of conventional cotton textiles. This makes the *Flexwarm®* panels "invisible" in comparison with heating panels made of carbon fibers, which are frequently used inside smart clothing for cold protection. Actually, the heated layer itself and the whole garment can be twisted or rolled (in the suitcase, for example) without any risk of damage.

The *Flexwarm®* jacket is light in weight (250 g) and has four heating zones: one on the chest, one on the back and two on the arms. The built-in semi gloves can be used for an additional increment of the thermophysiological comfort. The heating level is controlled by the wearer and can be increased up to 65 °C.

Snowballs (https://www.snowballsunderwear.com/)

Snowballs™ is cooling underwear for men, aiming to protect and even increase the male fertility. It is known that one of the reasons for male reproductive problems is the high temperature of the testes. A significant augmentation of the sperm count and sperm concentration has been established after scrotal cooling during nights (Jung, Schill, & Schuppe, 2005). However, the inventors of Snowballs™ are clear in their suggestion the men to use this cooling underwear as a supplementary help of medical treatment.

This wearable cooling item uses 100% organic cotton yarn, knitted in a soft fabric. The cooling ability comes from carboxymethyl cellulose sodium gel, which is freezable and inserted in a biodegradable packaging, shaped ergonomically. This is the *SnowWedge*™ package, which has to be inserted in the *Snowballs*™ underwear. The freezing of the *SnowWedge*™ lasts no longer than 60 min. The scrotal cooling itself has to be prolonged to one hour and the producers suggest this local cooling to be done in specific cases: after an intensive activity, a workout or in the bed.

Intel (http://intel.com) and Chromat (http://chromat.co)

The sportswear design company Chromat together with Intel has developed two clothing items: a bra and a dress, which can change their shape in response with the temperature or stress level in the body of the wearer. Actually, the new technology for production of "responsive items", keeps track on the body's metabolism, which is very susceptible to temperature change, production of adrenaline hormone or hormones from the thyroid gland.

The two garments are equipped with the Intel's *Curie Module*, which involves 32-bit microcontroller (Intel Quark), gyroscope and accelerometer, working with 6-axis motion sensor, digital signal processing hub, Bluetooth low energy component, 80 kB Ram and 384 kb flash memory. 3D-printed frames of carbon fibers with shape memory alloys are used in the design.

Carbon fibers are far more flexible than metal wires, traditionally used for old-fashioned heating systems in clothing. The panels of carbon fiber do not need external wires, which are usually the main drawback of the garments that create controlled microenvironment around the body. In the case of the *Chromat Adrenaline Dress,* the increment of the sweating or breathing rate provokes activation of the carbon mesh and it expands, giving room for higher air transfer and heat release from the body to the environment. The method of approach is similar in the *Chromat Aeros Sports Bra*, but instead of expanding, the carbon mesh opens vents to increase the ventilation rate between the skin and the surrounding air. Certainly, the vents are closed after the skin is cooled down. The two approaches support body's thermoregulation, helping the thermophysiological comfort of the wearer during stressful activity or workout.

Hohenstein Institute in Bönnigheim (http://www.hohenstein.de)

The researchers from Hohenstein Institute in Bönnigheim have developed a portable cooling vest that can help people to avoid brain damage when suffering from cardiac arrest. It is a new method for hypothermia treatment that provokes the human thermoregulatory system to react so as to save the brain. This can help in several cases, as every year 375000 persons suffer a cardiac arrest only in Europe (Nikolov & Cunningham, 2003). The cardiac arrest involves heartbeat and breathing stopping, which require immediate reanimation, otherwise the chances of survival lessens with ten percent per second. Even after a successful reanimation that has restored the vital functions, a brain damage frequently occurs due to the lack of oxygen supplied.

The cooling vest aims to very fast cool down the body, as the cooling can protect the tissues of the brain against neurological damage during periods of lack of blood flow and oxygen supply. The cooling vest may decrease the core body temperature to 34 °C and even lower without power and everywhere. The principle used is based on a hollow textile fabric, called *cooling pad*, which is absolutely impermeable. The cooling pads are built-in into a vest and connected to a metal container with the mineral zeolite. When necessary, a valve is opened, and the water inside the cooling pads is filled with zeolite under pressure, which cools the pads to nearly freezing temperature. This non-invasive surface cooling of the body provokes thermophysiological reactions in the skin and core body that cools the brain and helps it to survive during a period of oxygen lack. The most important for this wearable device is that its application does not require medical knowledge and can be implemented as a survival kit at any public place, giving a better chance to patients with cardiac arrest to overcome the brain-related effects of the health incident.

Outlast Technologies Inc. (www.outlast.com)

Outlast Technologies Inc. is world famous with the application of phase change materials (PCMs) in several commodities. Outlast's products are used in clothing for outdoor activities and sports, footwear, military uniforms, upholstery, packaging, etc.

Phase Change Materials (PCMs) are any kind of materials, which can experience the process of change from one state to another, i.e. from liquid to solid and back. They absorb, store or discharge heat in accordance with the changes in the surrounding temperature.

PCMs, used in the production of intelligent textiles, are mostly paraffin from different types: heneicosane, eicosane, hexadecane, heptadecane, octadecane, nonadecane, which are applied in different combinations. They oscillate between solid and liquid phases, absorbing heat during the transformation from a solid to a liquid phase and discharging heat when going back to the solid state.

Although nowadays it is easy to explain the physical phenomenon of PCMs, the primary idea is related with a very big leap in the thinking what textiles and clothing can do. The first application of PCMs was in the space suits for astronauts for the US Space Program. In 2003 namely Outlast Technologies Inc. was selected as a Certified Space Technology and later on, in 2005, the company was inducted into the Space Technology Hall of Fame.

PCMs are very interesting as a source for heat storage because at least for two reasons. The first one is that the heat released during the phase change processes is very high: traditional textile materials absorb about 1 KJ per 1 kg of heat while increasing its temperature by 1°C. In the same conditions, paraffin PCMs absorb approximately 200 KJ per 1 kg of heat during the melting process, which is approx. 200 times higher.

The second reason is that there are no temperature changes during the process of heat absorption, storage, and discharge. During the melting process, the temperature of PCMs and the temperature of the surrounding air remain constant (30-34°C). The same is valid for the process of PCMSs crystallization.

In order to prevent the dissolution during the liquid phase, the paraffin has to be "closed" in small, impenetrable, hard capsules, made from plastics with dimensions of the order of microns. These are the *Thermocules* microcapsules, developed by Outlast Technologies Inc.

The *Thermocules* are incorporated in fibers and textiles of different types: woven, knitted, nonwovens, aiming to help body's thermoregulation. There are various ways for adding the PCM-microcapsules to the textiles. The first one is encapsulation during the wet spinning process of synthetic fibers. PCM-microcapsules with round, square or triangular shapes are added to an acrylic polymer before the wet spinning. As a result, the microcapsules are permanently fixed within the fiber structure. Fabrics made from such fibers have a softer hand, greater stretch, high air-permeability, and breathability.

Another solution for adding the PCM-microcapsules to the textiles is to add them in acrylic or polyurethane coating compound, which is applied to the fabric surface. Once the coating is dried, it forms a bond with the fabric. PCM-microcapsules can be also mixed into water-blown polyurethane foam, which is consequently applied to the fabric during lamination. The result is a fabric between two plies of a flexible thermoplastic film with PCMs.

The interaction between the intelligent clothing and accessories with PCMs and the thermoregulatory system of the body depends on the body's activity and the temperature of the environment. The following situations can be distinguished:

- The temperature of the environment is higher than the melting temperature of the PCM (warm environment). The PCM in the microcapsules starts to absorb energy, moving from a solid to a liquid state. This process is accompanied by short-term cooling of the clothing item. The storage of heat stops when PCM is totally melted.
- The temperature of the body is higher than the melting temperature of the PCM (for example, due to active exercises and increased metabolism). If the heat, generated by the body, is higher than the possible heat release from the clothing system to the environment, the PCM starts to accumulate heat, melting. Thus PCM stores the superfluous heat for the future.
- The temperature of the environment is lower than the crystallization temperature of the PCM (cold environment). The PCM comes back to a solid state, releasing heat. The result is a momentary warming effect, which minimizes the body heat losses and helps the thermophysiological comfort.
- The temperature of the body is lower than the crystallization temperature of the PCM (i.e. during outdoor activities). Normally the temperature of the skin near the torso is about 33°C, the skin temperature of the feet is nearly 30 – 31 °c. These values are around the crystallization point of some PCMs, which start to change from a liquid to solid state, discharging heat.

SMART AND INTELLIGENT TEXTILES

EXO2 (http://exo2.co.uk)

ThermoKnit™ is a technology of EXO2 that is actually a smart knitted textile. The creators of the technology and the designers of heated gloves for outdoor activities claim that ThermoKnit™ overcomes the disadvantages of other smart gloves, used for heating the hands: the irregular spread of the heat. Indeed, there are many producers of heating gloves on the market, which use a similar method of approaching the task: they heat the palm and the tips of the fingers, but the efficient heat distribution around the hand is difficult. As a result, the individual does not have a noticeable feeling of improved thermophysiological comfort. In addition, the traditional wire-based heating systems have different shortcomings, like the risk of overheating or power breaking.

The ThermoKnit™ knitted layer operates at very low (3.7 V) voltage. It consists of FabRoc® polymer, extruded in high conductive yarns. Silver yarns are also applied as busbars, which provide the power effectively to the FabRoc® yarns. Due to the smart layer, the whole length of fingers and thumbs is heated, helping the vasodilatation of the blood vessels in the cold environment and heating the blood flow, circulating in the hands. The heating temperature can reach up to 46 oC.

The power is supplied by a lightweight battery, kept in a leather pouch, away from the hands of the wearer, to keep them free for other activities. The heating of the hands can be up to 3 hours and the battery needs additional 6 hours to be fully recharged.

The ThermoKnit™ conductive yarns can be knitted not only for the inner layer of gloves but to be also used for the production of mittens and socks. Other applications involve clothing design, upholstery textiles in cars and means of transport, blankets and bedding, as well upper layers of mattresses.

Volt Resistance (http://voltheat.com/)

Volt Resistance has patented their own heating technology for people, whose thermoregulatory system needs help to provide thermophysiological comfort in severe cold conditions or during prolonged exposure to low temperatures. The *Zero Layer*® heating technology is claimed to be more efficient than any other on the market, as there is no need the heating system to be an "invisible layer" in the clothing item. Usually, the heating panels are covered by additional, soft textile layer, which decreases the emission from the heating system to the body. In addition, this decreases the efficiency of the battery power and limits the time of the exposure in conditions of thermophysiological comfort.

Ultra fine stainless fibers are used for the production of very thin, imperceptible insulated woven fabric that is the heating panel. In addition, the panel is covered with a thermally bonded tape so as to be sealed and permanently protected during the exploitation cycle. Thus, the heating panel is resistant to mechanical loads and wearing and is designed to assure particular quantity of instant radiative heat, needed to protect the body from cold.

The *Zero Layer®* heating technology is designed to work with three types of power supply: 12 V, 7 V, and 3 V lithium rechargeable batteries. A different power source is integrated, depending on the type of the item and the cold protection that the product has to assure. The products, equipped with 12 V power systems can be powered also by the electrical system of different means of transport, using an adapter for temperature control.

An insulated jacket, equipped with *Zero Layer®* heating technology, involves three heating panels: one larger on the back and two smaller on the chests. Powered by 2900 mAh, 7.4 V lithium battery, it can assure the protection of the human body from cold for approx. 2 hours, releasing heat with temperature up to 75 °C.

Zero Layer® heating technology is applied also for production of heated slippers, which can be used both indoors and outdoors. Powered by two 3.7 V, 2000 mAh batteries, they release heat during a period of 3 – 10 hours. Four levels of heating are foreseen, adjusted via a temperature adapter. To ensure the feet protection from cold, additional insulation layers are applied in the design of the slippers: fleece lining from inside and upper shell of polyamide fabric with high resistance; footbed cushion from memory foam and rubber outsole.

Another cold protection item, designed by using *Zero Layer®* technology of Volt Resistance, is heated gloves, which can protect (heat) each finger and the thumb separately along its entire length, together with the palm and the upper side of the hand. The gloves involve breathable and waterproof fine membrane to assure the transfer of water vapor from the skin surface to the environment but to protect the hands from liquid penetration from outside in. Each glove is powered by 2900 mAh, 7.4 V batteries, which can guarantee heat release from three hours on the maximum level of heat release (approx. 65 °C) to up to twelve hours if minimal temperature settings are applied. The built-in microprocessor controller allows the temperature to be adjusted smoothly and 4 heating levels are available.

Centre for Wearable Sensors (http://jacobsschool.ucsd.edu/wearablesensors/)

The *ATTACH* (Adaptive Textiles Technology with Active Cooling and Heating) project has been funded with $2.6M to develop intelligent textile, which can help the thermoregulatory system of the body in any kind (warm or cold) indoor environment.

The basic aim is rather energy savings than human health and well-being, but in this situation the two targets are very close to each other. It is supposed that the local heating or cooling of the human body would be cheaper than the heating or cooling of a large room. It is even calculated that the energy costs in the indoor environment with the new intelligent textile would be cut by at least 15%.

The research idea behind the project is the intelligent fabric to change the temperature of the skin surface in accordance with the fluctuations of the room air temperature. The thickness of the fabric will increase with the decrement of the room air temperature, thus impeding the transfer of heat between the skin surface and the surrounding air. If the room air temperature increases, the textile layer will become thinner, so as to help the heat and air transfer from the body to the environment. These abilities of the fabric will be provided by built-in polymers, which shrink at high temperature inside the textile or expand at low temperature.

The comfortable skin temperature for greater part of humans is around 34 °C. The new intelligent textile would provide the required skin temperature as new, “intelligent part” of the human body, without the use of wires, batteries, and adapters for the temperature. Certainly, the heating of the body is not enough the people to assess the indoor environment as comfortable one. If a person feels unpleasant when breaths in cold air and the uncovered parts of the body (face, ears, and hands) are cold, the savings from the unheated room becomes meaningless, as the inhabitant is not in the state of thermophysiological comfort (suffering from local thermal discomfort).

As usual, the intelligent textile can be easily converted into a smart one, adding electronic components to it. These are the planes of the researchers, working on the *ATTACH* project. Additional thermoelectric (heating and cooling devices) will be incorporated in the intelligent textile so as to help the thermoregulatory system of the body. These particular spots of incorporation will be on the back side of the torso and underneath the feet, as these areas of the body can be overheated if the person is active indoors. Certainly, the thermoelectrics will need power, which is planned to be taken from batteries or biofuel – the human sweat itself. Hi-tech approach for printable wearable devices will be used to print the thermoelectrics, the batteries and the biofuel cells, so as they to be stretchable, thin and durable enough to be worn.

Kolon Glotech Inc. (http://www.kolonglotech.co.kr)

The wearable technology, invented by Kolon Glotech Inc. is *Heatex*®, smart fabric with an electric heating system. Actually, this is a commercial product, developed on the basis of the *Termed PrinTronix*® technology. This is an interdisciplinary high-tech achievement, created on the basis of textiles, printing technology, information technology and conductive materials. The *PrinTronix*® technology overcomes the problem with conventional clothing for cold protection, which has to consist of several layers, and air entrapped between them in order to provide the necessary thermal insulation. The result is a bulky clothing item, which is not very suitable for active performance outdoor (i.e. working activity or military services). The involvement of wired heating elements removes the bulkiness of the cold protective clothing but creates new problems with the weight of the item, the maintenance of the power circuits, etc. *Termed PrinTronix®* is flexible, fabric-based heating layer, which overcomes all these disadvantages.

Heatex® is a system of several layers, made of materials with different conductivity: upper textile layer, insulation layer, heating panel layer, and a substrate. The temperature range varies between 40 °C and 50 °C, powered by two battery packs with different weight. A high-tech "twist" of the smart *Heatex®* fabric is the possibility the heating release to be controlled via a smartphone remote control application. As a result, new supplements are added to the "heated textile": GPS tracking, location forwarding, etc.

The *Heatex®* fabric is used in the production of clothing items with a wide range of applications. It is a layer of different sports clothing, when the activity is very dynamic and the heat release from the body is quite high, despite the low outdoor temperature. Another application is in workwear, especially for activities in low outdoor temperatures, but also in artificial cold. *Heatex®* is also applied in military clothing, as a cold-proof system and in wearable accessories like heated gloves, hats, socks, and sleeping bags. Out of the range of wearable items, *Heatex®* can be used in car seats for additional improvement of the thermophysiological comfort of the passengers.

Soothsoft Innovations Worldwide Inc. (http://www.soothsoft.com)

Soothsoft Innovations Worldwide Inc. has designed a self-cooling pillow that can ascertain the thermophysiological comfort during sleep. The thermoregulation of the body is less active during sleep. Therefore in a hot indoor environment (over 22 °C) the contact of the body with the textiles for bedding can provoke unpleasant heat increment and sweat release, especially around the head and neck. *Chillow®* is a non-electric cushion that assures a "cold" spot for every pillow, using non-toxic materials. It can be used during sleep, but it can also be a wearable device with several applications.

The temperature of the *Chillow's®* surface is between 22 and 27 °C, which is at least 7 °C less than the temperature of the skin surface, regardless the temperature of the indoor air. Thus the skin surface in contact with the item feels refreshed, but not cold, warm or wet. Apart from helping the thermophysiological comfort, the Chillow® aims in reducing headaches, hot flashes, and fibromyalgia.

The Chillow® pad is only 2 cm thick and it starts its function once it is filled with water. Patented foam, which is inside the pillow, absorbs the water and converts it in a memory foam. After activation the product keeps working without any power and allergic reactions.

Addidas (http://www.global.adidas.com)

Climachill™ is an intelligent fabric that aims to keep athletes in thermophysiological comfort or closer to that state by applying a hi-tech cooling technology. A unique mixture of titanium blended *SubZero* yarns and 3D aluminum spheres are used for the production of a knitted textile, which is then used for designing of training wear, as well as outdoor and tennis outfit. The instant cooling effect helps the thermoregulatory system to decrease the body temperature during training and competitions, hence ensuring better performance and faster recovery from the physical strain.

The aluminum spheres are the core of the cooling technology: they are located so as to be in a contact with the warmest parts of the skin, cooling these spots down during the activity. The cooling capacity of the *Climachill*™ jersey is increased with one-third by the flat titanium yarns, used as an underlay of the right side fabric. When the body is in thermoneutral conditions (thermophysiological comfort) the *Climachill*™ clothing items do not react. When the core body and, consecutively – the skin surface – start to increase their temperature due to activity, the conductive elements in the Addidas' intelligent fabric begin to absorb heat from the warm body's spots and to transfer it to the environment. The result is that *Climachill*™ adds power to the sweating process, helping the body to release the excessive heat.

Columbia Sportswear Company (www.columbia.com)

Omni-Freeze® ZERO, the Cooling Technology of Columbia Sportswear Company, is another example of intelligent textiles, used to help the human thermophysiological comfort. The technology uses thousands of polymer rings – circular coolers that interact with the natural release of water vapor and sweat from the skin surface. Though the technology is closely guarded, it is supposed that the researchers, working on the new development, have shifted from the classical design solution to a new one. The classical design solution to deal with the sweat involves fast omitting of the liquid from the skin surface through the textile layer (or layers) to the environment. The new approach, applied in *Omni-Freeze® ZERO* fabric is to keep the sweat near the body and to work with it so as to increase its cooling effect. The new technology is applied in a wide range of clothing, accessories, and footwear, under the Columbia's brands.

Venture Heat (http://www.ventureheat.com/)

Venture Heat produce wearable heated blanket with sleeves (hoodie buddy). Wearable blankets and wearable sleeping bags are the most popular wearable textiles, which can help the body thermoregulatory system to ensure the human thermophysiological comfort. They create a microenvironment around the human body, which is "separated" from the surrounding air, being it indoor or outdoor environment. Okamoto-Mizuno, Tsuzuki, Ohshiro, & Mizuno (2005) have found that the use of heated blankets at low temperature (3 °C) and a relative humidity (from 50% to 80%) is very beneficial for the relaxation during sleep. Low extremities have been specified as the best zone for warming the body in case of relaxed posture. Heated blankets, however, can cause thermal stress in the case of higher air temperature. In such cases, the blanket must be switched off at night.

The wearable blankets of Venture Heat are based on their own micro alloy fibers technology for production of the heating panels. This wearable heater has one large panel for deep infrared heat on the chests, which helps the heating of the air layer, entrapped between the body and the textile layer. The infrared heat also provides additional features: it can be used for heating of the muscles before intensive activity and sports or can be used for relaxation and rehabilitation.

The heated blanket is plugged in at home, using an adapter to reduce the voltage to 12 V. The central FIR heating panel emits the necessary heat, but at the same time, it reduces the pain in the muscles, penetrating deep into the body. Four levels of heating can be applied and a built-in timer switches off the device after 30 minutes. This protects the person from unexpected overheating in case of deep sleep.

Penn State University (www.psu.edu)

Researchers from PSU have developed a new material that can be used in cooling devices for people, working in a hot environment (firefighters, founders) or people under physical heat stress (athletes). The new material is a nanowire array, which has reversible temperature change if exposed to an electric field (Zhang et al. 2016). Other, already known materials with similar electrocaloric effect, are not appropriate for embedment in clothing items due to their rigidity, their toxicity, or due to the high voltage, needed to power them.

The nanowire array can be transferred to any substrate, including a textile layer. The application of an electric field of 36 V leads to cooling of the array by 3 °C. The researchers claim that a battery pack with a weight of 500 g can cool the material for around two hours. The relatively low voltage is safety to wear and can be used in many application.

CONCLUSION

The wearable technologies, used for helping the thermophysiological comfort will increase their applications, becoming slowly smarter, smaller and cheaper. Along with this, the application of smart and intelligent clothing and textiles will be extended from aerospace and military clothing, as well as professional sportswear towards everyday clothing items, textiles and accessories faster than ever, due to the appearance of new players on the market that run small technology companies and seek fast money turnover. Cooling and heating devices for personal wear will be developed in close connection with smart electronics and hi-tech gadgets. The reason for these expectations is the very intensive research on the topic all over the world, the development of new materials and the interdisciplinary collaboration between researchers from different fields.

The wearable technologies, used for helping the human thermophysiological comfort, can maintain the microenvironment around the human body during resting and sedentary activity. At the same time, they can react to the changes in the temperature of the environment and body's metabolism, thus helping the active thermoregulation. The wearables can protect the body in cold and warm environmental conditions, decreasing the risks of discomfort, reduced performance and injuries. They also can make use of the thermoregulation mechanisms to help the human health and even save the life in other, not related to the thermophysiological comfort, health issues.

Last, but not least, wearable technologies, related to the body thermoregulation can lead to a paradigm shift: from protective to social wearables. They can collect information and send an emergency alert to the parents if their child in the school has a fever or a bedridden person at home has run a temperature or feels cold. The wearable devices may help the science, conducting real-time measurements of the temperature and providing information for large groups of people or by geographical regions.

REFERENCES

An, B. W., Gwak, E. J., Kim, K., Kim, Y. C., Jang, J., Kim, J. Y., & Park, J. U. (2016). Stretchable, transparent electrodes as wearable heaters using nanotrough networks of metallic glasses with superior mechanical properties and thermal stability. *Nano Letters*, *16*(1), 471–478. doi:10.1021/acs.nanolett.5b04134 PMID:26670378

Angelova, R. A. (2007). Maintaining the Workers Comfort and Safety in Extreme Temperatures Industrial Environment. In *Industrial Ventilation* (pp. 197–205). Pamporovo, Bulgaria: Avanguard.

Angelova, R. A. (2016). *Human Thermophysiological Comfort*. Boca Raton, FL: CRC Press.

Celcar, D. (2014). Subjective Evaluation of the Thermal Comfort of Clothing Evaluated in Cold Environment. *Glasnik Hemičara. Tehnologa I Ekologa Republike Srpske*, *10*(1), 65–71.

CEN/TR 16298. (2011). *Textiles and textile products - Smart textiles – Definitions, categorisation, applications and standardization needs*. European Committee for Standardization.

Choi, S., Park, J., Hyun, W., Kim, J., Kim, J., Lee, Y. B., & Kim, D. H. et al. (2015). Stretchable heater using ligand-exchanged silver nanowire nanocomposite for wearable articular thermotherapy. *ACS Nano*, *9*(6), 6626–6633. doi:10.1021/acsnano.5b02790 PMID:26027637

Díaz, M., & Becker, D. E. (2010). Thermoregulation: Physiological and Clinical Considerations during Sedation and General Anesthesia. *Anesthesia Progress*, *57*(1), 25–33. doi:10.2344/0003-3006-57.1.25 PMID:20331336

Foulger, S. H., & Gregory (Clemson), R. V. (2003). *Intelligent Textiles Based on Environmentally Responsive in Fibers*. National Textile Center Annual Report, Clemson University.

Gropper, S., & Smith, J. (2012). *Advanced nutrition and human metabolism*. Cengage Learning.

Harris, J. A., & Benedict, F. G. (1918). A biometric study of human basal metabolism. *Proceedings of the National Academy of Sciences of the United States of America*, *4*(12), 370–373. doi:10.1073/pnas.4.12.370 PMID:16576330

Havenith, G. (1999). Heat balance when wearing protective clothing. *The Annals of Occupational Hygiene*, *43*(5), 289–296. doi:10.1016/S0003-4878(99)00051-4 PMID:10481628

Havenith, G. (2002). Interaction of clothing and thermoregulation. *Exogenous Dermatology*, *1*(5), 221–230. doi:10.1159/000068802

Holmér, I. (2005). Protection against the cold. In R. Shishoo (Ed.), *Textiles in sport* (pp. 262–286). Cambridge, UK: Woodhead Publishing. doi:10.1533/9781845690885.4.262

Holmér, I. (2009). Evaluation of cold workplaces: An overview of standards for assessment of cold stress. *Industrial Health*, *47*(3), 228–234. doi:10.2486/indhealth.47.228 PMID:19531908

ISO 7730. (1995). *Moderate Thermal Environments – Determination of the PMV and PPD Indices and Specification of the Conditions for Thermal Comfort*. International Organization for Standardization.

Jung, A., Schill, W. B., & Schuppe, H. C. (2005). Improvement of semen quality by nocturnal scrotal cooling in oligozoospermic men with a history of testicular maldescent. *International Journal of Andrology*, *28*(2), 93–98. doi:10.1111/j.1365-2605.2004.00517.x PMID:15811070

Morrissey, M., & Rossi, R. M. (2013). Clothing systems for outdoor activities. *Textile Progress*, *45*(2-3), 145–181. doi:10.1080/00405167.2013.845540

Nesbitt, P., Lam, Y., & Thompson, L. (1999). Human metabolism of mammalian lignan precursors in raw and processed flaxseed. *The American Journal of Clinical Nutrition*, *69*(3), 549–555. PMID:10075344

Nielsen, B. (1978). Physiology of thermoregulation during swimming. In B. Eriksson & B. Furberg (Eds.), *Swimming Medicine IV* (pp. 297–303). Baltimore, MD: University Park Press.

Nikolov, N. M., & Cunningham, A. J. (2003). Mild therapeutic hypothermia to improve the neurologic outcome after cardiac arrest. *Survey of Anesthesiology*, *47*(4), 219–220. doi:10.1097/01.sa.0000087691.31092.12

Okamoto-Mizuno, K., Tsuzuki, K., Ohshiro, Y., & Mizuno, K. (2005). Effects of an electric blanket on sleep stages and body temperature in young men. *Ergonomics*, *48*(7), 749–757. doi:10.1080/00140130500120874 PMID:16076735

Park, S., & Jayaraman, S. (2003). Smart textiles: Wearable electronic systems. *MRS Bulletin*, *28*(8), 585–591. doi:10.1557/mrs2003.170

Parsons, K. (2003). *Human Thermal Environments. The Effects of Hot, Moderate and Cold Environments on Human Health, Comfort and Performance*. London, UK: Taylor and Francis.

Tarlochan, F., & Ramesh, S. (2005). Heat transfer model for predicting survival time in cold water immersion. *Biomedical Engineering: Applications. Basis and Communications*, *17*(4), 159–166. doi:10.4015/S1016237205000251

Van Langenhove, L., & Hertleer, C. (2004). Smart textiles in vehicles: A foresight. *Journal of Textiles and Apparel. Technology and Management*, *3*(4), 1–6.

KEY TERMS AND DEFINITIONS

Intelligent Clothing: Clothing items and accessories that do not involve electronic components and devices, but have specific technological performance.

Intelligent Textiles: Woven, knitted or non-woven textiles and their combinations that do not involve electronic components and devices, but have specific technological performance.

Smart Clothing: Clothing items and accessories that involve electronic components and devices in their structure and / or process information electronically.

Smart Textiles: Woven, knitted or non-woven textiles and their combinations that involve electronic components and devices in their structure and / or process information electronically.

Thermophysiological Comfort: A state of the human body, when the generated by the body heat is equal to the heat losses to the surroundings.

Thermoregulatory System: A system in the body, which is responsible for keeping the body temperature within certain limits.

Wearable Devices: Wearable technological items that involve computer components, electronics or intelligent materials, with practical functions and features.

Related References

To continue our tradition of advancing academic research, we have compiled a list of recommended IGI Global readings. These references will provide additional information and guidance to further enrich your knowledge and assist you with your own research and future publications.

Adeyemo, O. (2013). The nationwide health information network: A biometric approach to prevent medical identity theft. In *User-driven healthcare: Concepts, methodologies, tools, and applications* (pp. 1636–1649). Hershey, PA: IGI Global. doi:10.4018/978-1-4666-2770-3.ch081

Adler, M., & Henman, P. (2009). Justice beyond the courts: The implications of computerisation for procedural justice in social security. In A. Martínez & P. Abat (Eds.), *E-justice: Using information communication technologies in the court system* (pp. 65–86). Hershey, PA: IGI Global. doi:10.4018/978-1-59904-998-4.ch005

Aflalo, E., & Gabay, E. (2013). An information system for coping with student dropout. In L. Tomei (Ed.), *Learning tools and teaching approaches through ICT advancements* (pp. 176–187). Hershey, PA: IGI Global. doi:10.4018/978-1-4666-2017-9.ch016

Ahmed, M. A., Janssen, M., & van den Hoven, J. (2012). Value sensitive transfer (VST) of systems among countries: Towards a framework. *International Journal of Electronic Government Research*, *8*(1), 26–42. doi:10.4018/jegr.2012010102

Aikins, S. K. (2008). Issues and trends in internet-based citizen participation. In G. Garson & M. Khosrow-Pour (Eds.), *Handbook of research on public information technology* (pp. 31–40). Hershey, PA: IGI Global. doi:10.4018/978-1-59904-857-4.ch004

Aikins, S. K. (2009). A comparative study of municipal adoption of internet-based citizen participation. In C. Reddick (Ed.), *Handbook of research on strategies for local e-government adoption and implementation: Comparative studies* (pp. 206–230). Hershey, PA: IGI Global. doi:10.4018/978-1-60566-282-4.ch011

Aikins, S. K. (2012). Improving e-government project management: Best practices and critical success factors. In *Digital democracy: Concepts, methodologies, tools, and applications* (pp. 1314–1332). Hershey, PA: IGI Global. doi:10.4018/978-1-4666-1740-7.ch065

Akabawi, M. S. (2011). Ghabbour group ERP deployment: Learning from past technology failures. In E. Business Research and Case Center (Ed.), Cases on business and management in the MENA region: New trends and opportunities (pp. 177-203). Hershey, PA: IGI Global. doi:10.4018/978-1-60960-583-4.ch012

Akabawi, M. S. (2013). Ghabbour group ERP deployment: Learning from past technology failures. In *Industrial engineering: Concepts, methodologies, tools, and applications* (pp. 933–958). Hershey, PA: Engineering Science Reference. doi:10.4018/978-1-4666-1945-6.ch051

Akbulut, A. Y., & Motwani, J. (2008). Integration and information sharing in e-government. In G. Putnik & M. Cruz-Cunha (Eds.), *Encyclopedia of networked and virtual organizations* (pp. 729–734). Hershey, PA: IGI Global. doi:10.4018/978-1-59904-885-7.ch096

Akers, E. J. (2008). Technology diffusion in public administration. In G. Garson & M. Khosrow-Pour (Eds.), *Handbook of research on public information technology* (pp. 339–348). Hershey, PA: IGI Global. doi:10.4018/978-1-59904-857-4.ch033

Al-Shafi, S. (2008). Free wireless internet park services: An investigation of technology adoption in Qatar from a citizens perspective. *Journal of Cases on Information Technology, 10*(3), 21–34. doi:10.4018/jcit.2008070103

Al-Shafi, S., & Weerakkody, V. (2009). Implementing free wi-fi in public parks: An empirical study in Qatar. *International Journal of Electronic Government Research, 5*(3), 21–35. doi:10.4018/jegr.2009070102

Aladwani, A. M. (2002). Organizational actions, computer attitudes and end-user satisfaction in public organizations: An empirical study. In C. Snodgrass & E. Szewczak (Eds.), *Human factors in information systems* (pp. 153–168). Hershey, PA: IGI Global. doi:10.4018/978-1-931777-10-0.ch012

Aladwani, A. M. (2002). Organizational actions, computer attitudes, and end-user satisfaction in public organizations: An empirical study. *Journal of Organizational and End User Computing*, *14*(1), 42–49. doi:10.4018/joeuc.2002010104

Allen, B., Juillet, L., Paquet, G., & Roy, J. (2005). E-government and private-public partnerships: Relational challenges and strategic directions. In M. Khosrow-Pour (Ed.), *Practicing e-government: A global perspective* (pp. 364–382). Hershey, PA: IGI Global. doi:10.4018/978-1-59140-637-2.ch016

Alshawaf, A., & Knalil, O. E. (2008). IS success factors and IS organizational impact: Does ownership type matter in Kuwait? *International Journal of Enterprise Information Systems*, *4*(2), 13–33. doi:10.4018/jeis.2008040102

Ambali, A. R. (2009). Digital divide and its implication on Malaysian e-government: Policy initiatives. In H. Rahman (Ed.), *Social and political implications of data mining: Knowledge management in e-government* (pp. 267–287). Hershey, PA: IGI Global. doi:10.4018/978-1-60566-230-5.ch016

Amoretti, F. (2007). Digital international governance. In A. Anttiroiko & M. Malkia (Eds.), *Encyclopedia of digital government* (pp. 365–370). Hershey, PA: IGI Global. doi:10.4018/978-1-59140-789-8.ch056

Amoretti, F. (2008). Digital international governance. In A. Anttiroiko (Ed.), *Electronic government: Concepts, methodologies, tools, and applications* (pp. 688–696). Hershey, PA: IGI Global. doi:10.4018/978-1-59904-947-2.ch058

Amoretti, F. (2008). E-government at supranational level in the European Union. In A. Anttiroiko (Ed.), *Electronic government: Concepts, methodologies, tools, and applications* (pp. 1047–1055). Hershey, PA: IGI Global. doi:10.4018/978-1-59904-947-2.ch079

Amoretti, F. (2008). E-government regimes. In A. Anttiroiko (Ed.), *Electronic government: Concepts, methodologies, tools, and applications* (pp. 3846–3856). Hershey, PA: IGI Global. doi:10.4018/978-1-59904-947-2.ch280

Amoretti, F. (2009). Electronic constitution: A Braudelian perspective. In F. Amoretti (Ed.), *Electronic constitution: Social, cultural, and political implications* (pp. 1–19). Hershey, PA: IGI Global. doi:10.4018/978-1-60566-254-1.ch001

Amoretti, F., & Musella, F. (2009). Institutional isomorphism and new technologies. In M. Khosrow-Pour (Ed.), *Encyclopedia of information science and technology* (2nd ed.; pp. 2066–2071). Hershey, PA: IGI Global. doi:10.4018/978-1-60566-026-4.ch325

Andersen, K. V., & Henriksen, H. Z. (2007). E-government research: Capabilities, interaction, orientation, and values. In D. Norris (Ed.), *Current issues and trends in e-government research* (pp. 269–288). Hershey, PA: IGI Global. doi:10.4018/978-1-59904-283-1.ch013

Anderson, K. V., & Henriksen, H. Z. (2005). The first leg of e-government research: Domains and application areas 19982003. *International Journal of Electronic Government Research, 1*(4), 26–44. doi:10.4018/jegr.2005100102

Anttiroiko, A. (2009). Democratic e-governance. In M. Khosrow-Pour (Ed.), *Encyclopedia of information science and technology* (2nd ed.; pp. 990–995). Hershey, PA: IGI Global. doi:10.4018/978-1-60566-026-4.ch158

Baker, P. M., Bell, A., & Moon, N. W. (2009). Accessibility issues in municipal wireless networks. In C. Reddick (Ed.), *Handbook of research on strategies for local e-government adoption and implementation: Comparative studies* (pp. 569–588). Hershey, PA: IGI Global. doi:10.4018/978-1-60566-282-4.ch030

Becker, S. A., Keimer, R., & Muth, T. (2010). A case on university and community collaboration: The sci-tech entrepreneurial training services (ETS) program. In S. Becker & R. Niebuhr (Eds.), *Cases on technology innovation: Entrepreneurial successes and pitfalls* (pp. 68–90). Hershey, PA: IGI Global. doi:10.4018/978-1-61520-609-4.ch003

Becker, S. A., Keimer, R., & Muth, T. (2012). A case on university and community collaboration: The sci-tech entrepreneurial training services (ETS) program. In Regional development: Concepts, methodologies, tools, and applications (pp. 947-969). Hershey, PA: IGI Global. doi:10.4018/978-1-4666-0882-5.ch507

Bernardi, R. (2012). Information technology and resistance to public sector reforms: A case study in Kenya. In T. Papadopoulos & P. Kanellis (Eds.), *Public sector reform using information technologies: Transforming policy into practice* (pp. 59–78). Hershey, PA: IGI Global. doi:10.4018/978-1-60960-839-2.ch004

Bernardi, R. (2013). Information technology and resistance to public sector reforms: A case study in Kenya. In *User-driven healthcare: Concepts, methodologies, tools, and applications* (pp. 14–33). Hershey, PA: IGI Global. doi:10.4018/978-1-4666-2770-3.ch002

Bolívar, M. P., Pérez, M. D., & Hernández, A. M. (2012). Municipal e-government services in emerging economies: The Latin-American and Caribbean experiences. In Y. Chen & P. Chu (Eds.), *Electronic governance and cross-boundary collaboration: Innovations and advancing tools* (pp. 198–226). Hershey, PA: IGI Global. doi:10.4018/978-1-60960-753-1.ch011

Borycki, E. M., & Kushniruk, A. W. (2010). Use of clinical simulations to evaluate the impact of health information systems and ubiquitous computing devices upon health professional work. In S. Mohammed & J. Fiaidhi (Eds.), *Ubiquitous health and medical informatics: The ubiquity 2.0 trend and beyond* (pp. 552–573). Hershey, PA: IGI Global. doi:10.4018/978-1-61520-777-0.ch026

Borycki, E. M., & Kushniruk, A. W. (2011). Use of clinical simulations to evaluate the impact of health information systems and ubiquitous computing devices upon health professional work. In *Clinical technologies: Concepts, methodologies, tools and applications* (pp. 532–553). Hershey, PA: IGI Global. doi:10.4018/978-1-60960-561-2.ch220

Buchan, J. (2011). Developing a dynamic and responsive online learning environment: A case study of a large Australian university. In B. Czerkawski (Ed.), *Free and open source software for e-learning: Issues, successes and challenges* (pp. 92–109). Hershey, PA: IGI Global. doi:10.4018/978-1-61520-917-0.ch006

Buenger, A. W. (2008). Digital convergence and cybersecurity policy. In G. Garson & M. Khosrow-Pour (Eds.), *Handbook of research on public information technology* (pp. 395–405). Hershey, PA: IGI Global. doi:10.4018/978-1-59904-857-4.ch038

Burn, J. M., & Loch, K. D. (2002). The societal impact of world wide web - Key challenges for the 21st century. In A. Salehnia (Ed.), *Ethical issues of information systems* (pp. 88–106). Hershey, PA: IGI Global. doi:10.4018/978-1-931777-15-5.ch007

Burn, J. M., & Loch, K. D. (2003). The societal impact of the world wide web-Key challenges for the 21st century. In M. Khosrow-Pour (Ed.), *Advanced topics in information resources management* (Vol. 2, pp. 32–51). Hershey, PA: IGI Global. doi:10.4018/978-1-59140-062-2.ch002

Bwalya, K. J., Du Plessis, T., & Rensleigh, C. (2012). The "quicksilver initiatives" as a framework for e-government strategy design in developing economies. In K. Bwalya & S. Zulu (Eds.), *Handbook of research on e-government in emerging economies: Adoption, e-participation, and legal frameworks* (pp. 605–623). Hershey, PA: IGI Global. doi:10.4018/978-1-4666-0324-0.ch031

Cabotaje, C. E., & Alampay, E. A. (2013). Social media and citizen engagement: Two cases from the Philippines. In S. Saeed & C. Reddick (Eds.), *Human-centered system design for electronic governance* (pp. 225–238). Hershey, PA: IGI Global. doi:10.4018/978-1-4666-3640-8.ch013

Camillo, A., Di Pietro, L., Di Virgilio, F., & Franco, M. (2013). Work-groups conflict at PetroTech-Italy, S.R.L.: The influence of culture on conflict dynamics. In B. Christiansen, E. Turkina, & N. Williams (Eds.), *Cultural and technological influences on global business* (pp. 272–289). Hershey, PA: IGI Global. doi:10.4018/978-1-4666-3966-9.ch015

Capra, E., Francalanci, C., & Marinoni, C. (2008). Soft success factors for m-government. In A. Anttiroiko (Ed.), *Electronic government: Concepts, methodologies, tools, and applications* (pp. 1213–1233). Hershey, PA: IGI Global. doi:10.4018/978-1-59904-947-2.ch089

Cartelli, A. (2009). The implementation of practices with ICT as a new teaching-learning paradigm. In A. Cartelli & M. Palma (Eds.), *Encyclopedia of information communication technology* (pp. 413–417). Hershey, PA: IGI Global. doi:10.4018/978-1-59904-845-1.ch055

Charalabidis, Y., Lampathaki, F., & Askounis, D. (2010). Investigating the landscape in national interoperability frameworks. *International Journal of E-Services and Mobile Applications*, *2*(4), 28–41. doi:10.4018/jesma.2010100103

Charalabidis, Y., Lampathaki, F., & Askounis, D. (2012). Investigating the landscape in national interoperability frameworks. In A. Scupola (Ed.), *Innovative mobile platform developments for electronic services design and delivery* (pp. 218–231). Hershey, PA: IGI Global. doi:10.4018/978-1-4666-1568-7.ch013

Chen, I. (2005). Distance education associations. In C. Howard, J. Boettcher, L. Justice, K. Schenk, P. Rogers, & G. Berg (Eds.), *Encyclopedia of distance learning* (pp. 599–612). Hershey, PA: IGI Global. doi:10.4018/978-1-59140-555-9.ch087

Chen, I. (2008). Distance education associations. In L. Tomei (Ed.), *Online and distance learning: Concepts, methodologies, tools, and applications* (pp. 562–579). Hershey, PA: IGI Global. doi:10.4018/978-1-59904-935-9.ch048

Chen, Y. (2008). Managing IT outsourcing for digital government. In A. Anttiroiko (Ed.), *Electronic government: Concepts, methodologies, tools, and applications* (pp. 3107–3114). Hershey, PA: IGI Global. doi:10.4018/978-1-59904-947-2.ch229

Chen, Y., & Dimitrova, D. V. (2006). Electronic government and online engagement: Citizen interaction with government via web portals. *International Journal of Electronic Government Research*, *2*(1), 54–76. doi:10.4018/jegr.2006010104

Chen, Y., & Knepper, R. (2005). Digital government development strategies: Lessons for policy makers from a comparative perspective. In W. Huang, K. Siau, & K. Wei (Eds.), *Electronic government strategies and implementation* (pp. 394–420). Hershey, PA: IGI Global. doi:10.4018/978-1-59140-348-7.ch017

Chen, Y., & Knepper, R. (2008). Digital government development strategies: Lessons for policy makers from a comparative perspective. In H. Rahman (Ed.), *Developing successful ICT strategies: Competitive advantages in a global knowledge-driven society* (pp. 334–356). Hershey, PA: IGI Global. doi:10.4018/978-1-59904-654-9.ch017

Cherian, E. J., & Ryan, T. W. (2014). Incongruent needs: Why differences in the iron-triangle of priorities make health information technology adoption and use difficult. In C. El Morr (Ed.), *Research perspectives on the role of informatics in health policy and management* (pp. 209–221). Hershey, PA: IGI Global. doi:10.4018/978-1-4666-4321-5.ch012

Cho, H. J., & Hwang, S. (2010). Government 2.0 in Korea: Focusing on e-participation services. In C. Reddick (Ed.), *Politics, democracy and e-government: Participation and service delivery* (pp. 94–114). Hershey, PA: IGI Global. doi:10.4018/978-1-61520-933-0.ch006

Chorus, C., & Timmermans, H. (2010). Ubiquitous travel environments and travel control strategies: Prospects and challenges. In M. Wachowicz (Ed.), *Movement-aware applications for sustainable mobility: Technologies and approaches* (pp. 30–51). Hershey, PA: IGI Global. doi:10.4018/978-1-61520-769-5.ch003

Chuanshen, R. (2007). E-government construction and China's administrative litigation act. In A. Anttiroiko & M. Malkia (Eds.), *Encyclopedia of digital government* (pp. 507–510). Hershey, PA: IGI Global. doi:10.4018/978-1-59140-789-8.ch077

Ciaghi, A., & Villafiorita, A. (2012). Law modeling and BPR for public administration improvement. In K. Bwalya & S. Zulu (Eds.), *Handbook of research on e-government in emerging economies: Adoption, e-participation, and legal frameworks* (pp. 391–410). Hershey, PA: IGI Global. doi:10.4018/978-1-4666-0324-0.ch019

Ciaramitaro, B. L., & Skrocki, M. (2012). mHealth: Mobile healthcare. In B. Ciaramitaro (Ed.), Mobile technology consumption: Opportunities and challenges (pp. 99-109). Hershey, PA: IGI Global. doi:10.4018/978-1-61350-150-4.ch007

Comite, U. (2012). Innovative processes and managerial effectiveness of e-procurement in healthcare. In A. Manoharan & M. Holzer (Eds.), *Active citizen participation in e-government: A global perspective* (pp. 206–229). Hershey, PA: IGI Global. doi:10.4018/978-1-4666-0116-1.ch011

Cordella, A. (2013). E-government success: How to account for ICT, administrative rationalization, and institutional change. In J. Gil-Garcia (Ed.), *E-government success factors and measures: Theories, concepts, and methodologies* (pp. 40–51). Hershey, PA: IGI Global. doi:10.4018/978-1-4666-4058-0.ch003

Cropf, R. A. (2009). ICT and e-democracy. In M. Khosrow-Pour (Ed.), *Encyclopedia of information science and technology* (2nd ed.; pp. 1789–1793). Hershey, PA: IGI Global. doi:10.4018/978-1-60566-026-4.ch281

Cropf, R. A. (2009). The virtual public sphere. In M. Pagani (Ed.), *Encyclopedia of multimedia technology and networking* (2nd ed.; pp. 1525–1530). Hershey, PA: IGI Global. doi:10.4018/978-1-60566-014-1.ch206

D'Abundo, M. L. (2013). Electronic health record implementation in the United States healthcare industry: Making the process of change manageable. In V. Wang (Ed.), *Handbook of research on technologies for improving the 21st century workforce: Tools for lifelong learning* (pp. 272–286). Hershey, PA: IGI Global. doi:10.4018/978-1-4666-2181-7.ch018

Damurski, L. (2012). E-participation in urban planning: Online tools for citizen engagement in Poland and in Germany. *International Journal of E-Planning Research, 1*(3), 40–67. doi:10.4018/ijepr.2012070103

de Almeida, M. O. (2007). E-government strategy in Brazil: Increasing transparency and efficiency through e-government procurement. In M. Gascó-Hernandez (Ed.), *Latin America online: Cases, successes and pitfalls* (pp. 34–82). Hershey, PA: IGI Global. doi:10.4018/978-1-59140-974-8.ch002

de Juana Espinosa, S. (2008). Empirical study of the municipalitites' motivations for adopting online presence. In A. Anttiroiko (Ed.), *Electronic government: Concepts, methodologies, tools, and applications* (pp. 3593–3608). Hershey, PA: IGI Global. doi:10.4018/978-1-59904-947-2.ch262

de Souza Dias, D. (2002). Motivation for using information technology. In C. Snodgrass & E. Szewczak (Eds.), *Human factors in information systems* (pp. 55–60). Hershey, PA: IGI Global. doi:10.4018/978-1-931777-10-0.ch005

Demediuk, P. (2006). Government procurement ICT's impact on the sustainability of SMEs and regional communities. In S. Marshall, W. Taylor, & X. Yu (Eds.), *Encyclopedia of developing regional communities with information and communication technology* (pp. 321–324). Hershey, PA: IGI Global. doi:10.4018/978-1-59140-575-7.ch056

Devonshire, E., Forsyth, H., Reid, S., & Simpson, J. M. (2013). The challenges and opportunities of online postgraduate coursework programs in a traditional university context. In B. Tynan, J. Willems, & R. James (Eds.), *Outlooks and opportunities in blended and distance learning* (pp. 353–368). Hershey, PA: IGI Global. doi:10.4018/978-1-4666-4205-8.ch026

Di Cerbo, F., Scotto, M., Sillitti, A., Succi, G., & Vernazza, T. (2007). Toward a GNU/Linux distribution for corporate environments. In S. Sowe, I. Stamelos, & I. Samoladas (Eds.), *Emerging free and open source software practices* (pp. 215–236). Hershey, PA: IGI Global. doi:10.4018/978-1-59904-210-7.ch010

Diesner, J., & Carley, K. M. (2005). Revealing social structure from texts: Meta-matrix text analysis as a novel method for network text analysis. In V. Narayanan & D. Armstrong (Eds.), *Causal mapping for research in information technology* (pp. 81–108). Hershey, PA: IGI Global. doi:10.4018/978-1-59140-396-8.ch004

Dologite, D. G., Mockler, R. J., Bai, Q., & Viszhanyo, P. F. (2006). IS change agents in practice in a US-Chinese joint venture. In M. Hunter & F. Tan (Eds.), *Advanced topics in global information management* (Vol. 5, pp. 331–352). Hershey, PA: IGI Global. doi:10.4018/978-1-59140-923-6.ch015

Drnevich, P., Brush, T. H., & Luckock, G. T. (2011). Process and structural implications for IT-enabled outsourcing. *International Journal of Strategic Information Technology and Applications*, 2(4), 30–43. doi:10.4018/jsita.2011100103

Dwivedi, A. N. (2009). Handbook of research on information technology management and clinical data administration in healthcare (Vols. 1–2). Hershey, PA: IGI Global. doi:10.4018/978-1-60566-356-2

Elbeltagi, I., McBride, N., & Hardaker, G. (2006). Evaluating the factors affecting DSS usage by senior managers in local authorities in Egypt. In M. Hunter & F. Tan (Eds.), *Advanced topics in global information management* (Vol. 5, pp. 283–307). Hershey, PA: IGI Global. doi:10.4018/978-1-59140-923-6.ch013

Eom, S., & Fountain, J. E. (2013). Enhancing information services through public-private partnerships: Information technology knowledge transfer underlying structures to develop shared services in the U.S. and Korea. In J. Gil-Garcia (Ed.), *E-government success around the world: Cases, empirical studies, and practical recommendations* (pp. 15–40). Hershey, PA: IGI Global. doi:10.4018/978-1-4666-4173-0.ch002

Esteves, T., Leuenberger, D., & Van Leuven, N. (2012). Reaching citizen 2.0: How government uses social media to send public messages during times of calm and times of crisis. In K. Kloby & M. D'Agostino (Eds.), *Citizen 2.0: Public and governmental interaction through web 2.0 technologies* (pp. 250–268). Hershey, PA: IGI Global. doi:10.4018/978-1-4666-0318-9.ch013

Estevez, E., Fillottrani, P., Janowski, T., & Ojo, A. (2012). Government information sharing: A framework for policy formulation. In Y. Chen & P. Chu (Eds.), *Electronic governance and cross-boundary collaboration: Innovations and advancing tools* (pp. 23–55). Hershey, PA: IGI Global. doi:10.4018/978-1-60960-753-1.ch002

Ezz, I. E. (2008). E-governement emerging trends: Organizational challenges. In A. Anttiroiko (Ed.), *Electronic government: Concepts, methodologies, tools, and applications* (pp. 3721–3737). Hershey, PA: IGI Global. doi:10.4018/978-1-59904-947-2.ch269

Fabri, M. (2009). The Italian style of e-justice in a comparative perspective. In A. Martínez & P. Abat (Eds.), *E-justice: Using information communication technologies in the court system* (pp. 1–19). Hershey, PA: IGI Global. doi:10.4018/978-1-59904-998-4.ch001

Fagbe, T., & Adekola, O. D. (2010). Workplace safety and personnel well-being: The impact of information technology. *International Journal of Green Computing*, *1*(1), 28–33. doi:10.4018/jgc.2010010103

Fagbe, T., & Adekola, O. D. (2011). Workplace safety and personnel well-being: The impact of information technology. In *Global business: Concepts, methodologies, tools and applications* (pp. 1438–1444). Hershey, PA: IGI Global. doi:10.4018/978-1-60960-587-2.ch509

Farmer, L. (2008). Affective collaborative instruction with librarians. In S. Kelsey & K. St.Amant (Eds.), *Handbook of research on computer mediated communication* (pp. 15–24). Hershey, PA: IGI Global. doi:10.4018/978-1-59904-863-5.ch002

Favier, L., & Mekhantar, J. (2007). Use of OSS by local e-administration: The French situation. In K. St.Amant & B. Still (Eds.), *Handbook of research on open source software: Technological, economic, and social perspectives* (pp. 428–444). Hershey, PA: IGI Global. doi:10.4018/978-1-59140-999-1.ch033

Fernando, S. (2009). Issues of e-learning in third world countries. In M. Khosrow-Pour (Ed.), *Encyclopedia of information science and technology* (2nd ed.; pp. 2273–2277). Hershey, PA: IGI Global. doi:10.4018/978-1-60566-026-4.ch360

Filho, J. R., & dos Santos Junior, J. R. (2009). Local e-government in Brazil: Poor interaction and local politics as usual. In C. Reddick (Ed.), *Handbook of research on strategies for local e-government adoption and implementation: Comparative studies* (pp. 863–878). Hershey, PA: IGI Global. doi:10.4018/978-1-60566-282-4.ch045

Fletcher, P. D. (2004). Portals and policy: Implications of electronic access to U.S. federal government information services. In A. Pavlichev & G. Garson (Eds.), *Digital government: Principles and best practices* (pp. 52–62). Hershey, PA: IGI Global. doi:10.4018/978-1-59140-122-3.ch004

Fletcher, P. D. (2008). Portals and policy: Implications of electronic access to U.S. federal government information services. In A. Anttiroiko (Ed.), *Electronic government: Concepts, methodologies, tools, and applications* (pp. 3970–3979). Hershey, PA: IGI Global. doi:10.4018/978-1-59904-947-2.ch289

Forlano, L. (2004). The emergence of digital government: International perspectives. In A. Pavlichev & G. Garson (Eds.), *Digital government: Principles and best practices* (pp. 34–51). Hershey, PA: IGI Global. doi:10.4018/978-1-59140-122-3.ch003

Franzel, J. M., & Coursey, D. H. (2004). Government web portals: Management issues and the approaches of five states. In A. Pavlichev & G. Garson (Eds.), *Digital government: Principles and best practices* (pp. 63–77). Hershey, PA: IGI Global. doi:10.4018/978-1-59140-122-3.ch005

Gaivéo, J. M. (2013). Security of ICTs supporting healthcare activities. In M. Cruz-Cunha, I. Miranda, & P. Gonçalves (Eds.), *Handbook of research on ICTs for human-centered healthcare and social care services* (pp. 208–228). Hershey, PA: IGI Global. doi:10.4018/978-1-4666-3986-7.ch011

Garson, G. D. (1999). *Information technology and computer applications in public administration: Issues and trends*. Hershey, PA: IGI Global. doi:10.4018/978-1-87828-952-0

Garson, G. D. (2003). Toward an information technology research agenda for public administration. In G. Garson (Ed.), *Public information technology: Policy and management issues* (pp. 331–357). Hershey, PA: IGI Global. doi:10.4018/978-1-59140-060-8.ch014

Garson, G. D. (2004). The promise of digital government. In A. Pavlichev & G. Garson (Eds.), *Digital government: Principles and best practices* (pp. 2–15). Hershey, PA: IGI Global. doi:10.4018/978-1-59140-122-3.ch001

Garson, G. D. (2007). An information technology research agenda for public administration. In G. Garson (Ed.), *Modern public information technology systems: Issues and challenges* (pp. 365–392). Hershey, PA: IGI Global. doi:10.4018/978-1-59904-051-6.ch018

Gasco, M. (2007). Civil servants' resistance towards e-government development. In A. Anttiroiko & M. Malkia (Eds.), *Encyclopedia of digital government* (pp. 190–195). Hershey, PA: IGI Global. doi:10.4018/978-1-59140-789-8.ch028

Gasco, M. (2008). Civil servants' resistance towards e-government development. In A. Anttiroiko (Ed.), *Electronic government: Concepts, methodologies, tools, and applications* (pp. 2580–2588). Hershey, PA: IGI Global. doi:10.4018/978-1-59904-947-2.ch190

Ghere, R. K. (2010). Accountability and information technology enactment: Implications for social empowerment. In E. Ferro, Y. Dwivedi, J. Gil-Garcia, & M. Williams (Eds.), *Handbook of research on overcoming digital divides: Constructing an equitable and competitive information society* (pp. 515–532). Hershey, PA: IGI Global. doi:10.4018/978-1-60566-699-0.ch028

Gibson, I. W. (2012). Simulation modeling of healthcare delivery. In A. Kolker & P. Story (Eds.), *Management engineering for effective healthcare delivery: Principles and applications* (pp. 69–89). Hershey, PA: IGI Global. doi:10.4018/978-1-60960-872-9.ch003

Gil-Garcia, J. R. (2007). Exploring e-government benefits and success factors. In A. Anttiroiko & M. Malkia (Eds.), *Encyclopedia of digital government* (pp. 803–811). Hershey, PA: IGI Global. doi:10.4018/978-1-59140-789-8.ch122

Gil-Garcia, J. R., & González Miranda, F. (2010). E-government and opportunities for participation: The case of the Mexican state web portals. In C. Reddick (Ed.), *Politics, democracy and e-government: Participation and service delivery* (pp. 56–74). Hershey, PA: IGI Global. doi:10.4018/978-1-61520-933-0.ch004

Goldfinch, S. (2012). Public trust in government, trust in e-government, and use of e-government. In Z. Yan (Ed.), *Encyclopedia of cyber behavior* (pp. 987–995). Hershey, PA: IGI Global. doi:10.4018/978-1-4666-0315-8.ch081

Goodyear, M. (2012). Organizational change contributions to e-government project transitions. In S. Aikins (Ed.), *Managing e-government projects: Concepts, issues, and best practices* (pp. 1–21). Hershey, PA: IGI Global. doi:10.4018/978-1-4666-0086-7.ch001

Gordon, S., & Mulligan, P. (2003). Strategic models for the delivery of personal financial services: The role of infocracy. In S. Gordon (Ed.), *Computing information technology: The human side* (pp. 220–232). Hershey, PA: IGI Global. doi:10.4018/978-1-93177-752-0.ch014

Gordon, T. F. (2007). Legal knowledge systems. In A. Anttiroiko & M. Malkia (Eds.), *Encyclopedia of digital government* (pp. 1161–1166). Hershey, PA: IGI Global. doi:10.4018/978-1-59140-789-8.ch175

Graham, J. E., & Semich, G. W. (2008). Integrating technology to transform pedagogy: Revisiting the progress of the three phase TUI model for faculty development. In L. Tomei (Ed.), *Adapting information and communication technologies for effective education* (pp. 1–12). Hershey, PA: IGI Global. doi:10.4018/978-1-59904-922-9.ch001

Grandinetti, L., & Pisacane, O. (2012). Web services for healthcare management. In D. Prakash Vidyarthi (Ed.), *Technologies and protocols for the future of internet design: Reinventing the web* (pp. 60–94). Hershey, PA: IGI Global. doi:10.4018/978-1-4666-0203-8.ch004

Groenewegen, P., & Wagenaar, F. P. (2008). VO as an alternative to hierarchy in the Dutch police sector. In G. Putnik & M. Cruz-Cunha (Eds.), *Encyclopedia of networked and virtual organizations* (pp. 1851–1857). Hershey, PA: IGI Global. doi:10.4018/978-1-59904-885-7.ch245

Gronlund, A. (2001). Building an infrastructure to manage electronic services. In S. Dasgupta (Ed.), *Managing internet and intranet technologies in organizations: Challenges and opportunities* (pp. 71–103). Hershey, PA: IGI Global. doi:10.4018/978-1-878289-95-7.ch006

Gronlund, A. (2002). Introduction to electronic government: Design, applications and management. In Å. Grönlund (Ed.), *Electronic government: Design, applications and management* (pp. 1–21). Hershey, PA: IGI Global. doi:10.4018/978-1-930708-19-8.ch001

Gupta, A., Woosley, R., Crk, I., & Sarnikar, S. (2009). An information technology architecture for drug effectiveness reporting and post-marketing surveillance. In J. Tan (Ed.), *Medical informatics: Concepts, methodologies, tools, and applications* (pp. 631–646). Hershey, PA: IGI Global. doi:10.4018/978-1-60566-050-9.ch047

Hallin, A., & Lundevall, K. (2007). mCity: User focused development of mobile services within the city of Stockholm. In I. Kushchu (Ed.), Mobile government: An emerging direction in e-government (pp. 12-29). Hershey, PA: IGI Global. doi:10.4018/978-1-59140-884-0.ch002

Hallin, A., & Lundevall, K. (2009). mCity: User focused development of mobile services within the city of Stockholm. In S. Clarke (Ed.), Evolutionary concepts in end user productivity and performance: Applications for organizational progress (pp. 268-280). Hershey, PA: IGI Global. doi:10.4018/978-1-60566-136-0.ch017

Hallin, A., & Lundevall, K. (2009). mCity: User focused development of mobile services within the city of Stockholm. In D. Taniar (Ed.), Mobile computing: Concepts, methodologies, tools, and applications (pp. 3455-3467). Hershey, PA: IGI Global. doi:10.4018/978-1-60566-054-7.ch253

Hanson, A. (2005). Overcoming barriers in the planning of a virtual library. In M. Khosrow-Pour (Ed.), *Encyclopedia of information science and technology* (pp. 2255–2259). Hershey, PA: IGI Global. doi:10.4018/978-1-59140-553-5.ch397

Haque, A. (2008). Information technology and surveillance: Implications for public administration in a new word order. In T. Loendorf & G. Garson (Eds.), *Patriotic information systems* (pp. 177–185). Hershey, PA: IGI Global. doi:10.4018/978-1-59904-594-8.ch008

Hauck, R. V., Thatcher, S. M., & Weisband, S. P. (2012). Temporal aspects of information technology use: Increasing shift work effectiveness. In J. Wang (Ed.), *Advancing the service sector with evolving technologies: Techniques and principles* (pp. 87–104). Hershey, PA: IGI Global. doi:10.4018/978-1-4666-0044-7.ch006

Hawk, S., & Witt, T. (2006). Telecommunications courses in information systems programs. *International Journal of Information and Communication Technology Education*, *2*(1), 79–92. doi:10.4018/jicte.2006010107

Helms, M. M., Moore, R., & Ahmadi, M. (2009). Information technology (IT) and the healthcare industry: A SWOT analysis. In J. Tan (Ed.), *Medical informatics: Concepts, methodologies, tools, and applications* (pp. 134–152). Hershey, PA: IGI Global. doi:10.4018/978-1-60566-050-9.ch012

Hendrickson, S. M., & Young, M. E. (2014). Electronic records management at a federally funded research and development center. In J. Krueger (Ed.), *Cases on electronic records and resource management implementation in diverse environments* (pp. 334–350). Hershey, PA: IGI Global. doi:10.4018/978-1-4666-4466-3.ch020

Henman, P. (2010). Social policy and information communication technologies. In J. Martin & L. Hawkins (Eds.), *Information communication technologies for human services education and delivery: Concepts and cases* (pp. 215–229). Hershey, PA: IGI Global. doi:10.4018/978-1-60566-735-5.ch014

Hismanoglu, M. (2011). Important issues in online education: E-pedagogy and marketing. In U. Demiray & S. Sever (Eds.), *Marketing online education programs: Frameworks for promotion and communication* (pp. 184–209). Hershey, PA: IGI Global. doi:10.4018/978-1-60960-074-7.ch012

Ho, K. K. (2008). The e-government development, IT strategies, and portals of the Hong Kong SAR government. In A. Anttiroiko (Ed.), *Electronic government: Concepts, methodologies, tools, and applications* (pp. 715–733). Hershey, PA: IGI Global. doi:10.4018/978-1-59904-947-2.ch060

Holden, S. H. (2003). The evolution of information technology management at the federal level: Implications for public administration. In G. Garson (Ed.), *Public information technology: Policy and management issues* (pp. 53–73). Hershey, PA: IGI Global. doi:10.4018/978-1-59140-060-8.ch003

Holden, S. H. (2007). The evolution of federal information technology management literature: Does IT finally matter? In G. Garson (Ed.), *Modern public information technology systems: Issues and challenges* (pp. 17–34). Hershey, PA: IGI Global. doi:10.4018/978-1-59904-051-6.ch002

Holland, J. W. (2009). Automation of American criminal justice. In M. Khosrow-Pour (Ed.), *Encyclopedia of information science and technology* (2nd ed.; pp. 300–302). Hershey, PA: IGI Global. doi:10.4018/978-1-60566-026-4.ch051

Holloway, K. (2013). Fair use, copyright, and academic integrity in an online academic environment. In *Digital rights management: Concepts, methodologies, tools, and applications* (pp. 917–928). Hershey, PA: IGI Global. doi:10.4018/978-1-4666-2136-7.ch044

Horiuchi, C. (2005). E-government databases. In L. Rivero, J. Doorn, & V. Ferraggine (Eds.), *Encyclopedia of database technologies and applications* (pp. 206–210). Hershey, PA: IGI Global. doi:10.4018/978-1-59140-560-3.ch035

Horiuchi, C. (2006). Creating IS quality in government settings. In E. Duggan & J. Reichgelt (Eds.), *Measuring information systems delivery quality* (pp. 311–327). Hershey, PA: IGI Global. doi:10.4018/978-1-59140-857-4.ch014

Hsiao, N., Chu, P., & Lee, C. (2012). Impact of e-governance on businesses: Model development and case study. In *Digital democracy: Concepts, methodologies, tools, and applications* (pp. 1407–1425). Hershey, PA: IGI Global. doi:10.4018/978-1-4666-1740-7.ch070

Huang, T., & Lee, C. (2010). Evaluating the impact of e-government on citizens: Cost-benefit analysis. In C. Reddick (Ed.), *Citizens and e-government: Evaluating policy and management* (pp. 37–52). Hershey, PA: IGI Global. doi:10.4018/978-1-61520-931-6.ch003

Hunter, M. G., Diochon, M., Pugsley, D., & Wright, B. (2002). Unique challenges for small business adoption of information technology: The case of the Nova Scotia ten. In S. Burgess (Ed.), *Managing information technology in small business: Challenges and solutions* (pp. 98–117). Hershey, PA: IGI Global. doi:10.4018/978-1-930708-35-8.ch006

Hurskainen, J. (2003). Integration of business systems and applications in merger and alliance: Case metso automation. In T. Reponen (Ed.), *Information technology enabled global customer service* (pp. 207–225). Hershey, PA: IGI Global. doi:10.4018/978-1-59140-048-6.ch012

Iazzolino, G., & Pietrantonio, R. (2011). The soveria.it project: A best practice of e-government in southern Italy. In D. Piaggesi, K. Sund, & W. Castelnovo (Eds.), *Global strategy and practice of e-governance: Examples from around the world* (pp. 34–56). Hershey, PA: IGI Global. doi:10.4018/978-1-60960-489-9.ch003

Imran, A., & Gregor, S. (2012). A process model for successful e-government adoption in the least developed countries: A case of Bangladesh. In F. Tan (Ed.), *International comparisons of information communication technologies: Advancing applications* (pp. 321–350). Hershey, PA: IGI Global. doi:10.4018/978-1-61350-480-2.ch014

Inoue, Y., & Bell, S. T. (2005). Electronic/digital government innovation, and publishing trends with IT. In M. Khosrow-Pour (Ed.), *Encyclopedia of information science and technology* (pp. 1018–1023). Hershey, PA: IGI Global. doi:10.4018/978-1-59140-553-5.ch180

Islam, M. M., & Ehsan, M. (2013). Understanding e-governance: A theoretical approach. In M. Islam & M. Ehsan (Eds.), *From government to e-governance: Public administration in the digital age* (pp. 38–49). Hershey, PA: IGI Global. doi:10.4018/978-1-4666-1909-8.ch003

Jaeger, B. (2009). E-government and e-democracy in the making. In M. Khosrow-Pour (Ed.), *Encyclopedia of information science and technology* (2nd ed.; pp. 1318–1322). Hershey, PA: IGI Global. doi:10.4018/978-1-60566-026-4.ch208

Jain, R. B. (2007). Revamping the administrative structure and processes in India for online diplomacy. In A. Anttiroiko & M. Malkia (Eds.), *Encyclopedia of digital government* (pp. 1418–1423). Hershey, PA: IGI Global. doi:10.4018/978-1-59140-789-8.ch217

Jain, R. B. (2008). Revamping the administrative structure and processes in India for online diplomacy. In A. Anttiroiko (Ed.), *Electronic government: Concepts, methodologies, tools, and applications* (pp. 3142–3149). Hershey, PA: IGI Global. doi:10.4018/978-1-59904-947-2.ch233

Jauhiainen, J. S., & Inkinen, T. (2009). E-governance and the information society in periphery. In C. Reddick (Ed.), *Handbook of research on strategies for local e-government adoption and implementation: Comparative studies* (pp. 497–514). Hershey, PA: IGI Global. doi:10.4018/978-1-60566-282-4.ch026

Jensen, M. J. (2009). Electronic democracy and citizen influence in government. In C. Reddick (Ed.), *Handbook of research on strategies for local e-government adoption and implementation: Comparative studies* (pp. 288–305). Hershey, PA: IGI Global. doi:10.4018/978-1-60566-282-4.ch015

Jiao, Y., Hurson, A. R., Potok, T. E., & Beckerman, B. G. (2009). Integrating mobile-based systems with healthcare databases. In J. Erickson (Ed.), *Database technologies: Concepts, methodologies, tools, and applications* (pp. 484–504). Hershey, PA: IGI Global. doi:10.4018/978-1-60566-058-5.ch031

Joia, L. A. (2002). A systematic model to integrate information technology into metabusinesses: A case study in the engineering realms. In F. Tan (Ed.), *Advanced topics in global information management* (Vol. 1, pp. 250–267). Hershey, PA: IGI Global. doi:10.4018/978-1-930708-43-3.ch016

Jones, T. H., & Song, I. (2000). Binary equivalents of ternary relationships in entity-relationship modeling: A logical decomposition approach. *Journal of Database Management, 11*(2), 12–19. doi:10.4018/jdm.2000040102

Juana-Espinosa, S. D. (2007). Empirical study of the municipalitites' motivations for adopting online presence. In L. Al-Hakim (Ed.), *Global e-government: Theory, applications and benchmarking* (pp. 261–279). Hershey, PA: IGI Global. doi:10.4018/978-1-59904-027-1.ch015

Jun, K., & Weare, C. (2012). Bridging from e-government practice to e-government research: Past trends and future directions. In K. Bwalya & S. Zulu (Eds.), *Handbook of research on e-government in emerging economies: Adoption, e-participation, and legal frameworks* (pp. 263–289). Hershey, PA: IGI Global. doi:10.4018/978-1-4666-0324-0.ch013

Junqueira, A., Diniz, E. H., & Fernandez, M. (2010). Electronic government implementation projects with multiple agencies: Analysis of the electronic invoice project under PMBOK framework. In J. Cordoba-Pachon & A. Ochoa-Arias (Eds.), *Systems thinking and e-participation: ICT in the governance of society* (pp. 135–153). Hershey, PA: IGI Global. doi:10.4018/978-1-60566-860-4.ch009

Juntunen, A. (2009). Joint service development with the local authorities. In C. Reddick (Ed.), *Handbook of research on strategies for local e-government adoption and implementation: Comparative studies* (pp. 902–920). Hershey, PA: IGI Global. doi:10.4018/978-1-60566-282-4.ch047

Kamel, S. (2001). *Using DSS for crisis management*. Hershey, PA: IGI Global. doi:10.4018/978-1-87828-961-2.ch020

Kamel, S. (2006). DSS for strategic decision making. In M. Khosrow-Pour (Ed.), *Cases on information technology and organizational politics & culture* (pp. 230–246). Hershey, PA: IGI Global. doi:10.4018/978-1-59904-411-8.ch015

Kamel, S. (2009). The software industry in Egypt as a potential contributor to economic growth. In M. Khosrow-Pour (Ed.), *Encyclopedia of information science and technology* (2nd ed.; pp. 3531–3537). Hershey, PA: IGI Global. doi:10.4018/978-1-60566-026-4.ch562

Kamel, S., & Hussein, M. (2008). Xceed: Pioneering the contact center industry in Egypt. *Journal of Cases on Information Technology*, *10*(1), 67–91. doi:10.4018/jcit.2008010105

Kamel, S., & Wahba, K. (2003). The use of a hybrid model in web-based education: "The Global campus project. In A. Aggarwal (Ed.), *Web-based education: Learning from experience* (pp. 331–346). Hershey, PA: IGI Global. doi:10.4018/978-1-59140-102-5.ch020

Kardaras, D. K., & Papathanassiou, E. A. (2008). An exploratory study of the e-government services in Greece. In G. Garson & M. Khosrow-Pour (Eds.), *Handbook of research on public information technology* (pp. 162–174). Hershey, PA: IGI Global. doi:10.4018/978-1-59904-857-4.ch016

Kassahun, A. E., Molla, A., & Sarkar, P. (2012). Government process reengineering: What we know and what we need to know. In *Digital democracy: Concepts, methodologies, tools, and applications* (pp. 1730–1752). Hershey, PA: IGI Global. doi:10.4018/978-1-4666-1740-7.ch086

Khan, B. (2005). Technological issues. In B. Khan (Ed.), *Managing e-learning strategies: Design, delivery, implementation and evaluation* (pp. 154–180). Hershey, PA: IGI Global. doi:10.4018/978-1-59140-634-1.ch004

Khasawneh, A., Bsoul, M., Obeidat, I., & Al Azzam, I. (2012). Technology fears: A study of e-commerce loyalty perception by Jordanian customers. In J. Wang (Ed.), *Advancing the service sector with evolving technologies: Techniques and principles* (pp. 158–165). Hershey, PA: IGI Global. doi:10.4018/978-1-4666-0044-7.ch010

Khatibi, V., & Montazer, G. A. (2012). E-research methodology. In A. Juan, T. Daradoumis, M. Roca, S. Grasman, & J. Faulin (Eds.), *Collaborative and distributed e-research: Innovations in technologies, strategies and applications* (pp. 62–81). Hershey, PA: IGI Global. doi:10.4018/978-1-4666-0125-3.ch003

Kidd, T. (2011). The dragon in the school's backyard: A review of literature on the uses of technology in urban schools. In L. Tomei (Ed.), *Online courses and ICT in education: Emerging practices and applications* (pp. 242–257). Hershey, PA: IGI Global. doi:10.4018/978-1-60960-150-8.ch019

Kidd, T. T. (2010). My experience tells the story: Exploring technology adoption from a qualitative perspective - A pilot study. In H. Song & T. Kidd (Eds.), *Handbook of research on human performance and instructional technology* (pp. 247–262). Hershey, PA: IGI Global. doi:10.4018/978-1-60566-782-9.ch015

Kieley, B., Lane, G., Paquet, G., & Roy, J. (2002). e-Government in Canada: Services online or public service renewal? In Å. Grönlund (Ed.), Electronic government: Design, applications and management (pp. 340-355). Hershey, PA: IGI Global. doi:10.4018/978-1-930708-19-8.ch016

Kim, P. (2012). "Stay out of the way! My kid is video blogging through a phone!": A lesson learned from math tutoring social media for children in underserved communities. In *Wireless technologies: Concepts, methodologies, tools and applications* (pp. 1415–1428). Hershey, PA: IGI Global. doi:10.4018/978-1-61350-101-6.ch517

Kirlidog, M. (2010). Financial aspects of national ICT strategies. In S. Kamel (Ed.), *E-strategies for technological diffusion and adoption: National ICT approaches for socioeconomic development* (pp. 277–292). Hershey, PA: IGI Global. doi:10.4018/978-1-60566-388-3.ch016

Kisielnicki, J. (2006). Transfer of information and knowledge in the project management. In E. Coakes & S. Clarke (Eds.), *Encyclopedia of communities of practice in information and knowledge management* (pp. 544–551). Hershey, PA: IGI Global. doi:10.4018/978-1-59140-556-6.ch091

Kittner, M., & Van Slyke, C. (2006). Reorganizing information technology services in an academic environment. In M. Khosrow-Pour (Ed.), *Cases on the human side of information technology* (pp. 49–66). Hershey, PA: IGI Global. doi:10.4018/978-1-59904-405-7.ch004

Knoell, H. D. (2008). Semi virtual workplaces in German financial service enterprises. In P. Zemliansky & K. St.Amant (Eds.), *Handbook of research on virtual workplaces and the new nature of business practices* (pp. 570–581). Hershey, PA: IGI Global. doi:10.4018/978-1-59904-893-2.ch041

Koh, S. L., & Maguire, S. (2009). Competing in the age of information technology in a developing economy: Experiences of an Indian bank. In S. Koh & S. Maguire (Eds.), *Information and communication technologies management in turbulent business environments* (pp. 326–350). Hershey, PA: IGI Global. doi:10.4018/978-1-60566-424-8.ch018

Kollmann, T., & Häsel, M. (2009). Competence of information technology professionals in internet-based ventures. In I. Lee (Ed.), *Electronic business: Concepts, methodologies, tools, and applications* (pp. 1905–1919). Hershey, PA: IGI Global. doi:10.4018/978-1-60566-056-1.ch118

Kollmann, T., & Häsel, M. (2009). Competence of information technology professionals in internet-based ventures. In A. Cater-Steel (Ed.), *Information technology governance and service management: Frameworks and adaptations* (pp. 239–253). Hershey, PA: IGI Global. doi:10.4018/978-1-60566-008-0.ch013

Kollmann, T., & Häsel, M. (2010). Competence of information technology professionals in internet-based ventures. In *Electronic services: Concepts, methodologies, tools and applications* (pp. 1551–1565). Hershey, PA: IGI Global. doi:10.4018/978-1-61520-967-5.ch094

Kraemer, K., & King, J. L. (2006). Information technology and administrative reform: Will e-government be different? *International Journal of Electronic Government Research, 2*(1), 1–20. doi:10.4018/jegr.2006010101

Kraemer, K., & King, J. L. (2008). Information technology and administrative reform: Will e-government be different? In D. Norris (Ed.), *E-government research: Policy and management* (pp. 1–20). Hershey, PA: IGI Global. doi:10.4018/978-1-59904-913-7.ch001

Lampathaki, F., Tsiakaliaris, C., Stasis, A., & Charalabidis, Y. (2011). National interoperability frameworks: The way forward. In Y. Charalabidis (Ed.), *Interoperability in digital public services and administration: Bridging e-government and e-business* (pp. 1–24). Hershey, PA: IGI Global. doi:10.4018/978-1-61520-887-6.ch001

Lan, Z., & Scott, C. R. (1996). The relative importance of computer-mediated information versus conventional non-computer-mediated information in public managerial decision making. *Information Resources Management Journal, 9*(1), 27–0. doi:10.4018/irmj.1996010103

Law, W. (2004). *Public sector data management in a developing economy*. Hershey, PA: IGI Global. doi:10.4018/978-1-59140-259-6.ch034

Law, W. K. (2005). Information resources development challenges in a cross-cultural environment. In M. Khosrow-Pour (Ed.), *Encyclopedia of information science and technology* (pp. 1476–1481). Hershey, PA: IGI Global. doi:10.4018/978-1-59140-553-5.ch259

Law, W. K. (2009). Cross-cultural challenges for information resources management. In M. Khosrow-Pour (Ed.), *Encyclopedia of information science and technology* (2nd ed.; pp. 840–846). Hershey, PA: IGI Global. doi:10.4018/978-1-60566-026-4.ch136

Law, W. K. (2011). Cross-cultural challenges for information resources management. In *Global business: Concepts, methodologies, tools and applications* (pp. 1924–1932). Hershey, PA: IGI Global. doi:10.4018/978-1-60960-587-2.ch704

Malkia, M., & Savolainen, R. (2004). eTransformation in government, politics and society: Conceptual framework and introduction. In M. Malkia, A. Anttiroiko, & R. Savolainen (Eds.), eTransformation in governance: New directions in government and politics (pp. 1-21). Hershey, PA: IGI Global. doi:10.4018/978-1-59140-130-8.ch001

Mandujano, S. (2011). Network manageability security. In D. Kar & M. Syed (Eds.), *Network security, administration and management: Advancing technology and practice* (pp. 158–181). Hershey, PA: IGI Global. doi:10.4018/978-1-60960-777-7.ch009

Marich, M. J., Schooley, B. L., & Horan, T. A. (2012). A normative enterprise architecture for guiding end-to-end emergency response decision support. In M. Jennex (Ed.), *Managing crises and disasters with emerging technologies: Advancements* (pp. 71–87). Hershey, PA: IGI Global. doi:10.4018/978-1-4666-0167-3.ch006

Markov, R., & Okujava, S. (2008). Costs, benefits, and risks of e-government portals. In G. Putnik & M. Cruz-Cunha (Eds.), *Encyclopedia of networked and virtual organizations* (pp. 354–363). Hershey, PA: IGI Global. doi:10.4018/978-1-59904-885-7.ch047

Martin, N., & Rice, J. (2013). Evaluating and designing electronic government for the future: Observations and insights from Australia. In V. Weerakkody (Ed.), *E-government services design, adoption, and evaluation* (pp. 238–258). Hershey, PA: IGI Global. doi:10.4018/978-1-4666-2458-0.ch014

i. Martinez, A. C. (2008). Accessing administration's information via internet in Spain. In F. Tan (Ed.), *Global information technologies: Concepts, methodologies, tools, and applications* (pp. 2558–2573). Hershey, PA: IGI Global. doi:10.4018/978-1-59904-939-7.ch186

Mbarika, V. W., Meso, P. N., & Musa, P. F. (2006). A disconnect in stakeholders' perceptions from emerging realities of teledensity growth in Africa's least developed countries. In M. Hunter & F. Tan (Eds.), *Advanced topics in global information management* (Vol. 5, pp. 263–282). Hershey, PA: IGI Global. doi:10.4018/978-1-59140-923-6.ch012

Mbarika, V. W., Meso, P. N., & Musa, P. F. (2008). A disconnect in stakeholders' perceptions from emerging realities of teledensity growth in Africa's least developed countries. In F. Tan (Ed.), *Global information technologies: Concepts, methodologies, tools, and applications* (pp. 2948–2962). Hershey, PA: IGI Global. doi:10.4018/978-1-59904-939-7.ch209

Means, T., Olson, E., & Spooner, J. (2013). Discovering ways that don't work on the road to success: Strengths and weaknesses revealed by an active learning studio classroom project. In A. Benson, J. Moore, & S. Williams van Rooij (Eds.), *Cases on educational technology planning, design, and implementation: A project management perspective* (pp. 94–113). Hershey, PA: IGI Global. doi:10.4018/978-1-4666-4237-9.ch006

Melitski, J., Holzer, M., Kim, S., Kim, C., & Rho, S. (2008). Digital government worldwide: An e-government assessment of municipal web sites. In G. Garson & M. Khosrow-Pour (Eds.), *Handbook of research on public information technology* (pp. 790–804). Hershey, PA: IGI Global. doi:10.4018/978-1-59904-857-4.ch069

Memmola, M., Palumbo, G., & Rossini, M. (2009). Web & RFID technology: New frontiers in costing and process management for rehabilitation medicine. In L. Al-Hakim & M. Memmola (Eds.), *Business web strategy: Design, alignment, and application* (pp. 145–169). Hershey, PA: IGI Global. doi:10.4018/978-1-60566-024-0.ch008

Meng, Z., Fahong, Z., & Lei, L. (2008). Information technology and environment. In Y. Kurihara, S. Takaya, H. Harui, & H. Kamae (Eds.), *Information technology and economic development* (pp. 201–212). Hershey, PA: IGI Global. doi:10.4018/978-1-59904-579-5.ch014

Mentzingen de Moraes, A. J., Ferneda, E., Costa, I., & Spinola, M. D. (2011). Practical approach for implementation of governance process in IT: Information technology areas. In N. Shi & G. Silvius (Eds.), *Enterprise IT governance, business value and performance measurement* (pp. 19–40). Hershey, PA: IGI Global. doi:10.4018/978-1-60566-346-3.ch002

Merwin, G. A. Jr, McDonald, J. S., & Odera, L. C. (2008). Economic development: Government's cutting edge in IT. In M. Raisinghani (Ed.), *Handbook of research on global information technology management in the digital economy* (pp. 1–37). Hershey, PA: IGI Global. doi:10.4018/978-1-59904-875-8.ch001

Meso, P., & Duncan, N. (2002). Can national information infrastructures enhance social development in the least developed countries? An empirical investigation. In M. Dadashzadeh (Ed.), *Information technology management in developing countries* (pp. 23–51). Hershey, PA: IGI Global. doi:10.4018/978-1-931777-03-2.ch002

Meso, P. N., & Duncan, N. B. (2002). Can national information infrastructures enhance social development in the least developed countries? In F. Tan (Ed.), *Advanced topics in global information management* (Vol. 1, pp. 207–226). Hershey, PA: IGI Global. doi:10.4018/978-1-930708-43-3.ch014

Middleton, M. (2008). Evaluation of e-government web sites. In G. Garson & M. Khosrow-Pour (Eds.), *Handbook of research on public information technology* (pp. 699–710). Hershey, PA: IGI Global. doi:10.4018/978-1-59904-857-4.ch063

Mingers, J. (2010). Pluralism, realism, and truth: The keys to knowledge in information systems research. In D. Paradice (Ed.), *Emerging systems approaches in information technologies: Concepts, theories, and applications* (pp. 86–98). Hershey, PA: IGI Global. doi:10.4018/978-1-60566-976-2.ch006

Mital, K. M. (2012). ICT, unique identity and inclusive growth: An Indian perspective. In A. Manoharan & M. Holzer (Eds.), *E-governance and civic engagement: Factors and determinants of e-democracy* (pp. 584–612). Hershey, PA: IGI Global. doi:10.4018/978-1-61350-083-5.ch029

Mizell, A. P. (2008). Helping close the digital divide for financially disadvantaged seniors. In F. Tan (Ed.), *Global information technologies: Concepts, methodologies, tools, and applications* (pp. 2396–2402). Hershey, PA: IGI Global. doi:10.4018/978-1-59904-939-7.ch173

Molinari, F., Wills, C., Koumpis, A., & Moumtzi, V. (2011). A citizen-centric platform to support networking in the area of e-democracy. In H. Rahman (Ed.), *Cases on adoption, diffusion and evaluation of global e-governance systems: Impact at the grass roots* (pp. 282–302). Hershey, PA: IGI Global. doi:10.4018/978-1-61692-814-8.ch014

Molinari, F., Wills, C., Koumpis, A., & Moumtzi, V. (2013). A citizen-centric platform to support networking in the area of e-democracy. In H. Rahman (Ed.), *Cases on progressions and challenges in ICT utilization for citizen-centric governance* (pp. 265–297). Hershey, PA: IGI Global. doi:10.4018/978-1-4666-2071-1.ch013

Monteverde, F. (2010). The process of e-government public policy inclusion in the governmental agenda: A framework for assessment and case study. In J. Cordoba-Pachon & A. Ochoa-Arias (Eds.), *Systems thinking and e-participation: ICT in the governance of society* (pp. 233–245). Hershey, PA: IGI Global. doi:10.4018/978-1-60566-860-4.ch015

Moodley, S. (2008). Deconstructing the South African government's ICT for development discourse. In A. Anttiroiko (Ed.), *Electronic government: Concepts, methodologies, tools, and applications* (pp. 622–631). Hershey, PA: IGI Global. doi:10.4018/978-1-59904-947-2.ch053

Moodley, S. (2008). Deconstructing the South African government's ICT for development discourse. In C. Van Slyke (Ed.), *Information communication technologies: Concepts, methodologies, tools, and applications* (pp. 816–825). Hershey, PA: IGI Global. doi:10.4018/978-1-59904-949-6.ch052

Mora, M., Cervantes-Perez, F., Gelman-Muravchik, O., Forgionne, G. A., & Mejia-Olvera, M. (2003). DMSS implementation research: A conceptual analysis of the contributions and limitations of the factor-based and stage-based streams. In G. Forgionne, J. Gupta, & M. Mora (Eds.), *Decision-making support systems: Achievements and challenges for the new decade* (pp. 331–356). Hershey, PA: IGI Global. doi:10.4018/978-1-59140-045-5.ch020

Mörtberg, C., & Elovaara, P. (2010). Attaching people and technology: Between e and government. In S. Booth, S. Goodman, & G. Kirkup (Eds.), *Gender issues in learning and working with information technology: Social constructs and cultural contexts* (pp. 83–98). Hershey, PA: IGI Global. doi:10.4018/978-1-61520-813-5.ch005

Murphy, J., Harper, E., Devine, E. C., Burke, L. J., & Hook, M. L. (2011). Case study: Lessons learned when embedding evidence-based knowledge in a nurse care planning and documentation system. In A. Cashin & R. Cook (Eds.), *Evidence-based practice in nursing informatics: Concepts and applications* (pp. 174–190). Hershey, PA: IGI Global. doi:10.4018/978-1-60960-034-1.ch014

Mutula, S. M. (2013). E-government's role in poverty alleviation: Case study of South Africa. In H. Rahman (Ed.), *Cases on progressions and challenges in ICT utilization for citizen-centric governance* (pp. 44–68). Hershey, PA: IGI Global. doi:10.4018/978-1-4666-2071-1.ch003

Nath, R., & Angeles, R. (2005). Relationships between supply characteristics and buyer-supplier coupling in e-procurement: An empirical analysis. *International Journal of E-Business Research, 1*(2), 40–55. doi:10.4018/jebr.2005040103

Nissen, M. E. (2006). Application cases in government. In M. Nissen (Ed.), *Harnessing knowledge dynamics: Principled organizational knowing & learning* (pp. 152–181). Hershey, PA: IGI Global. doi:10.4018/978-1-59140-773-7.ch008

Norris, D. F. (2003). Leading-edge information technologies and American local governments. In G. Garson (Ed.), *Public information technology: Policy and management issues* (pp. 139–169). Hershey, PA: IGI Global. doi:10.4018/978-1-59140-060-8.ch007

Norris, D. F. (2008). Information technology among U.S. local governments. In G. Garson & M. Khosrow-Pour (Eds.), *Handbook of research on public information technology* (pp. 132–144). Hershey, PA: IGI Global. doi:10.4018/978-1-59904-857-4.ch013

Northrop, A. (1999). The challenge of teaching information technology in public administration graduate programs. In G. Garson (Ed.), *Information technology and computer applications in public administration: Issues and trends* (pp. 1–22). Hershey, PA: IGI Global. doi:10.4018/978-1-87828-952-0.ch001

Northrop, A. (2003). Information technology and public administration: The view from the profession. In G. Garson (Ed.), *Public information technology: Policy and management issues* (pp. 1–19). Hershey, PA: IGI Global. doi:10.4018/978-1-59140-060-8.ch001

Northrop, A. (2007). Lip service? How PA journals and textbooks view information technology. In G. Garson (Ed.), *Modern public information technology systems: Issues and challenges* (pp. 1–16). Hershey, PA: IGI Global. doi:10.4018/978-1-59904-051-6.ch001

Null, E. (2013). Legal and political barriers to municipal networks in the United States. In A. Abdelaal (Ed.), *Social and economic effects of community wireless networks and infrastructures* (pp. 27–56). Hershey, PA: IGI Global. doi:10.4018/978-1-4666-2997-4.ch003

Okunoye, A., Frolick, M., & Crable, E. (2006). ERP implementation in higher education: An account of pre-implementation and implementation phases. *Journal of Cases on Information Technology*, *8*(2), 110–132. doi:10.4018/jcit.2006040106

Olasina, G. (2012). A review of egovernment services in Nigeria. In A. Tella & A. Issa (Eds.), *Library and information science in developing countries: Contemporary issues* (pp. 205–221). Hershey, PA: IGI Global. doi:10.4018/978-1-61350-335-5.ch015

Orgeron, C. P. (2008). A model for reengineering IT job classes in state government. In G. Garson & M. Khosrow-Pour (Eds.), *Handbook of research on public information technology* (pp. 735–746). Hershey, PA: IGI Global. doi:10.4018/978-1-59904-857-4.ch066

Owsinski, J. W., & Pielak, A. M. (2011). Local authority websites in rural areas: Measuring quality and functionality, and assessing the role. In Z. Andreopoulou, B. Manos, N. Polman, & D. Viaggi (Eds.), *Agricultural and environmental informatics, governance and management: Emerging research applications* (pp. 39–60). Hershey, PA: IGI Global. doi:10.4018/978-1-60960-621-3.ch003

Owsiński, J. W., Pielak, A. M., Sęp, K., & Stańczak, J. (2014). Local web-based networks in rural municipalities: Extension, density, and meaning. In Z. Andreopoulou, V. Samathrakis, S. Louca, & M. Vlachopoulou (Eds.), *E-innovation for sustainable development of rural resources during global economic crisis* (pp. 126–151). Hershey, PA: IGI Global. doi:10.4018/978-1-4666-4550-9.ch011

Pagani, M., & Pasinetti, C. (2008). Technical and functional quality in the development of t-government services. In A. Anttiroiko (Ed.), *Electronic government: Concepts, methodologies, tools, and applications* (pp. 2943–2965). Hershey, PA: IGI Global. doi:10.4018/978-1-59904-947-2.ch220

Pani, A. K., & Agrahari, A. (2005). On e-markets in emerging economy: An Indian experience. In M. Khosrow-Pour (Ed.), *Advanced topics in electronic commerce* (Vol. 1, pp. 287–299). Hershey, PA: IGI Global. doi:10.4018/978-1-59140-819-2.ch015

Papadopoulos, T., Angelopoulos, S., & Kitsios, F. (2011). A strategic approach to e-health interoperability using e-government frameworks. In A. Lazakidou, K. Siassiakos, & K. Ioannou (Eds.), *Wireless technologies for ambient assisted living and healthcare: Systems and applications* (pp. 213–229). Hershey, PA: IGI Global. doi:10.4018/978-1-61520-805-0.ch012

Papadopoulos, T., Angelopoulos, S., & Kitsios, F. (2013). A strategic approach to e-health interoperability using e-government frameworks. In *User-driven healthcare: Concepts, methodologies, tools, and applications* (pp. 791–807). Hershey, PA: IGI Global. doi:10.4018/978-1-4666-2770-3.ch039

Papaleo, G., Chiarella, D., Aiello, M., & Caviglione, L. (2012). Analysis, development and deployment of statistical anomaly detection techniques for real e-mail traffic. In T. Chou (Ed.), *Information assurance and security technologies for risk assessment and threat management: Advances* (pp. 47–71). Hershey, PA: IGI Global. doi:10.4018/978-1-61350-507-6.ch003

Papp, R. (2003). Information technology & FDA compliance in the pharmaceutical industry. In M. Khosrow-Pour (Ed.), *Annals of cases on information technology* (Vol. 5, pp. 262–273). Hershey, PA: IGI Global. doi:10.4018/978-1-59140-061-5.ch017

Parsons, T. W. (2007). Developing a knowledge management portal. In A. Tatnall (Ed.), *Encyclopedia of portal technologies and applications* (pp. 223–227). Hershey, PA: IGI Global. doi:10.4018/978-1-59140-989-2.ch039

Passaris, C. E. (2007). Immigration and digital government. In A. Anttiroiko & M. Malkia (Eds.), *Encyclopedia of digital government* (pp. 988–994). Hershey, PA: IGI Global. doi:10.4018/978-1-59140-789-8.ch148

Pavlichev, A. (2004). The e-government challenge for public administration. In A. Pavlichev & G. Garson (Eds.), *Digital government: Principles and best practices* (pp. 276–290). Hershey, PA: IGI Global. doi:10.4018/978-1-59140-122-3.ch018

Penrod, J. I., & Harbor, A. F. (2000). Designing and implementing a learning organization-oriented information technology planning and management process. In L. Petrides (Ed.), *Case studies on information technology in higher education: Implications for policy and practice* (pp. 7–19). Hershey, PA: IGI Global. doi:10.4018/978-1-878289-74-2.ch001

Planas-Silva, M. D., & Joseph, R. C. (2011). Perspectives on the adoption of electronic resources for use in clinical trials. In M. Guah (Ed.), *Healthcare delivery reform and new technologies: Organizational initiatives* (pp. 19–28). Hershey, PA: IGI Global. doi:10.4018/978-1-60960-183-6.ch002

Pomazalová, N., & Rejman, S. (2013). The rationale behind implementation of new electronic tools for electronic public procurement. In N. Pomazalová (Ed.), *Public sector transformation processes and internet public procurement: Decision support systems* (pp. 85–117). Hershey, PA: Engineering Science Reference. doi:10.4018/978-1-4666-2665-2.ch006

Postorino, M. N. (2012). City competitiveness and airport: Information science perspective. In M. Bulu (Ed.), *City competitiveness and improving urban subsystems: Technologies and applications* (pp. 61–83). Hershey, PA: IGI Global. doi:10.4018/978-1-61350-174-0.ch004

Poupa, C. (2002). Electronic government in Switzerland: Priorities for 2001-2005 - Electronic voting and federal portal. In Å. Grönlund (Ed.), *Electronic government: Design, applications and management* (pp. 356–369). Hershey, PA: IGI Global. doi:10.4018/978-1-930708-19-8.ch017

Powell, S. R. (2010). Interdisciplinarity in telecommunications and networking. In *Networking and telecommunications: Concepts, methodologies, tools and applications* (pp. 33–40). Hershey, PA: IGI Global. doi:10.4018/978-1-60566-986-1.ch004

Priya, P. S., & Mathiyalagan, N. (2011). A study of the implementation status of two e-governance projects in land revenue administration in India. In M. Shareef, V. Kumar, U. Kumar, & Y. Dwivedi (Eds.), *Stakeholder adoption of e-government services: Driving and resisting factors* (pp. 214–230). Hershey, PA: IGI Global. doi:10.4018/978-1-60960-601-5.ch011

Prysby, C., & Prysby, N. (2000). Electronic mail, employee privacy and the workplace. In L. Janczewski (Ed.), *Internet and intranet security management: Risks and solutions* (pp. 251–270). Hershey, PA: IGI Global. doi:10.4018/978-1-878289-71-1.ch009

Prysby, C. L., & Prysby, N. D. (2003). Electronic mail in the public workplace: Issues of privacy and public disclosure. In G. Garson (Ed.), *Public information technology: Policy and management issues* (pp. 271–298). Hershey, PA: IGI Global. doi:10.4018/978-1-59140-060-8.ch012

Prysby, C. L., & Prysby, N. D. (2007). You have mail, but who is reading it? Issues of e-mail in the public workplace. In G. Garson (Ed.), *Modern public information technology systems: Issues and challenges* (pp. 312–336). Hershey, PA: IGI Global. doi:10.4018/978-1-59904-051-6.ch016

Radl, A., & Chen, Y. (2005). Computer security in electronic government: A state-local education information system. *International Journal of Electronic Government Research*, *1*(1), 79–99. doi:10.4018/jegr.2005010105

Rahman, H. (2008). Information dynamics in developing countries. In C. Van Slyke (Ed.), *Information communication technologies: Concepts, methodologies, tools, and applications* (pp. 104–114). Hershey, PA: IGI Global. doi:10.4018/978-1-59904-949-6.ch008

Ramanathan, J. (2009). Adaptive IT architecture as a catalyst for network capability in government. In P. Saha (Ed.), *Advances in government enterprise architecture* (pp. 149–172). Hershey, PA: IGI Global. doi:10.4018/978-1-60566-068-4.ch007

Ramos, I., & Berry, D. M. (2006). Social construction of information technology supporting work. In M. Khosrow-Pour (Ed.), *Cases on information technology: Lessons learned* (Vol. 7, pp. 36–52). Hershey, PA: IGI Global. doi:10.4018/978-1-59140-673-0.ch003

Ray, D., Gulla, U., Gupta, M. P., & Dash, S. S. (2009). Interoperability and constituents of interoperable systems in public sector. In V. Weerakkody, M. Janssen, & Y. Dwivedi (Eds.), *Handbook of research on ICT-enabled transformational government: A global perspective* (pp. 175–195). Hershey, PA: IGI Global. doi:10.4018/978-1-60566-390-6.ch010

Reddick, C. G. (2007). E-government and creating a citizen-centric government: A study of federal government CIOs. In G. Garson (Ed.), *Modern public information technology systems: Issues and challenges* (pp. 143–165). Hershey, PA: IGI Global. doi:10.4018/978-1-59904-051-6.ch008

Reddick, C. G. (2010). Citizen-centric e-government. In C. Reddick (Ed.), *Homeland security preparedness and information systems: Strategies for managing public policy* (pp. 45–75). Hershey, PA: IGI Global. doi:10.4018/978-1-60566-834-5.ch002

Reddick, C. G. (2010). E-government and creating a citizen-centric government: A study of federal government CIOs. In C. Reddick (Ed.), *Homeland security preparedness and information systems: Strategies for managing public policy* (pp. 230–250). Hershey, PA: IGI Global. doi:10.4018/978-1-60566-834-5.ch012

Reddick, C. G. (2010). Perceived effectiveness of e-government and its usage in city governments: Survey evidence from information technology directors. In C. Reddick (Ed.), *Homeland security preparedness and information systems: Strategies for managing public policy* (pp. 213–229). Hershey, PA: IGI Global. doi:10.4018/978-1-60566-834-5.ch011

Reddick, C. G. (2012). Customer relationship management adoption in local governments in the United States. In S. Chhabra & M. Kumar (Eds.), *Strategic enterprise resource planning models for e-government: Applications and methodologies* (pp. 111–124). Hershey, PA: IGI Global. doi:10.4018/978-1-60960-863-7.ch008

Reeder, F. S., & Pandy, S. M. (2008). Identifying effective funding models for e-government. In A. Anttiroiko (Ed.), *Electronic government: Concepts, methodologies, tools, and applications* (pp. 1108–1138). Hershey, PA: IGI Global. doi:10.4018/978-1-59904-947-2.ch083

Riesco, D., Acosta, E., & Montejano, G. (2003). An extension to a UML activity graph from workflow. In L. Favre (Ed.), *UML and the unified process* (pp. 294–314). Hershey, PA: IGI Global. doi:10.4018/978-1-93177-744-5.ch015

Ritzhaupt, A. D., & Gill, T. G. (2008). A hybrid and novel approach to teaching computer programming in MIS curriculum. In S. Negash, M. Whitman, A. Woszczynski, K. Hoganson, & H. Mattord (Eds.), *Handbook of distance learning for real-time and asynchronous information technology education* (pp. 259–281). Hershey, PA: IGI Global. doi:10.4018/978-1-59904-964-9.ch014

Roche, E. M. (1993). International computing and the international regime. *Journal of Global Information Management, 1*(2), 33–44. doi:10.4018/jgim.1993040103

Rocheleau, B. (2007). Politics, accountability, and information management. In G. Garson (Ed.), *Modern public information technology systems: Issues and challenges* (pp. 35–71). Hershey, PA: IGI Global. doi:10.4018/978-1-59904-051-6.ch003

Rodrigues Filho, J. (2010). E-government in Brazil: Reinforcing dominant institutions or reducing citizenship? In C. Reddick (Ed.), *Politics, democracy and e-government: Participation and service delivery* (pp. 347–362). Hershey, PA: IGI Global. doi:10.4018/978-1-61520-933-0.ch021

Rodriguez, S. R., & Thorp, D. A. (2013). eLearning for industry: A case study of the project management process. In A. Benson, J. Moore, & S. Williams van Rooij (Eds.), Cases on educational technology planning, design, and implementation: A project management perspective (pp. 319-342). Hershey, PA: IGI Global. doi:10.4018/978-1-4666-4237-9.ch017

Roman, A. V. (2013). Delineating three dimensions of e-government success: Security, functionality, and transformation. In J. Gil-Garcia (Ed.), *E-government success factors and measures: Theories, concepts, and methodologies* (pp. 171–192). Hershey, PA: IGI Global. doi:10.4018/978-1-4666-4058-0.ch010

Ross, S. C., Tyran, C. K., & Auer, D. J. (2008). Up in smoke: Rebuilding after an IT disaster. In H. Nemati (Ed.), *Information security and ethics: Concepts, methodologies, tools, and applications* (pp. 3659–3675). Hershey, PA: IGI Global. doi:10.4018/978-1-59904-937-3.ch248

Ross, S. C., Tyran, C. K., Auer, D. J., Junell, J. M., & Williams, T. G. (2005). Up in smoke: Rebuilding after an IT disaster. *Journal of Cases on Information Technology, 7*(2), 31–49. doi:10.4018/jcit.2005040103

Roy, J. (2008). Security, sovereignty, and continental interoperability: Canada's elusive balance. In T. Loendorf & G. Garson (Eds.), *Patriotic information systems* (pp. 153–176). Hershey, PA: IGI Global. doi:10.4018/978-1-59904-594-8.ch007

Rubeck, R. F., & Miller, G. A. (2009). vGOV: Remote video access to government services. In A. Scupola (Ed.), Cases on managing e-services (pp. 253-268). Hershey, PA: IGI Global. doi:10.4018/978-1-60566-064-6.ch017

Saekow, A., & Boonmee, C. (2011). The challenges of implementing e-government interoperability in Thailand: Case of official electronic correspondence letters exchange across government departments. In Y. Charalabidis (Ed.), *Interoperability in digital public services and administration: Bridging e-government and e-business* (pp. 40–61). Hershey, PA: IGI Global. doi:10.4018/978-1-61520-887-6.ch003

Saekow, A., & Boonmee, C. (2012). The challenges of implementing e-government interoperability in Thailand: Case of official electronic correspondence letters exchange across government departments. In *Digital democracy: Concepts, methodologies, tools, and applications* (pp. 1883–1905). Hershey, PA: IGI Global. doi:10.4018/978-1-4666-1740-7.ch094

Sagsan, M., & Medeni, T. (2012). Understanding "knowledge management (KM) paradigms" from social media perspective: An empirical study on discussion group for KM at professional networking site. In M. Cruz-Cunha, P. Gonçalves, N. Lopes, E. Miranda, & G. Putnik (Eds.), *Handbook of research on business social networking: Organizational, managerial, and technological dimensions* (pp. 738–755). Hershey, PA: IGI Global. doi:10.4018/978-1-61350-168-9.ch039

Sahi, G., & Madan, S. (2013). Information security threats in ERP enabled e-governance: Challenges and solutions. In *Enterprise resource planning: Concepts, methodologies, tools, and applications* (pp. 825–837). Hershey, PA: IGI Global. doi:10.4018/978-1-4666-4153-2.ch048

Sanford, C., & Bhattacherjee, A. (2008). IT implementation in a developing country municipality: A sociocognitive analysis. *International Journal of Technology and Human Interaction, 4*(3), 68–93. doi:10.4018/jthi.2008070104

Schelin, S. H. (2003). E-government: An overview. In G. Garson (Ed.), *Public information technology: Policy and management issues* (pp. 120–138). Hershey, PA: IGI Global. doi:10.4018/978-1-59140-060-8.ch006

Schelin, S. H. (2004). Training for digital government. In A. Pavlichev & G. Garson (Eds.), *Digital government: Principles and best practices* (pp. 263–275). Hershey, PA: IGI Global. doi:10.4018/978-1-59140-122-3.ch017

Schelin, S. H. (2007). E-government: An overview. In G. Garson (Ed.), *Modern public information technology systems: Issues and challenges* (pp. 110–126). Hershey, PA: IGI Global. doi:10.4018/978-1-59904-051-6.ch006

Schelin, S. H., & Garson, G. (2004). Theoretical justification of critical success factors. In G. Garson & S. Schelin (Eds.), *IT solutions series: Humanizing information technology: Advice from experts* (pp. 4–15). Hershey, PA: IGI Global. doi:10.4018/978-1-59140-245-9.ch002

Scime, A. (2002). Information systems and computer science model curricula: A comparative look. In M. Dadashzadeh, A. Saber, & S. Saber (Eds.), *Information technology education in the new millennium* (pp. 146–158). Hershey, PA: IGI Global. doi:10.4018/978-1-931777-05-6.ch018

Scime, A. (2009). Computing curriculum analysis and development. In M. Khosrow-Pour (Ed.), *Encyclopedia of information science and technology* (2nd ed.; pp. 667–671). Hershey, PA: IGI Global. doi:10.4018/978-1-60566-026-4.ch108

Scime, A., & Wania, C. (2008). Computing curricula: A comparison of models. In C. Van Slyke (Ed.), *Information communication technologies: Concepts, methodologies, tools, and applications* (pp. 1270–1283). Hershey, PA: IGI Global. doi:10.4018/978-1-59904-949-6.ch088

Seidman, S. B. (2009). An international perspective on professional software engineering credentials. In H. Ellis, S. Demurjian, & J. Naveda (Eds.), *Software engineering: Effective teaching and learning approaches and practices* (pp. 351–361). Hershey, PA: IGI Global. doi:10.4018/978-1-60566-102-5.ch018

Seifert, J. W. (2007). E-government act of 2002 in the United States. In A. Anttiroiko & M. Malkia (Eds.), *Encyclopedia of digital government* (pp. 476–481). Hershey, PA: IGI Global. doi:10.4018/978-1-59140-789-8.ch072

Seifert, J. W., & Relyea, H. C. (2008). E-government act of 2002 in the United States. In A. Anttiroiko (Ed.), *Electronic government: Concepts, methodologies, tools, and applications* (pp. 154–161). Hershey, PA: IGI Global. doi:10.4018/978-1-59904-947-2.ch013

Seufert, S. (2002). E-learning business models: Framework and best practice examples. In M. Raisinghani (Ed.), *Cases on worldwide e-commerce: Theory in action* (pp. 70–94). Hershey, PA: IGI Global. doi:10.4018/978-1-930708-27-3.ch004

Shareef, M. A., & Archer, N. (2012). E-government service development. In M. Shareef, N. Archer, & S. Dutta (Eds.), *E-government service maturity and development: Cultural, organizational and technological perspectives* (pp. 1–14). Hershey, PA: IGI Global. doi:10.4018/978-1-60960-848-4.ch001

Shareef, M. A., & Archer, N. (2012). E-government initiatives: Review studies on different countries. In M. Shareef, N. Archer, & S. Dutta (Eds.), *E-government service maturity and development: Cultural, organizational and technological perspectives* (pp. 40–76). Hershey, PA: IGI Global. doi:10.4018/978-1-60960-848-4.ch003

Shareef, M. A., Kumar, U., & Kumar, V. (2011). E-government development: Performance evaluation parameters. In M. Shareef, V. Kumar, U. Kumar, & Y. Dwivedi (Eds.), *Stakeholder adoption of e-government services: Driving and resisting factors* (pp. 197–213). Hershey, PA: IGI Global. doi:10.4018/978-1-60960-601-5.ch010

Shareef, M. A., Kumar, U., Kumar, V., & Niktash, M. (2012). Electronic-government vision: Case studies for objectives, strategies, and initiatives. In M. Shareef, N. Archer, & S. Dutta (Eds.), *E-government service maturity and development: Cultural, organizational and technological perspectives* (pp. 15–39). Hershey, PA: IGI Global. doi:10.4018/978-1-60960-848-4.ch002

Shukla, P., Kumar, A., & Anu Kumar, P. B. (2013). Impact of national culture on business continuity management system implementation. *International Journal of Risk and Contingency Management*, *2*(3), 23–36. doi:10.4018/ijrcm.2013070102

Shulman, S. W. (2007). The federal docket management system and the prospect for digital democracy in U S rulemaking. In G. Garson (Ed.), *Modern public information technology systems: Issues and challenges* (pp. 166–184). Hershey, PA: IGI Global. doi:10.4018/978-1-59904-051-6.ch009

Simonovic, S. (2007). Problems of offline government in e-Serbia. In A. Anttiroiko & M. Malkia (Eds.), *Encyclopedia of digital government* (pp. 1342–1351). Hershey, PA: IGI Global. doi:10.4018/978-1-59140-789-8.ch205

Simonovic, S. (2008). Problems of offline government in e-Serbia. In A. Anttiroiko (Ed.), *Electronic government: Concepts, methodologies, tools, and applications* (pp. 2929–2942). Hershey, PA: IGI Global. doi:10.4018/978-1-59904-947-2.ch219

Singh, A. M. (2005). Information systems and technology in South Africa. In M. Khosrow-Pour (Ed.), *Encyclopedia of information science and technology* (pp. 1497–1502). Hershey, PA: IGI Global. doi:10.4018/978-1-59140-553-5.ch263

Singh, S., & Naidoo, G. (2005). Towards an e-government solution: A South African perspective. In W. Huang, K. Siau, & K. Wei (Eds.), *Electronic government strategies and implementation* (pp. 325–353). Hershey, PA: IGI Global. doi:10.4018/978-1-59140-348-7.ch014

Snoke, R., & Underwood, A. (2002). Generic attributes of IS graduates: An analysis of Australian views. In F. Tan (Ed.), *Advanced topics in global information management* (Vol. 1, pp. 370–384). Hershey, PA: IGI Global. doi:10.4018/978-1-930708-43-3.ch023

Sommer, L. (2006). Revealing unseen organizations in higher education: A study framework and application example. In A. Metcalfe (Ed.), *Knowledge management and higher education: A critical analysis* (pp. 115–146). Hershey, PA: IGI Global. doi:10.4018/978-1-59140-509-2.ch007

Song, H., Kidd, T., & Owens, E. (2011). Examining technological disparities and instructional practices in English language arts classroom: Implications for school leadership and teacher training. In L. Tomei (Ed.), *Online courses and ICT in education: Emerging practices and applications* (pp. 258–274). Hershey, PA: IGI Global. doi:10.4018/978-1-60960-150-8.ch020

Speaker, P. J., & Kleist, V. F. (2003). Using information technology to meet electronic commerce and MIS education demands. In A. Aggarwal (Ed.), *Web-based education: Learning from experience* (pp. 280–291). Hershey, PA: IGI Global. doi:10.4018/978-1-59140-102-5.ch017

Spitler, V. K. (2007). Learning to use IT in the workplace: Mechanisms and masters. In M. Mahmood (Ed.), *Contemporary issues in end user computing* (pp. 292–323). Hershey, PA: IGI Global. doi:10.4018/978-1-59140-926-7.ch013

Stellefson, M. (2011). Considerations for marketing distance education courses in health education: Five important questions to examine before development. In U. Demiray & S. Sever (Eds.), *Marketing online education programs: Frameworks for promotion and communication* (pp. 222–234). Hershey, PA: IGI Global. doi:10.4018/978-1-60960-074-7.ch014

Straub, D. W., & Loch, K. D. (2006). Creating and developing a program of global research. *Journal of Global Information Management, 14*(2), 1–28. doi:10.4018/jgim.2006040101

Straub, D. W., Loch, K. D., & Hill, C. E. (2002). Transfer of information technology to the Arab world: A test of cultural influence modeling. In M. Dadashzadeh (Ed.), *Information technology management in developing countries* (pp. 92–134). Hershey, PA: IGI Global. doi:10.4018/978-1-931777-03-2.ch005

Straub, D. W., Loch, K. D., & Hill, C. E. (2003). Transfer of information technology to the Arab world: A test of cultural influence modeling. In F. Tan (Ed.), *Advanced topics in global information management* (Vol. 2, pp. 141–172). Hershey, PA: IGI Global. doi:10.4018/978-1-59140-064-6.ch009

Suki, N. M., Ramayah, T., Ming, M. K., & Suki, N. M. (2013). Factors enhancing employed job seekers intentions to use social networking sites as a job search tool. In A. Mesquita (Ed.), *User perception and influencing factors of technology in everyday life* (pp. 265–281). Hershey, PA: IGI Global. doi:10.4018/978-1-4666-1954-8.ch018

Suomi, R. (2006). Introducing electronic patient records to hospitals: Innovation adoption paths. In T. Spil & R. Schuring (Eds.), *E-health systems diffusion and use: The innovation, the user and the use IT model* (pp. 128–146). Hershey, PA: IGI Global. doi:10.4018/978-1-59140-423-1.ch008

Swim, J., & Barker, L. (2012). Pathways into a gendered occupation: Brazilian women in IT. *International Journal of Social and Organizational Dynamics in IT, 2*(4), 34–51. doi:10.4018/ijsodit.2012100103

Tarafdar, M., & Vaidya, S. D. (2006). Adoption and implementation of IT in developing nations: Experiences from two public sector enterprises in India. In M. Khosrow-Pour (Ed.), *Cases on information technology planning, design and implementation* (pp. 208–233). Hershey, PA: IGI Global. doi:10.4018/978-1-59904-408-8.ch013

Tarafdar, M., & Vaidya, S. D. (2008). Adoption and implementation of IT in developing nations: Experiences from two public sector enterprises in India. In G. Garson & M. Khosrow-Pour (Eds.), *Handbook of research on public information technology* (pp. 905–924). Hershey, PA: IGI Global. doi:10.4018/978-1-59904-857-4.ch076

Thesing, Z. (2007). Zarina thesing, pumpkin patch. In M. Hunter (Ed.), *Contemporary chief information officers: Management experiences* (pp. 83–94). Hershey, PA: IGI Global. doi:10.4018/978-1-59904-078-3.ch007

Thomas, J. C. (2004). Public involvement in public administration in the information age: Speculations on the effects of technology. In M. Malkia, A. Anttiroiko, & R. Savolainen (Eds.), *eTransformation in governance: New directions in government and politics* (pp. 67–84). Hershey, PA: IGI Global. doi:10.4018/978-1-59140-130-8.ch004

Treiblmaier, H., & Chong, S. (2013). Trust and perceived risk of personal information as antecedents of online information disclosure: Results from three countries. In F. Tan (Ed.), *Global diffusion and adoption of technologies for knowledge and information sharing* (pp. 341–361). Hershey, PA: IGI Global. doi:10.4018/978-1-4666-2142-8.ch015

van Grembergen, W., & de Haes, S. (2008). IT governance in practice: Six case studies. In W. van Grembergen & S. De Haes (Eds.), *Implementing information technology governance: Models, practices and cases* (pp. 125–237). Hershey, PA: IGI Global. doi:10.4018/978-1-59904-924-3.ch004

van Os, G., Homburg, V., & Bekkers, V. (2013). Contingencies and convergence in European social security: ICT coordination in the back office of the welfare state. In M. Cruz-Cunha, I. Miranda, & P. Gonçalves (Eds.), *Handbook of research on ICTs and management systems for improving efficiency in healthcare and social care* (pp. 268–287). Hershey, PA: IGI Global. doi:10.4018/978-1-4666-3990-4.ch013

Velloso, A. B., Gassenferth, W., & Machado, M. A. (2012). Evaluating IBMEC-RJ's intranet usability using fuzzy logic. In M. Cruz-Cunha, P. Gonçalves, N. Lopes, E. Miranda, & G. Putnik (Eds.), *Handbook of research on business social networking: Organizational, managerial, and technological dimensions* (pp. 185–205). Hershey, PA: IGI Global. doi:10.4018/978-1-61350-168-9.ch010

Villablanca, A. C., Baxi, H., & Anderson, K. (2009). Novel data interface for evaluating cardiovascular outcomes in women. In A. Dwivedi (Ed.), *Handbook of research on information technology management and clinical data administration in healthcare* (pp. 34–53). Hershey, PA: IGI Global. doi:10.4018/978-1-60566-356-2.ch003

Villablanca, A. C., Baxi, H., & Anderson, K. (2011). Novel data interface for evaluating cardiovascular outcomes in women. In *Clinical technologies: Concepts, methodologies, tools and applications* (pp. 2094–2113). Hershey, PA: IGI Global. doi:10.4018/978-1-60960-561-2.ch806

Virkar, S. (2011). Information and communication technologies in administrative reform for development: Exploring the case of property tax systems in Karnataka, India. In J. Steyn, J. Van Belle, & E. Mansilla (Eds.), *ICTs for global development and sustainability: Practice and applications* (pp. 127–149). Hershey, PA: IGI Global. doi:10.4018/978-1-61520-997-2.ch006

Virkar, S. (2013). Designing and implementing e-government projects: Actors, influences, and fields of play. In S. Saeed & C. Reddick (Eds.), *Human-centered system design for electronic governance* (pp. 88–110). Hershey, PA: IGI Global. doi:10.4018/978-1-4666-3640-8.ch007

Wallace, A. (2009). E-justice: An Australian perspective. In A. Martínez & P. Abat (Eds.), *E-justice: Using information communication technologies in the court system* (pp. 204–228). Hershey, PA: IGI Global. doi:10.4018/978-1-59904-998-4.ch014

Wang, G. (2012). E-democratic administration and bureaucratic responsiveness: A primary study of bureaucrats' perceptions of the civil service e-mail box in Taiwan. In K. Kloby & M. D'Agostino (Eds.), *Citizen 2.0: Public and governmental interaction through web 2.0 technologies* (pp. 146–173). Hershey, PA: IGI Global. doi:10.4018/978-1-4666-0318-9.ch009

Wangpipatwong, S., Chutimaskul, W., & Papasratorn, B. (2011). Quality enhancing the continued use of e-government web sites: Evidence from e-citizens of Thailand. In V. Weerakkody (Ed.), *Applied technology integration in governmental organizations: New e-government research* (pp. 20–36). Hershey, PA: IGI Global. doi:10.4018/978-1-60960-162-1.ch002

Wedemeijer, L. (2006). Long-term evolution of a conceptual schema at a life insurance company. In M. Khosrow-Pour (Ed.), *Cases on database technologies and applications* (pp. 202–226). Hershey, PA: IGI Global. doi:10.4018/978-1-59904-399-9.ch012

Whybrow, E. (2008). Digital access, ICT fluency, and the economically disadvantages: Approaches to minimize the digital divide. In F. Tan (Ed.), *Global information technologies: Concepts, methodologies, tools, and applications* (pp. 1409–1422). Hershey, PA: IGI Global. doi:10.4018/978-1-59904-939-7.ch102

Whybrow, E. (2008). Digital access, ICT fluency, and the economically disadvantages: Approaches to minimize the digital divide. In C. Van Slyke (Ed.), *Information communication technologies: Concepts, methodologies, tools, and applications* (pp. 764–777). Hershey, PA: IGI Global. doi:10.4018/978-1-59904-949-6.ch049

Wickramasinghe, N., & Geisler, E. (2010). Key considerations for the adoption and implementation of knowledge management in healthcare operations. In M. Saito, N. Wickramasinghe, M. Fuji, & E. Geisler (Eds.), *Redesigning innovative healthcare operation and the role of knowledge management* (pp. 125–142). Hershey, PA: IGI Global. doi:10.4018/978-1-60566-284-8.ch009

Wickramasinghe, N., & Geisler, E. (2012). Key considerations for the adoption and implementation of knowledge management in healthcare operations. In *Organizational learning and knowledge: Concepts, methodologies, tools and applications* (pp. 1316–1328). Hershey, PA: IGI Global. doi:10.4018/978-1-60960-783-8.ch405

Wickramasinghe, N., & Goldberg, S. (2007). A framework for delivering m-health excellence. In L. Al-Hakim (Ed.), *Web mobile-based applications for healthcare management* (pp. 36–61). Hershey, PA: IGI Global. doi:10.4018/978-1-59140-658-7.ch002

Wickramasinghe, N., & Goldberg, S. (2008). Critical success factors for delivering m-health excellence. In N. Wickramasinghe & E. Geisler (Eds.), *Encyclopedia of healthcare information systems* (pp. 339–351). Hershey, PA: IGI Global. doi:10.4018/978-1-59904-889-5.ch045

Wyld, D. (2009). Radio frequency identification (RFID) technology. In J. Symonds, J. Ayoade, & D. Parry (Eds.), *Auto-identification and ubiquitous computing applications* (pp. 279–293). Hershey, PA: IGI Global. doi:10.4018/978-1-60566-298-5.ch017

Yaghmaei, F. (2010). Understanding computerised information systems usage in community health. In J. Rodrigues (Ed.), *Health information systems: Concepts, methodologies, tools, and applications* (pp. 1388–1399). Hershey, PA: IGI Global. doi:10.4018/978-1-60566-988-5.ch088

Yee, G., El-Khatib, K., Korba, L., Patrick, A. S., Song, R., & Xu, Y. (2005). Privacy and trust in e-government. In W. Huang, K. Siau, & K. Wei (Eds.), *Electronic government strategies and implementation* (pp. 145–190). Hershey, PA: IGI Global. doi:10.4018/978-1-59140-348-7.ch007

Yeh, S., & Chu, P. (2010). Evaluation of e-government services: A citizen-centric approach to citizen e-complaint services. In C. Reddick (Ed.), *Citizens and e-government: Evaluating policy and management* (pp. 400–417). Hershey, PA: IGI Global. doi:10.4018/978-1-61520-931-6.ch022

Young-Jin, S., & Seang-tae, K. (2008). E-government concepts, measures, and best practices. In A. Anttiroiko (Ed.), *Electronic government: Concepts, methodologies, tools, and applications* (pp. 32–57). Hershey, PA: IGI Global. doi:10.4018/978-1-59904-947-2.ch004

Yun, H. J., & Opheim, C. (2012). New technology communication in American state governments: The impact on citizen participation. In K. Bwalya & S. Zulu (Eds.), *Handbook of research on e-government in emerging economies: Adoption, e-participation, and legal frameworks* (pp. 573–590). Hershey, PA: IGI Global. doi:10.4018/978-1-4666-0324-0.ch029

Zhang, N., Guo, X., Chen, G., & Chau, P. Y. (2011). User evaluation of e-government systems: A Chinese cultural perspective. In F. Tan (Ed.), *International enterprises and global information technologies: Advancing management practices* (pp. 63–84). Hershey, PA: IGI Global. doi:10.4018/978-1-60960-605-3.ch004

Zuo, Y., & Hu, W. (2011). Trust-based information risk management in a supply chain network. In J. Wang (Ed.), *Supply chain optimization, management and integration: Emerging applications* (pp. 181–196). Hershey, PA: IGI Global. doi:10.4018/978-1-60960-135-5.ch013

Compilation of References

Accessibility. (2015, December 5). In *Wikipedia, the free encyclopedia*. Retrieved from https://en.wikipedia.org/w/index.php?title=Accessibility&oldid=693827449

Ackerman, E. (2015). 4-D Light Field Displays are exactly what Virtual Reality needs. *IEEE Spectrum*.

Acquaviva, A., Benini, L., & Riccó, B. (2001, December). Energy characterization of embedded real-time operating systems. *SIGARCH Comput. Archit. News*, *29*(5), 13–18. doi:10.1145/563647.563652

Adeyemo, O. (2013). The nationwide health information network: A biometric approach to prevent medical identity theft. In *User-driven healthcare: Concepts, methodologies, tools, and applications* (pp. 1636–1649). Hershey, PA: IGI Global. doi:10.4018/978-1-4666-2770-3.ch081

Adler, M., & Henman, P. (2009). Justice beyond the courts: The implications of computerisation for procedural justice in social security. In A. Martínez & P. Abat (Eds.), *E-justice: Using information communication technologies in the court system* (pp. 65–86). Hershey, PA: IGI Global. doi:10.4018/978-1-59904-998-4.ch005

Aflalo, E., & Gabay, E. (2013). An information system for coping with student dropout. In L. Tomei (Ed.), *Learning tools and teaching approaches through ICT advancements* (pp. 176–187). Hershey, PA: IGI Global. doi:10.4018/978-1-4666-2017-9.ch016

Afyf, A., Bellarbi, L., Riouch, F., Achour, A., Errachid, A., & Sennouni, M. A. (2015). Antenna for Wireless Body Area Network (WBAN) Applications Flexible Miniaturized UWB CPW II- shaped Slot. In *IEEE Third International Workshop on RFID And Adaptive Wireless Sensor Networks* (pp. 52–56).

Ahmed, M. A., Janssen, M., & van den Hoven, J. (2012). Value sensitive transfer (VST) of systems among countries: Towards a framework. *International Journal of Electronic Government Research*, *8*(1), 26–42. doi:10.4018/jegr.2012010102

Aikins, S. K. (2008). Issues and trends in internet-based citizen participation. In G. Garson & M. Khosrow-Pour (Eds.), *Handbook of research on public information technology* (pp. 31–40). Hershey, PA: IGI Global. doi:10.4018/978-1-59904-857-4.ch004

Aikins, S. K. (2009). A comparative study of municipal adoption of internet-based citizen participation. In C. Reddick (Ed.), *Handbook of research on strategies for local e-government adoption and implementation: Comparative studies* (pp. 206–230). Hershey, PA: IGI Global. doi:10.4018/978-1-60566-282-4.ch011

Aikins, S. K. (2012). Improving e-government project management: Best practices and critical success factors. In *Digital democracy: Concepts, methodologies, tools, and applications* (pp. 1314–1332). Hershey, PA: IGI Global. doi:10.4018/978-1-4666-1740-7.ch065

Akabawi, M. S. (2011). Ghabbour group ERP deployment: Learning from past technology failures. In E. Business Research and Case Center (Ed.), Cases on business and management in the MENA region: New trends and opportunities (pp. 177-203). Hershey, PA: IGI Global. doi:10.4018/978-1-60960-583-4.ch012

Akabawi, M. S. (2013). Ghabbour group ERP deployment: Learning from past technology failures. In *Industrial engineering: Concepts, methodologies, tools, and applications* (pp. 933–958). Hershey, PA: Engineering Science Reference. doi:10.4018/978-1-4666-1945-6.ch051

Akbulut, A. Y., & Motwani, J. (2008). Integration and information sharing in e-government. In G. Putnik & M. Cruz-Cunha (Eds.), *Encyclopedia of networked and virtual organizations* (pp. 729–734). Hershey, PA: IGI Global. doi:10.4018/978-1-59904-885-7.ch096

Akers, E. J. (2008). Technology diffusion in public administration. In G. Garson & M. Khosrow-Pour (Eds.), *Handbook of research on public information technology* (pp. 339–348). Hershey, PA: IGI Global. doi:10.4018/978-1-59904-857-4.ch033

Aladwani, A. M. (2002). Organizational actions, computer attitudes and end-user satisfaction in public organizations: An empirical study. In C. Snodgrass & E. Szewczak (Eds.), *Human factors in information systems* (pp. 153–168). Hershey, PA: IGI Global. doi:10.4018/978-1-931777-10-0.ch012

Aladwani, A. M. (2002). Organizational actions, computer attitudes, and end-user satisfaction in public organizations: An empirical study. *Journal of Organizational and End User Computing*, *14*(1), 42–49. doi:10.4018/joeuc.2002010104

Albinali, F., Intille, S., Haskell, W., & Rosenberger, M. (2010). Using wearable activity type detection to improve physical activity energy expenditure estimation. *Proceedings of the 12th ACM International Conference on Ubiquitous Computing, Ubicomp* '10. doi:10.1145/1864349.1864396

Alipour, S., Parvaresh, F., Ghajari, H., & Kimball, D. F. (2010). Propagation Characteristics for a 60 GHz Wireless Body Area Network (WBAN). In *The 2010 Military Communications Conference - Unclassified Program - Waveforms and Signal Processing Track* (pp. 719–723).

Allen, B., Juillet, L., Paquet, G., & Roy, J. (2005). E-government and private-public partnerships: Relational challenges and strategic directions. In M. Khosrow-Pour (Ed.), *Practicing e-government: A global perspective* (pp. 364–382). Hershey, PA: IGI Global. doi:10.4018/978-1-59140-637-2.ch016

Almeida, S. (2012, January). Análise epidemiológica do Acidente Vascular Cerebral no Brasil. *Revista Neurociências*, *20*, 481–482. doi:10.4181/RNC.2012.20.483ed.2p

Alomainy, A., Hao, Y., & Pasveer, F. (2007). Numerical and experimental evaluation of a compact sensor antenna for healthcare devices. *IEEE Transactions on Biomedical Circuits and Systems*, *1*(4), 242–249. doi:10.1109/TBCAS.2007.913127 PMID:23852005

Alomainy, A., Sani, A., Rahman, A., Santas, J. G., & Hao, Y. (2009). Transient characteristics of wearable antennas and radio propagation channels for ultrawideband body-centric wireless communications. *IEEE Transactions on Antennas and Propagation*, *57*(4 part 1), 875–884. doi:10.1109/TAP.2009.2014588

Al-Rahayfeh, A., & Faezipour, M. (2013). Enhanced frame rate for real-time eye tracking using circular Hough Transform. In *IEEE Long Island Systems, Applications and Technology Conference* (pp. 1–6). Long Island, NY: IEEE. doi:10.1109/LISAT.2013.6578214

Al-Shafi, S. (2008). Free wireless internet park services: An investigation of technology adoption in Qatar from a citizens perspective. *Journal of Cases on Information Technology*, *10*(3), 21–34. doi:10.4018/jcit.2008070103

Al-Shafi, S., & Weerakkody, V. (2009). Implementing free wi-fi in public parks: An empirical study in Qatar. *International Journal of Electronic Government Research*, *5*(3), 21–35. doi:10.4018/jegr.2009070102

Alshawaf, A., & Knalil, O. E. (2008). IS success factors and IS organizational impact: Does ownership type matter in Kuwait? *International Journal of Enterprise Information Systems*, *4*(2), 13–33. doi:10.4018/jeis.2008040102

Altun K. & Barshan B. (2012). Pedestrian dead reckoning employing simultaneous activity recognition cues. *Measurement Science and Technology*, *23*(2), 1-20.

Amal, A., Larbi, B., & Anouar, A. (2016). Miniaturized Wideband Flexible CPW Antenna with Hexagonal Ring Slots for Early Breast Cancer Detection. In S. S. M. Singapore (Ed.), *Ubiquitous Networking* (pp. 211–222). Springer. doi:10.1007/978-981-287-990-5_17

Ambali, A. R. (2009). Digital divide and its implication on Malaysian e-government: Policy initiatives. In H. Rahman (Ed.), *Social and political implications of data mining: Knowledge management in e-government* (pp. 267–287). Hershey, PA: IGI Global. doi:10.4018/978-1-60566-230-5.ch016

Amoretti, F. (2007). Digital international governance. In A. Anttiroiko & M. Malkia (Eds.), *Encyclopedia of digital government* (pp. 365–370). Hershey, PA: IGI Global. doi:10.4018/978-1-59140-789-8.ch056

Amoretti, F. (2008). Digital international governance. In A. Anttiroiko (Ed.), *Electronic government: Concepts, methodologies, tools, and applications* (pp. 688–696). Hershey, PA: IGI Global. doi:10.4018/978-1-59904-947-2.ch058

Amoretti, F. (2008). E-government at supranational level in the European Union. In A. Anttiroiko (Ed.), *Electronic government: Concepts, methodologies, tools, and applications* (pp. 1047–1055). Hershey, PA: IGI Global. doi:10.4018/978-1-59904-947-2.ch079

Amoretti, F. (2008). E-government regimes. In A. Anttiroiko (Ed.), *Electronic government: Concepts, methodologies, tools, and applications* (pp. 3846–3856). Hershey, PA: IGI Global. doi:10.4018/978-1-59904-947-2.ch280

Amoretti, F. (2009). Electronic constitution: A Braudelian perspective. In F. Amoretti (Ed.), *Electronic constitution: Social, cultural, and political implications* (pp. 1–19). Hershey, PA: IGI Global. doi:10.4018/978-1-60566-254-1.ch001

Amoretti, F., & Musella, F. (2009). Institutional isomorphism and new technologies. In M. Khosrow-Pour (Ed.), *Encyclopedia of information science and technology* (2nd ed.; pp. 2066–2071). Hershey, PA: IGI Global. doi:10.4018/978-1-60566-026-4.ch325

Amorim, V. J. P., Delabrida, S., & Oliveira, R. A. (2016). A Constraint-Driven Assessment of Operating Systems for Wearable Devices. *Proceedings of the 6th Brazilian Symposium on Computing Systems Engineering (SBESC)*, 150-155. doi:10.1109/SBESC.2016.030

An, B. W., Gwak, E. J., Kim, K., Kim, Y. C., Jang, J., Kim, J. Y., & Park, J. U. (2016). Stretchable, transparent electrodes as wearable heaters using nanotrough networks of metallic glasses with superior mechanical properties and thermal stability. *Nano Letters*, *16*(1), 471–478. doi:10.1021/acs.nanolett.5b04134 PMID:26670378

Andersen, K. V., & Henriksen, H. Z. (2007). E-government research: Capabilities, interaction, orientation, and values. In D. Norris (Ed.), *Current issues and trends in e-government research* (pp. 269–288). Hershey, PA: IGI Global. doi:10.4018/978-1-59904-283-1.ch013

Anderson, K. V., & Henriksen, H. Z. (2005). The first leg of e-government research: Domains and application areas 19982003. *International Journal of Electronic Government Research*, *1*(4), 26–44. doi:10.4018/jegr.2005100102

Andreão, R., Pereira, J. G., & Calvi, C. Z. (2006). TeleCardio: Telecardiologia a Serviço de Pacientes Hospitalizados em Domicílio. In Congresso da Sociedade Brasileira de Informática em Saúde (vol. 1, pp. 1267-1272). Florianópolis, SC.

Angelova, R. A. (2007). Maintaining the Workers Comfort and Safety in Extreme Temperatures Industrial Environment. In *Industrial Ventilation* (pp. 197–205). Pamporovo, Bulgaria: Avanguard.

Angelova, R. A. (2016). *Human Thermophysiological Comfort*. Boca Raton, FL: CRC Press.

Anttiroiko, A. (2009). Democratic e-governance. In M. Khosrow-Pour (Ed.), *Encyclopedia of information science and technology* (2nd ed.; pp. 990–995). Hershey, PA: IGI Global. doi:10.4018/978-1-60566-026-4.ch158

Apple. (2016, September). *Watchos*. Retrieved September 19, 2016, from https://developer.apple.com/watchos/

Arias, O., Wurm, J., Hoang, K., & Jin, Y. (2015, April). Privacy and security in internet of things and wearable devices. *IEEE Transactions on Multi-Scale Computing Systems*, *1*(2), 99–109. doi:10.1109/TMSCS.2015.2498605

Azariadi, D., Tsoutsouras, V., Xydis, S., & Soudris, D. (2016). Ecg signal analysis and arrhythmia detection on iot wearable medical devices. *5th International Conference on Modern Circuits and Systems Technologies (MOCAST)*, 1–4. doi:10.1109/MOCAST.2016.7495143

B., I., & Batchelor, J. C. (2008). A dual band belt antenna. In *International Workshop on Antenna Technology: Small Antennas and Novel Metamaterial* (pp. 374– 377).

Baimbetov, Y., Khalil, I., Steinbauer, M., & Anderst-Kotsis, G. (2015). Using Big Data for Emotionally Intelligent Mobile Services through Multi-Modal Emotion Recognition. In *Inclusive Smart Cities and e-Health* (pp. 127–138). Springer. doi:10.1007/978-3-319-19312-0_11

Baker, P. M., Bell, A., & Moon, N. W. (2009). Accessibility issues in municipal wireless networks. In C. Reddick (Ed.), *Handbook of research on strategies for local e-government adoption and implementation: Comparative studies* (pp. 569–588). Hershey, PA: IGI Global. doi:10.4018/978-1-60566-282-4.ch030

Balanis, C. A. (2005). *Antenna Theory Analysis and Design* (3rd ed.). John Wiley Sons, Inc.

Becker, M., Werkman, E., Anastasopoulos, M., & Kleinberger, T. (2006). Approaching Ambient Intelligent Home Care Systems. *Pervasive Health Conference and Workshops*, 1–10.

Becker, S. A., Keimer, R., & Muth, T. (2012). A case on university and community collaboration: The sci-tech entrepreneurial training services (ETS) program. In Regional development: Concepts, methodologies, tools, and applications (pp. 947-969). Hershey, PA: IGI Global. doi:10.4018/978-1-4666-0882-5.ch507

Becker, S. A., Keimer, R., & Muth, T. (2010). A case on university and community collaboration: The sci-tech entrepreneurial training services (ETS) program. In S. Becker & R. Niebuhr (Eds.), *Cases on technology innovation: Entrepreneurial successes and pitfalls* (pp. 68–90). Hershey, PA: IGI Global. doi:10.4018/978-1-61520-609-4.ch003

Benini, L., Bruni, D., Macii, A., Macii, E., & Poncino, M. (2003). Discharge current steering for battery lifetime optimization. *IEEE Transactions on Computers*, *52*(8), 985–995. doi:10.1109/TC.2003.1223633

Bernardi, R. (2012). Information technology and resistance to public sector reforms: A case study in Kenya. In T. Papadopoulos & P. Kanellis (Eds.), *Public sector reform using information technologies: Transforming policy into practice* (pp. 59–78). Hershey, PA: IGI Global. doi:10.4018/978-1-60960-839-2.ch004

Bernardi, R. (2013). Information technology and resistance to public sector reforms: A case study in Kenya. In *User-driven healthcare: Concepts, methodologies, tools, and applications* (pp. 14–33). Hershey, PA: IGI Global. doi:10.4018/978-1-4666-2770-3.ch002

Bickerstaffe, J. (2015). *Energy harvesting*. Sagentia. Retrieved June 11, 2016, from http://www.sagentia.com/resources/white-papers/2011/energy-harvesting.aspx

Billinghurst, A. C. M., & Lee, G. (2014). A Survey of Augmented Reality. *Foundations and Trends in Human-Computer Interaction*, *8*(2), 73-272.

Bishop, C. M. C. C. M. (2006). *Pattern recognition and machine learning* (Vol. 4). Springer.

Biswas, S., & Quwaider, M. (2013). Modeling energy harvesting sensors using accelerometer in body sensor networks. *8th International Conference on Body Area Networks (BODYNETS)*. doi:10.4108/icst.bodynets.2013.253588

Bodor, R., Jackson, B., & Papanikolopoulos, N. (2003). Vision-based human tracking and activity recognition. *Proceedings of the 11th Mediterranean Conference on Control and Automation*.

Bohn, J., Coroamă, V., Langheinrich, M., Mattern, F., & Rohs, M. (2004). Living in a world of smart everyday objects – Social, Economical and Ethical implications. *Human and Ecological Risk Assessment, 10*(5), 763 – 785. DOI: 10.1080/10807030490513793

Boisseau, S., Despesse, G., & Seddik, B. A. (2012). *Electrostatic Conversion for Vibration Energy Harvesting, Small-Scale Energy Harvesting*. Intech.

Bokhari, S. A., Zürcher, J. F., Mosig, J. R., & Gardiol, F. E. (1996). A small microstrip patch antenna with a convenient tuning option. *IEEE Transactions on Antennas and Propagation*, *44*(11), 1521–1528. doi:10.1109/8.542077

Bolívar, M. P., Pérez, M. D., & Hernández, A. M. (2012). Municipal e-government services in emerging economies: The Latin-American and Caribbean experiences. In Y. Chen & P. Chu (Eds.), *Electronic governance and cross-boundary collaboration: Innovations and advancing tools* (pp. 198–226). Hershey, PA: IGI Global. doi:10.4018/978-1-60960-753-1.ch011

Borycki, E. M., & Kushniruk, A. W. (2010). Use of clinical simulations to evaluate the impact of health information systems and ubiquitous computing devices upon health professional work. In S. Mohammed & J. Fiaidhi (Eds.), *Ubiquitous health and medical informatics: The ubiquity 2.0 trend and beyond* (pp. 552–573). Hershey, PA: IGI Global. doi:10.4018/978-1-61520-777-0.ch026

Borycki, E. M., & Kushniruk, A. W. (2011). Use of clinical simulations to evaluate the impact of health information systems and ubiquitous computing devices upon health professional work. In *Clinical technologies: Concepts, methodologies, tools and applications* (pp. 532–553). Hershey, PA: IGI Global. doi:10.4018/978-1-60960-561-2.ch220

Bradski, G. R. (1998). Real time face and object tracking as a component of a perceptual user interface. In *Fourth IEEE Workshop on Applications of Computer Vision Proceedings* (pp. 214-219). Princeton, NJ: IEEE. doi:10.1109/ACV.1998.732882

Brown, W. C. (1984). The history of power transmission by radio waves. *IEEE Transactions on Microwave Theory and Techniques*, *32*(9), 1230–1242. doi:10.1109/TMTT.1984.1132833

Bsching, F., Kulau, U., Gietzelt, M., & Wolf, L. (2012). Comparison and validation of capacitive accelerometers for health care applications. *Computer Methods and Programs in Biomedicine*, *106*(2), 79–88. doi:10.1016/j.cmpb.2011.10.009 PMID:22153570

Buchan, J. (2011). Developing a dynamic and responsive online learning environment: A case study of a large Australian university. In B. Czerkawski (Ed.), *Free and open source software for e-learning: Issues, successes and challenges* (pp. 92–109). Hershey, PA: IGI Global. doi:10.4018/978-1-61520-917-0.ch006

Buenger, A. W. (2008). Digital convergence and cybersecurity policy. In G. Garson & M. Khosrow-Pour (Eds.), *Handbook of research on public information technology* (pp. 395–405). Hershey, PA: IGI Global. doi:10.4018/978-1-59904-857-4.ch038

Buildroot. (2016, September). *Buildroot project.* Retrieved September 21, 2016 from https://buildroot.org/

Burn, J. M., & Loch, K. D. (2002). The societal impact of world wide web - Key challenges for the 21st century. In A. Salehnia (Ed.), *Ethical issues of information systems* (pp. 88–106). Hershey, PA: IGI Global. doi:10.4018/978-1-931777-15-5.ch007

Burn, J. M., & Loch, K. D. (2003). The societal impact of the world wide web-Key challenges for the 21st century. In M. Khosrow-Pour (Ed.), *Advanced topics in information resources management* (Vol. 2, pp. 32–51). Hershey, PA: IGI Global. doi:10.4018/978-1-59140-062-2.ch002

Bwalya, K. J., Du Plessis, T., & Rensleigh, C. (2012). The "quicksilver initiatives" as a framework for e-government strategy design in developing economies. In K. Bwalya & S. Zulu (Eds.), *Handbook of research on e-government in emerging economies: Adoption, e-participation, and legal frameworks* (pp. 605–623). Hershey, PA: IGI Global. doi:10.4018/978-1-4666-0324-0.ch031

Cabotaje, C. E., & Alampay, E. A. (2013). Social media and citizen engagement: Two cases from the Philippines. In S. Saeed & C. Reddick (Eds.), *Human-centered system design for electronic governance* (pp. 225–238). Hershey, PA: IGI Global. doi:10.4018/978-1-4666-3640-8.ch013

Camillo, A., Di Pietro, L., Di Virgilio, F., & Franco, M. (2013). Work-groups conflict at PetroTech-Italy, S.R.L.: The influence of culture on conflict dynamics. In B. Christiansen, E. Turkina, & N. Williams (Eds.), *Cultural and technological influences on global business* (pp. 272–289). Hershey, PA: IGI Global. doi:10.4018/978-1-4666-3966-9.ch015

Capra, E., Francalanci, C., & Marinoni, C. (2008). Soft success factors for m-government. In A. Anttiroiko (Ed.), *Electronic government: Concepts, methodologies, tools, and applications* (pp. 1213–1233). Hershey, PA: IGI Global. doi:10.4018/978-1-59904-947-2.ch089

Carmo, J. P., Dias, N., Mendes, P. M., Couto, C., & Correia, J. H. (2006). Low-power 2.4-GHz RF transceiver for wireless EEG module plug-and-play. In *13th IEEE International Conference on Electronics, Circuits and Systems* (pp. 1144–1147). doi:10.1109/ICECS.2006.379642

Cartelli, A. (2009). The implementation of practices with ICT as a new teaching-learning paradigm. In A. Cartelli & M. Palma (Eds.), *Encyclopedia of information communication technology* (pp. 413–417). Hershey, PA: IGI Global. doi:10.4018/978-1-59904-845-1.ch055

Carter, J., Saberin, J., Shah, T., Sai Ananthanarayanan, P. R., & Furse, C. (2010). Inexpensive fabric antenna for off-body wireless sensor communication. In *IEEE International Symposium on Antennas and Propagation* (pp. 9–12). http://doi.org/ doi:10.1109/APS.2010.5561753

Castro, A. G., Normann, I., Hois, J., & Kutz, O. (2008). Ontologizing Metadata for Assistive Technologies-The OASIS Repository. In *Ontologies in Interactive Systems, 2008. ONTORACT'08. First International Workshop on* (pp. 57–62). IEEE. Retrieved from http://ieeexplore.ieee.org/xpls/abs_all.jsp?arnumber=4756196

Castro, L., Branisso, H., Figueiredo, E., Nascimento, F., Rocha, A., & Carvalho, H. (2004). HandMed: an integrated mobile system for automatic capture of symptoms. In *IX Congresso Brasileiro de Informatica em Saúde* (*vol. 9*, pp. 1-6). Ribeirão Preto, SP: UNIFESP.

Casula, G., Montisci, G., & Mazzarella, G. (2013). A wideband PET inkjet-printed antenna for UHF RFID. *Antennas and Wireless Propagation Letters*, *12*, 1400–1403. doi:10.1109/LAWP.2013.2287307

CDC. (n.d.). *Important Facts about Falls | Home and Recreational Safety | CDC Injury Center*. Retrieved May 9, 2016, from http://www.cdc.gov/HomeandRecreationalSafety/Falls/adultfalls.html

Celcar, D. (2014). Subjective Evaluation of the Thermal Comfort of Clothing Evaluated in Cold Environment. *Glasnik Hemičara. Tehnologa I Ekologa Republike Srpske*, *10*(1), 65–71.

CEN/TR 16298. (2011). *Textiles and textile products - Smart textiles – Definitions, categorisation, applications and standardization needs*. European Committee for Standardization.

Chae, S. H., Ju, S., Choi, Y., Jun, S., Park, S. M., Lee, S., & Ji, C. H. et al. (2013). Electromagnetic vibration energy harvester using springless proof mass and ferrofluid as a lubricant. *Journal of Physics: Conference Series*, *476*(1).

Chandran, A. R., Conway, G. A., & Scanlon, W. G. (2008). Compact Low Profile Patch Antenna For Medical Body Area Netowrks at 868 MHz. In *IEEE Antennas and Propagation Society International Symposium* (pp. 8–11).

Charalabidis, Y., Lampathaki, F., & Askounis, D. (2010). Investigating the landscape in national interoperability frameworks. *International Journal of E-Services and Mobile Applications*, *2*(4), 28–41. doi:10.4018/jesma.2010100103

Charalabidis, Y., Lampathaki, F., & Askounis, D. (2012). Investigating the landscape in national interoperability frameworks. In A. Scupola (Ed.), *Innovative mobile platform developments for electronic services design and delivery* (pp. 218–231). Hershey, PA: IGI Global. doi:10.4018/978-1-4666-1568-7.ch013

Chaves, M. L. F. (2000). Acidente vascular encefálico: conceituação e fatores de risco. *Revista Brasileira de Hipertensão*, 372–382.

Chen, X. L., Kuster, N., Tan, Y. C., & Chavannes, N. (2012). Body effects on the GPS antenna of a wearable tracking device. In *6th European Conference on Antennas and Propagation, EuCAP* (pp. 3313–3316). http://doi.org/ doi:10.1109/EuCAP.2012.6205889

Chen, I. (2005). Distance education associations. In C. Howard, J. Boettcher, L. Justice, K. Schenk, P. Rogers, & G. Berg (Eds.), *Encyclopedia of distance learning* (pp. 599–612). Hershey, PA: IGI Global. doi:10.4018/978-1-59140-555-9.ch087

Chen, I. (2008). Distance education associations. In L. Tomei (Ed.), *Online and distance learning: Concepts, methodologies, tools, and applications* (pp. 562–579). Hershey, PA: IGI Global. doi:10.4018/978-1-59904-935-9.ch048

Chen, J. B., Endo, Y., Chan, K., Mazières, D., Dias, A., Seltzer, M., & Smith, M. D. (1996, February). The measured performance of personal computer operating systems. *ACM Transactions on Computer Systems, 14*(1), 3–40. doi:10.1145/225535.225536

Chen, Y. (2008). Managing IT outsourcing for digital government. In A. Anttiroiko (Ed.), *Electronic government: Concepts, methodologies, tools, and applications* (pp. 3107–3114). Hershey, PA: IGI Global. doi:10.4018/978-1-59904-947-2.ch229

Chen, Y., & Dimitrova, D. V. (2006). Electronic government and online engagement: Citizen interaction with government via web portals. *International Journal of Electronic Government Research, 2*(1), 54–76. doi:10.4018/jegr.2006010104

Chen, Y., & Knepper, R. (2005). Digital government development strategies: Lessons for policy makers from a comparative perspective. In W. Huang, K. Siau, & K. Wei (Eds.), *Electronic government strategies and implementation* (pp. 394–420). Hershey, PA: IGI Global. doi:10.4018/978-1-59140-348-7.ch017

Chen, Y., & Knepper, R. (2008). Digital government development strategies: Lessons for policy makers from a comparative perspective. In H. Rahman (Ed.), *Developing successful ICT strategies: Competitive advantages in a global knowledge-driven society* (pp. 334–356). Hershey, PA: IGI Global. doi:10.4018/978-1-59904-654-9.ch017

Chen, Z. N. (2007). *Antennas for Portable Devices* (1st ed.). Wiley-VCH. doi:10.1002/9780470319642

Cherian, E. J., & Ryan, T. W. (2014). Incongruent needs: Why differences in the iron-triangle of priorities make health information technology adoption and use difficult. In C. El Morr (Ed.), *Research perspectives on the role of informatics in health policy and management* (pp. 209–221). Hershey, PA: IGI Global. doi:10.4018/978-1-4666-4321-5.ch012

Chipara, O., Lu, C., Bailey, T. C., & Roman, G. (2010). Reliable clinical monitoring using wireless sensor networks: Experiences in a step-down hospital unit. *Proceedings of the 8th ACM Conference on Embedded Networked Sensor Systems, SenSys '10*. doi:10.1145/1869983.1869999

Cho, M. H., Lim, J. S., & Lee, C. H. (2009). ertos: The low-power realtime operating system for wearable computers. *IEEE 13th International Symposium on Consumer Electronics*, 1015–1019. doi:10.1109/ISCE.2009.5156939

Cho, H. J., & Hwang, S. (2010). Government 2.0 in Korea: Focusing on e-participation services. In C. Reddick (Ed.), *Politics, democracy and e-government: Participation and service delivery* (pp. 94–114). Hershey, PA: IGI Global. doi:10.4018/978-1-61520-933-0.ch006

Choi, S., Park, J., Hyun, W., Kim, J., Kim, J., Lee, Y. B., & Kim, D. H. et al. (2015). Stretchable heater using ligand-exchanged silver nanowire nanocomposite for wearable articular thermotherapy. *ACS Nano*, *9*(6), 6626–6633. doi:10.1021/acsnano.5b02790 PMID:26027637

Cho, M. H., & Lee, C. H. (2010, August). A low-power real-time operating system for arc (actual remote control) wearable device. *IEEE Transactions on Consumer Electronics*, *56*(3), 1602–1609. doi:10.1109/TCE.2010.5606303

Chorus, C., & Timmermans, H. (2010). Ubiquitous travel environments and travel control strategies: Prospects and challenges. In M. Wachowicz (Ed.), *Movement-aware applications for sustainable mobility: Technologies and approaches* (pp. 30–51). Hershey, PA: IGI Global. doi:10.4018/978-1-61520-769-5.ch003

Chuanshen, R. (2007). E-government construction and China's administrative litigation act. In A. Anttiroiko & M. Malkia (Eds.), *Encyclopedia of digital government* (pp. 507–510). Hershey, PA: IGI Global. doi:10.4018/978-1-59140-789-8.ch077

Chu, L. J. (1948). Small antennas. *Journal of Applied Physics*, *19*(12), 1163–1175. doi:10.1063/1.1715038

Ciaghi, A., & Villafiorita, A. (2012). Law modeling and BPR for public administration improvement. In K. Bwalya & S. Zulu (Eds.), *Handbook of research on e-government in emerging economies: Adoption, e-participation, and legal frameworks* (pp. 391–410). Hershey, PA: IGI Global. doi:10.4018/978-1-4666-0324-0.ch019

Ciaramitaro, B. L., & Skrocki, M. (2012). mHealth: Mobile healthcare. In B. Ciaramitaro (Ed.), Mobile technology consumption: Opportunities and challenges (pp. 99-109). Hershey, PA: IGI Global. doi:10.4018/978-1-61350-150-4.ch007

Cibin, C., Leuchtmann, P., Gimersky, M., Vahldieck, R., & Moscibroda, S. (2004). A flexible wearable antenna. In IEEE Antennas and Propagation Society (pp. 3589–3592). doi:10.1109/APS.2004.1330122

Cibin, C., Leuchtmann, P., Gimersky, M., & Vahldieck, R. (2004). Modified E-Shaped Pifa Antenna for Wearable Systems. In *URSI International Symposium on Electromagnetic Theory (URSI EMTS)* (pp. 873–875).

City. (2016, June 12). In *Wikipedia, the free encyclopedia*. Retrieved from https://en.wikipedia.org/w/index.php?title=City&oldid=724986747

Comite, U. (2012). Innovative processes and managerial effectiveness of e-procurement in healthcare. In A. Manoharan & M. Holzer (Eds.), *Active citizen participation in e-government: A global perspective* (pp. 206–229). Hershey, PA: IGI Global. doi:10.4018/978-1-4666-0116-1.ch011

Connor, S. O. (2015). *Wearables at work: the new frontier at Employee Surveillance*. Available online: http://www.ft.com/intl/cms/s/2/d7eee768-0b65-11e5-994d-00144feabdc0.html#axzz45rtn27V6

Constantinou, C., Nechayev, Y., Wu, X., & Hall, P. (2012). Body-area Propagation at 60 GHz. In Loughborough Antennas & Propagation Conference (pp. 1–4).

Cook, B., & Shamim, A. (2013). Utilizing wideband AMC structures for high-gain inkjet-printed antennas on lossy paper substrate. *IEEE Antennas and Wireless Propagation Letters*, *12*, 76–79. doi:10.1109/LAWP.2013.2240251

Cordella, A. (2013). E-government success: How to account for ICT, administrative rationalization, and institutional change. In J. Gil-Garcia (Ed.), *E-government success factors and measures: Theories, concepts, and methodologies* (pp. 40–51). Hershey, PA: IGI Global. doi:10.4018/978-1-4666-4058-0.ch003

Corner, A. A. (2013). Body-Worn Antennas Making a Splash : Lifejacket-Integrated Antennas for Global Search and Rescue Satellite System. *IEEE Antennas and Propagation Magazine*, *55*(2).

Cropf, R. A. (2009). ICT and e-democracy. In M. Khosrow-Pour (Ed.), *Encyclopedia of information science and technology* (2nd ed.; pp. 1789–1793). Hershey, PA: IGI Global. doi:10.4018/978-1-60566-026-4.ch281

Cropf, R. A. (2009). The virtual public sphere. In M. Pagani (Ed.), *Encyclopedia of multimedia technology and networking* (2nd ed.; pp. 1525–1530). Hershey, PA: IGI Global. doi:10.4018/978-1-60566-014-1.ch206

Cumbler, E., Anderson, T., Neumann, R., Jones, W. J., & Brega, K. (2010). Stroke alert program improves recognition and evaluation time of in-hospital ischemic stroke. *Journal of Stroke and Cerebrovascular Diseases*, *19*(6), 494–496. doi:10.1016/j.jstrokecerebrovasdis.2009.09.007 PMID:20538480

Curone, D., Dudnik, G., Loriga, G., Luprano, J., Magenes, G., Paradiso, R., … Bonfiglio, A. (2007). Smart Garments for Safety Improvement of Emergency/Disaster Operators. In *29th Annual International Conference of the IEEE Engineering in Medicine and Biology Society (EMBS)* (pp. 3962–3965). doi:10.1109/IEMBS.2007.4353201

Curto, S., McEvoy, P., Bao, X., & Ammann, M. J. (2009). Compact patch antenna for electromagnetic interaction with human tissue at 434 MHz. *IEEE Transactions on Antennas and Propagation*, *57*(9), 2564–2571. doi:10.1109/TAP.2009.2027040

D'Abundo, M. L. (2013). Electronic health record implementation in the United States healthcare industry: Making the process of change manageable. In V. Wang (Ed.), *Handbook of research on technologies for improving the 21st century workforce: Tools for lifelong learning* (pp. 272–286). Hershey, PA: IGI Global. doi:10.4018/978-1-4666-2181-7.ch018

D'Angelo, T., Delabrida, S., Oliveira, R. A. R., & Loureiro, A. A. (2016). Towards a Low-Cost Augmented Reality Head-Mounted Display with Real-Time Eye Center Location Capability. In *2016 Brazilian Symposium on Computing Systems Engineering* (pp. 24-31). João Pessoa, PB: IEEE. doi:10.1109/SBESC.2016.013

Damurski, L. (2012). E-participation in urban planning: Online tools for citizen engagement in Poland and in Germany. *International Journal of E-Planning Research*, *1*(3), 40–67. doi:10.4018/ijepr.2012070103

DAusilio, A. (2012). Arduino: A low-cost multipurpose lab equipment. *Behavior Research Methods*, *44*(2), 305–313. doi:10.3758/s13428-011-0163-z PMID:22037977

de Almeida, M. O. (2007). E-government strategy in Brazil: Increasing transparency and efficiency through e-government procurement. In M. Gascó-Hernandez (Ed.), *Latin America online: Cases, successes and pitfalls* (pp. 34–82). Hershey, PA: IGI Global. doi:10.4018/978-1-59140-974-8.ch002

de Juana Espinosa, S. (2008). Empirical study of the municipalitites' motivations for adopting online presence. In A. Anttiroiko (Ed.), *Electronic government: Concepts, methodologies, tools, and applications* (pp. 3593–3608). Hershey, PA: IGI Global. doi:10.4018/978-1-59904-947-2.ch262

de Oliveira Neto, J. S., & Kofuji, S. T. (2016). Inclusive Smart City: an exploratory study. In M. Antona & C. Stephanidis (Eds.), *Universal Access in Human-Computer Interaction. Access to Learning, Health and Well-Being*. Springer International Publishing. doi:10.1007/978-3-319-40244-4_44

de Souza Dias, D. (2002). Motivation for using information technology. In C. Snodgrass & E. Szewczak (Eds.), *Human factors in information systems* (pp. 55–60). Hershey, PA: IGI Global. doi:10.4018/978-1-931777-10-0.ch005

Declercq, F., Couckuyt, I., Rogier, H., & Dhaene, T. (2010). Complex permittivity characterization of textile materials by means of surrogate modeling. In *IEEE Antennas and Propagation Society International Symposium* (pp. 1–4).

DeJean, G., Bairavasubramanian, R., Thompson, D., Ponchak, G. M. T., & Papapolymerou, J. (2005). Liquid crystal polymer (LCP): A new organic material for the development of multilayer dual-frequency dual-polarization flexible antenna arrays. *IEEE Antennas and Wireless Propagation Letters*, *4*(1), 22–26. doi:10.1109/LAWP.2004.841626

Delabrida, S., D'Angelo, T., Oliveira, R. A. R., & Loureiro, A. A. (2016). Building wearables for geology: An operating system approach. *SIGOPS Operating Systems Review*, *50*(1), 31-45.

Delabrida, S., D'Angelo, T., Oliveira, R. A. R., & Loureiro, A. A. (2015). Towards a wearable device for monitoring ecological environments. In *2015 Brazilian Symposium on Computing Systems Engineering* (pp. 148-153). Foz do Iguaçu, PR: IEEE. doi:10.1109/SBESC.2015.35

Delabrida, S., DAngelo, T., Oliveira, R. A., & Loureiro, A. A. (2016, March). Building wearables for geology: An operating system approach. *SIGOPS Oper. Syst. Rev.*, *50*(1), 31–45. doi:10.1145/2903267.2903275

Demediuk, P. (2006). Government procurement ICT's impact on the sustainability of SMEs and regional communities. In S. Marshall, W. Taylor, & X. Yu (Eds.), *Encyclopedia of developing regional communities with information and communication technology* (pp. 321–324). Hershey, PA: IGI Global. doi:10.4018/978-1-59140-575-7.ch056

Devonshire, E., Forsyth, H., Reid, S., & Simpson, J. M. (2013). The challenges and opportunities of online postgraduate coursework programs in a traditional university context. In B. Tynan, J. Willems, & R. James (Eds.), *Outlooks and opportunities in blended and distance learning* (pp. 353–368). Hershey, PA: IGI Global. doi:10.4018/978-1-4666-4205-8.ch026

Di Cerbo, F., Scotto, M., Sillitti, A., Succi, G., & Vernazza, T. (2007). Toward a GNU/Linux distribution for corporate environments. In S. Sowe, I. Stamelos, & I. Samoladas (Eds.), *Emerging free and open source software practices* (pp. 215–236). Hershey, PA: IGI Global. doi:10.4018/978-1-59904-210-7.ch010

Díaz, M., & Becker, D. E. (2010). Thermoregulation: Physiological and Clinical Considerations during Sedation and General Anesthesia. *Anesthesia Progress*, *57*(1), 25–33. doi:10.2344/0003-3006-57.1.25 PMID:20331336

Diesner, J., & Carley, K. M. (2005). Revealing social structure from texts: Meta-matrix text analysis as a novel method for network text analysis. In V. Narayanan & D. Armstrong (Eds.), *Causal mapping for research in information technology* (pp. 81–108). Hershey, PA: IGI Global. doi:10.4018/978-1-59140-396-8.ch004

Direct Frame Buffer. (2016, September). *Direct frame buffer library*. Retrieved September 21, 2016 from http://elinux.org/DirectFB

Dolan, B. (2012). *Wearable Devices, a $6B market by 2016*. Market Report by IMS. Available for online at: http://mobihealthnews.com/18194/report-wearable-devices-a-6b-market-by-2016

Dologite, D. G., Mockler, R. J., Bai, Q., & Viszhanyo, P. F. (2006). IS change agents in practice in a US-Chinese joint venture. In M. Hunter & F. Tan (Eds.), *Advanced topics in global information management* (Vol. 5, pp. 331–352). Hershey, PA: IGI Global. doi:10.4018/978-1-59140-923-6.ch015

Donnan, G. A., Fisher, M., Macleod, M., & Davis, S. M. (2008). Stroke. *Lancet*, *371*(9624), 1612–1623. doi:10.1016/S0140-6736(08)60694-7 PMID:18468545

Doukas, C., & Maglogiannis, I. (2011). Managing wearable sensor data through Cloud Computing. *Proceedings of the Third International Conference on Cloud Computing Technology and Science*, 440 – 445. doi:10.1109/CloudCom.2011.65

Drnevich, P., Brush, T. H., & Luckock, G. T. (2011). Process and structural implications for IT-enabled outsourcing. *International Journal of Strategic Information Technology and Applications*, *2*(4), 30–43. doi:10.4018/jsita.2011100103

Dunne, L. E., & Smyth, B. (2007). Psychophysical Elements of Wearability. In *Proceedings of the SIGCHI Conference on Human Factors in Computing Systems* (pp. 299–302). New York, NY: ACM. http://doi.org/ doi:10.1145/1240624.1240674

Dwivedi, A. N. (2009). Handbook of research on information technology management and clinical data administration in healthcare (Vols. 1–2). Hershey, PA: IGI Global. doi:10.4018/978-1-60566-356-2

Elbeltagi, I., McBride, N., & Hardaker, G. (2006). Evaluating the factors affecting DSS usage by senior managers in local authorities in Egypt. In M. Hunter & F. Tan (Eds.), *Advanced topics in global information management* (Vol. 5, pp. 283–307). Hershey, PA: IGI Global. doi:10.4018/978-1-59140-923-6.ch013

Eluf Neto, J., Lotufo, P. A., & de Lólio, C. A. (1990). Tratamento da hipertensão e declínio da mortalidade por acidentes vasculares cerebrais. *Revista de Saude Publica, 24*(4), 332–336. doi:10.1590/S0034-89101990000400013 PMID:2103653

Elvin, N., & Erturk, A. (2013). *Advances in Energy Harvesting Methods.* Springer Link. doi:10.1007/978-1-4614-5705-3

Engineers, R. T. (2016, September). *Freertos.* Retrieved September 21, 2016 from http://www.freertos.org/

Eom, S., & Fountain, J. E. (2013). Enhancing information services through public-private partnerships: Information technology knowledge transfer underlying structures to develop shared services in the U.S. and Korea. In J. Gil-Garcia (Ed.), *E-government success around the world: Cases, empirical studies, and practical recommendations* (pp. 15–40). Hershey, PA: IGI Global. doi:10.4018/978-1-4666-4173-0.ch002

Esteves, T., Leuenberger, D., & Van Leuven, N. (2012). Reaching citizen 2.0: How government uses social media to send public messages during times of calm and times of crisis. In K. Kloby & M. D'Agostino (Eds.), *Citizen 2.0: Public and governmental interaction through web 2.0 technologies* (pp. 250–268). Hershey, PA: IGI Global. doi:10.4018/978-1-4666-0318-9.ch013

Estevez, E., Fillottrani, P., Janowski, T., & Ojo, A. (2012). Government information sharing: A framework for policy formulation. In Y. Chen & P. Chu (Eds.), *Electronic governance and cross-boundary collaboration: Innovations and advancing tools* (pp. 23–55). Hershey, PA: IGI Global. doi:10.4018/978-1-60960-753-1.ch002

Ezz, I. E. (2008). E-governement emerging trends: Organizational challenges. In A. Anttiroiko (Ed.), *Electronic government: Concepts, methodologies, tools, and applications* (pp. 3721–3737). Hershey, PA: IGI Global. doi:10.4018/978-1-59904-947-2.ch269

Fabri, M. (2009). The Italian style of e-justice in a comparative perspective. In A. Martínez & P. Abat (Eds.), *E-justice: Using information communication technologies in the court system* (pp. 1–19). Hershey, PA: IGI Global. doi:10.4018/978-1-59904-998-4.ch001

Faceli, K., Lorena, A. C., Gama, J. A., & Carvalho, A. C. P. L. F. (2011). *Inteligência artificial: uma abordagem de aprendizado de máquina* (1st ed.). LTC.

Fagbe, T., & Adekola, O. D. (2010). Workplace safety and personnel well-being: The impact of information technology. *International Journal of Green Computing*, *1*(1), 28–33. doi:10.4018/jgc.2010010103

Fagbe, T., & Adekola, O. D. (2011). Workplace safety and personnel well-being: The impact of information technology. In *Global business: Concepts, methodologies, tools and applications* (pp. 1438–1444). Hershey, PA: IGI Global. doi:10.4018/978-1-60960-587-2.ch509

Falcão, I. V., de Carvalho, E. M. F., Barreto, K. M. L., Lessa, F. J. D., & Leite, V. M. M. (2004, March). Acidente vascular cerebral precoce: Implicações para adultos em idade produtiva atendidos pelo Sistema Único de Saúde. *Revista Brasileira de Saú de Materno Infantil*, *4*(1), 95–101. doi:10.1590/S1519-38292004000100009

Farmer, L. (2008). Affective collaborative instruction with librarians. In S. Kelsey & K. St.Amant (Eds.), *Handbook of research on computer mediated communication* (pp. 15–24). Hershey, PA: IGI Global. doi:10.4018/978-1-59904-863-5.ch002

Favier, L., & Mekhantar, J. (2007). Use of OSS by local e-administration: The French situation. In K. St.Amant & B. Still (Eds.), *Handbook of research on open source software: Technological, economic, and social perspectives* (pp. 428–444). Hershey, PA: IGI Global. doi:10.4018/978-1-59140-999-1.ch033

Federal Communications Commission (FCC). (2014). Retrieved from http: //transition.fcc.gov/bureaus/engineering_technology/orders/2002/fcc02048.pdf

Fernando, S. (2009). Issues of e-learning in third world countries. In M. Khosrow-Pour (Ed.), *Encyclopedia of information science and technology* (2nd ed.; pp. 2273–2277). Hershey, PA: IGI Global. doi:10.4018/978-1-60566-026-4.ch360

Filho, J. R., & dos Santos Junior, J. R. (2009). Local e-government in Brazil: Poor interaction and local politics as usual. In C. Reddick (Ed.), *Handbook of research on strategies for local e-government adoption and implementation: Comparative studies* (pp. 863–878). Hershey, PA: IGI Global. doi:10.4018/978-1-60566-282-4.ch045

Firoozy, N., & Shirazi, M. (2011). Planar Inverted-F Antenna (PIFA) Design Dissection for Cellular Communication Application. *Journal of Electromagnetic Analysis and Applications*, *03*(10), 406–411. doi:10.4236/jemaa.2011.310064

Fitbit. (2016, September). *Fitbit activity trackers*. Retrieved September 20, 2016 from https://www.fitbit.com/

Fletcher, P. D. (2004). Portals and policy: Implications of electronic access to U.S. federal government information services. In A. Pavlichev & G. Garson (Eds.), *Digital government: Principles and best practices* (pp. 52–62). Hershey, PA: IGI Global. doi:10.4018/978-1-59140-122-3.ch004

Fletcher, P. D. (2008). Portals and policy: Implications of electronic access to U.S. federal government information services. In A. Anttiroiko (Ed.), *Electronic government: Concepts, methodologies, tools, and applications* (pp. 3970–3979). Hershey, PA: IGI Global. doi:10.4018/978-1-59904-947-2.ch289

Fletcher, R. R., Dobson, K., Goodwin, M. S., Eydgahi, H., Wilder-Smith, O., Fernholz, D., & Picard, R. W. et al. (2010). iCalm: Wearable sensor and network architecture for wirelessly communicating and logging autonomic activity. *Information Technology in Biomedicine. IEEE Transactions on, 14*(2), 215–223.

Forlano, L. (2004). The emergence of digital government: International perspectives. In A. Pavlichev & G. Garson (Eds.), *Digital government: Principles and best practices* (pp. 34–51). Hershey, PA: IGI Global. doi:10.4018/978-1-59140-122-3.ch003

Foth, M., Choi, J. H., & Satchell, C. (2011). Urban Informatics. In *Proceedings of the ACM 2011 Conference on Computer Supported Cooperative Work* (pp. 1–8). New York, NY: ACM. http:// doi.org/ doi:<ALIGNMENT.qj></ALIGNMENT>10.1145/1958824.1958826

Foulger, S. H., & Gregory (Clemson), R. V. (2003). *Intelligent Textiles Based on Environmentally Responsive in Fibers.* National Textile Center Annual Report, Clemson University.

Fraile, J. A., Bajo, J., Corchado, J. M., & Abraham, A. (2010). Applying wearable solutions in dependent environments. *Information Technology in Biomedicine. IEEE Transactions on, 14*(6), 1459–1467.

Franzel, J. M., & Coursey, D. H. (2004). Government web portals: Management issues and the approaches of five states. In A. Pavlichev & G. Garson (Eds.), *Digital government: Principles and best practices* (pp. 63–77). Hershey, PA: IGI Global. doi:10.4018/978-1-59140-122-3.ch005

Fuchs, H., State, A., Dunn, D., & Keller, K. (2015). *Converting commodity head-mounted displays for optical see-through augmented reality. Project Report.* Department of Computer Science, University of North Carolina at Chapel Hill.

Fujimoto, K. (2008). Mobile Antenna Systems Handbook. Artech House.

Gaetano, D., McEvoy, P., Ammann, M. J., Browne, J. E., Keating, L., & Horgan, F. (2013). Footwear antennas for body area telemetry. *IEEE Transactions on Antennas and Propagation, 61*(10), 4908–4916. doi:10.1109/TAP.2013.2272451

Gaivéo, J. M. (2013). Security of ICTs supporting healthcare activities. In M. Cruz-Cunha, I. Miranda, & P. Gonçalves (Eds.), *Handbook of research on ICTs for human-centered healthcare and social care services* (pp. 208–228). Hershey, PA: IGI Global. doi:10.4018/978-1-4666-3986-7.ch011

Garson, G. D. (1999). *Information technology and computer applications in public administration: Issues and trends.* Hershey, PA: IGI Global. doi:10.4018/978-1-87828-952-0

Garson, G. D. (2003). Toward an information technology research agenda for public administration. In G. Garson (Ed.), *Public information technology: Policy and management issues* (pp. 331–357). Hershey, PA: IGI Global. doi:10.4018/978-1-59140-060-8.ch014

Garson, G. D. (2004). The promise of digital government. In A. Pavlichev & G. Garson (Eds.), *Digital government: Principles and best practices* (pp. 2–15). Hershey, PA: IGI Global. doi:10.4018/978-1-59140-122-3.ch001

Garson, G. D. (2007). An information technology research agenda for public administration. In G. Garson (Ed.), *Modern public information technology systems: Issues and challenges* (pp. 365–392). Hershey, PA: IGI Global. doi:10.4018/978-1-59904-051-6.ch018

Gasco, M. (2007). Civil servants' resistance towards e-government development. In A. Anttiroiko & M. Malkia (Eds.), *Encyclopedia of digital government* (pp. 190–195). Hershey, PA: IGI Global. doi:10.4018/978-1-59140-789-8.ch028

Gasco, M. (2008). Civil servants' resistance towards e-government development. In A. Anttiroiko (Ed.), *Electronic government: Concepts, methodologies, tools, and applications* (pp. 2580–2588). Hershey, PA: IGI Global. doi:10.4018/978-1-59904-947-2.ch190

Gemperle, F., Kasabach, C., Stivoric, J., Bauer, M., & Martin, R. (1998). Design for wearability. In *Second International Symposium on Wearable Computers, 1998. Digest of Papers* (pp. 116–122). doi:10.1109/ISWC.1998.729537

Ghere, R. K. (2010). Accountability and information technology enactment: Implications for social empowerment. In E. Ferro, Y. Dwivedi, J. Gil-Garcia, & M. Williams (Eds.), *Handbook of research on overcoming digital divides: Constructing an equitable and competitive information society* (pp. 515–532). Hershey, PA: IGI Global. doi:10.4018/978-1-60566-699-0.ch028

Gibson, I. W. (2012). Simulation modeling of healthcare delivery. In A. Kolker & P. Story (Eds.), *Management engineering for effective healthcare delivery: Principles and applications* (pp. 69–89). Hershey, PA: IGI Global. doi:10.4018/978-1-60960-872-9.ch003

Gil-Garcia, J. R. (2007). Exploring e-government benefits and success factors. In A. Anttiroiko & M. Malkia (Eds.), *Encyclopedia of digital government* (pp. 803–811). Hershey, PA: IGI Global. doi:10.4018/978-1-59140-789-8.ch122

Gil-Garcia, J. R., & González Miranda, F. (2010). E-government and opportunities for participation: The case of the Mexican state web portals. In C. Reddick (Ed.), *Politics, democracy and e-government: Participation and service delivery* (pp. 56–74). Hershey, PA: IGI Global. doi:10.4018/978-1-61520-933-0.ch004

GNU. (2016, September). *The gnu c library (glibc)*. Retrieved September 21, 2016 from https://www.gnu.org/software/libc/

Goldfinch, S. (2012). Public trust in government, trust in e-government, and use of e-government. In Z. Yan (Ed.), *Encyclopedia of cyber behavior* (pp. 987–995). Hershey, PA: IGI Global. doi:10.4018/978-1-4666-0315-8.ch081

Goldstein, L. B., Adams, R., Alberts, M. J., Appel, L. J., Brass, L. M., Bushnell, C. D., & Sacco, R. L. et al. (2006, June). Primary prevention of ischemic stroke: a guideline from the American Heart Association / American Stroke Association Stroke Council: cosponsored by the Atherosclerotic Peripheral Vascular Disease Interdisciplinary Working Group; Cardiovascular Nursing Counc. *Stroke*, *113*(24).

Goncalves, F., Macedo, J., Nicolau, M. J., & Santos, A. (2013). Security architecture for Mobile E-Health applications in medication control. *Proceedings of the 21 International Conference on Software, Telecommunication and Computer Networks*, 1–8. doi:10.1109/SoftCOM.2013.6671901

Gonzalez, M. C., Hidalgo, C. A., & Barabasi, A.-L. (2008). Understanding individual human mobility patterns. *Nature*, *453*(7196), 779–782. doi:10.1038/nature06958 PMID:18528393

Goodyear, M. (2012). Organizational change contributions to e-government project transitions. In S. Aikins (Ed.), *Managing e-government projects: Concepts, issues, and best practices* (pp. 1–21). Hershey, PA: IGI Global. doi:10.4018/978-1-4666-0086-7.ch001

Google Android. (2016, September). *Android wear*. Retrieved September 19, 2016, from https://www.android.com/wear/

Google Brillo. (2016, September). *Brillo project*. Retrieved September 19, 2016, from https://developers.google.com/brillo/

Google Glass. (2016, September). *Glass*. Retrieved September 20, 2016 from https://www.google.com/glass/start/

Gordon, S., & Mulligan, P. (2003). Strategic models for the delivery of personal financial services: The role of infocracy. In S. Gordon (Ed.), *Computing information technology: The human side* (pp. 220–232). Hershey, PA: IGI Global. doi:10.4018/978-1-93177-752-0.ch014

Gordon, T. F. (2007). Legal knowledge systems. In A. Anttiroiko & M. Malkia (Eds.), *Encyclopedia of digital government* (pp. 1161–1166). Hershey, PA: IGI Global. doi:10.4018/978-1-59140-789-8.ch175

Gorlatova, M., Sarik, J., Grebla, G., Cong, M., Kymissis, I., & Zussman, G. (2014). Movers and shakers: Kinetic energy harvesting for the internet of things. *Performance Evaluation Review*, *42*(1), 407–419. doi:10.1145/2637364.2591986

Graham, J. E., & Semich, G. W. (2008). Integrating technology to transform pedagogy: Revisiting the progress of the three phase TUI model for faculty development. In L. Tomei (Ed.), *Adapting information and communication technologies for effective education* (pp. 1–12). Hershey, PA: IGI Global. doi:10.4018/978-1-59904-922-9.ch001

Grandinetti, L., & Pisacane, O. (2012). Web services for healthcare management. In D. Prakash Vidyarthi (Ed.), *Technologies and protocols for the future of internet design: Reinventing the web* (pp. 60–94). Hershey, PA: IGI Global. doi:10.4018/978-1-4666-0203-8.ch004

Green, C. (2015). *Wearable technology: Latest devices allow employers to track behaviour of their workers.* Available online: http://www.independent.co.uk/life-style/gadgets-and-tech/news/wearable-technology-latest-devices-allow-employers-to-track-behaviour-of-their-workers-10454342.html

Groenewegen, P., & Wagenaar, F. P. (2008). VO as an alternative to hierarchy in the Dutch police sector. In G. Putnik & M. Cruz-Cunha (Eds.), *Encyclopedia of networked and virtual organizations* (pp. 1851–1857). Hershey, PA: IGI Global. doi:10.4018/978-1-59904-885-7.ch245

Gronlund, A. (2001). Building an infrastructure to manage electronic services. In S. Dasgupta (Ed.), *Managing internet and intranet technologies in organizations: Challenges and opportunities* (pp. 71–103). Hershey, PA: IGI Global. doi:10.4018/978-1-878289-95-7.ch006

Gronlund, A. (2002). Introduction to electronic government: Design, applications and management. In Å. Grönlund (Ed.), *Electronic government: Design, applications and management* (pp. 1–21). Hershey, PA: IGI Global. doi:10.4018/978-1-930708-19-8.ch001

Gropper, S., & Smith, J. (2012). *Advanced nutrition and human metabolism.* Cengage Learning.

Gubbi, J., Buyya, R., Marusic, S., & Palaniswami, M. (2013). Internet of Things, A Vision, Architectural Elements and future directions. *Future Generation Computer Systems, 29*(7), 1645 – 1660. doi:10.1016/j.future.2013.01.010

Gupta, K., Lee, G. A., & Billinghurst, M. (2016). Do you see what I see? The effect of gaze tracking on task space remote collaboration. *IEEE Transactions on Visualization and Computer Graphics, 22*(11), 2413-2422.

Gupta, A., Woosley, R., Crk, I., & Sarnikar, S. (2009). An information technology architecture for drug effectiveness reporting and post-marketing surveillance. In J. Tan (Ed.), *Medical informatics: Concepts, methodologies, tools, and applications* (pp. 631–646). Hershey, PA: IGI Global. doi:10.4018/978-1-60566-050-9.ch047

Gustafsson, M., Sohl, C., & Kristensson, G. (2007). Physical limitations on antennas of arbitrary shape. In Royal Society a-Mathematical Physical and Engineering Sciences (Vol. 463, pp. 2589–2607). doi:10.1098/rspa.2007.1893

Hallin, A., & Lundevall, K. (2007). mCity: User focused development of mobile services within the city of Stockholm. In I. Kushchu (Ed.), Mobile government: An emerging direction in e-government (pp. 12-29). Hershey, PA: IGI Global. doi:10.4018/978-1-59140-884-0.ch002

Hallin, A., & Lundevall, K. (2009). mCity: User focused development of mobile services within the city of Stockholm. In D. Taniar (Ed.), Mobile computing: Concepts, methodologies, tools, and applications (pp. 3455-3467). Hershey, PA: IGI Global. doi:10.4018/978-1-60566-054-7.ch253

Hallin, A., & Lundevall, K. (2009). mCity: User focused development of mobile services within the city of Stockholm. In S. Clarke (Ed.), Evolutionary concepts in end user productivity and performance: Applications for organizational progress (pp. 268-280). Hershey, PA: IGI Global. doi:10.4018/978-1-60566-136-0.ch017

Hamilton, M. C. (2012). Recent advances in energy harvesting technology and techniques. *IECON 2012 - 38th Annual Conference on IEEE Industrial Electronics Society*. doi:10.1109/IECON.2012.6389019

Hanai, Y., Nishimura, J., & Kuroda, T. (2009). Haar-like filtering for human activity recognition using 3d accelerometer. In *Digital Signal Processing Workshop and 5th IEEE Signal Processing Education Workshop, 2009. DSP/SPE 2009. IEEE 13th* (pp. 675–678). IEEE. Retrieved from http://ieeexplore.ieee.org/xpls/abs_all.jsp?arnumber=4786008

Hansen, D. W., & Ji, Q. (2010). In the eye of the beholder: A survey of models for eyes and gaze. *IEEE Transactions on Pattern Analysis and Machine Intelligence*, *32*(3), 478-500.

Hanson, A. (2005). Overcoming barriers in the planning of a virtual library. In M. Khosrow-Pour (Ed.), *Encyclopedia of information science and technology* (pp. 2255–2259). Hershey, PA: IGI Global. doi:10.4018/978-1-59140-553-5.ch397

Hao, Y., Alomainy, A., Hall, P. S., Nechayev, Y. I., Parini, C. G., & Constantinou, C. C. (2012). *Antennas and Propagation for Body Centric Wireless Communications* (2nd ed.; pp. 38–41). Norwood, MA: Artech House, Inc.

Haque, A. (2008). Information technology and surveillance: Implications for public administration in a new word order. In T. Loendorf & G. Garson (Eds.), *Patriotic information systems* (pp. 177–185). Hershey, PA: IGI Global. doi:10.4018/978-1-59904-594-8.ch008

Harrington, R. F. (1960). Effect of antenna size on gain, bandwidth, and efficiency. *Journal of Research of the National Bureau of Standards, Section D. Radio Propagation*, *64D*(1), 1–12. doi:10.6028/jres.064D.003

Harris, J. A., & Benedict, F. G. (1918). A biometric study of human basal metabolism. *Proceedings of the National Academy of Sciences of the United States of America*, *4*(12), 370–373. doi:10.1073/pnas.4.12.370 PMID:16576330

Harrop, P., Hayward, J., Das, R., & Holland, G. (2015). *Wearable Technology 2015 2025: Technologies, Markets, Forecasts.* Retrieved June 11, 2016, from http://www.idtechex.com/research/reports/wearable-technology-2015-2025-technologies-markets-forecasts-000427.asp

Haskou, A., Ramadan, A., Al-Husseini, M., Kasem, F., Kabalan, K. Y., & ElHajj, A. (2012). A simple estimation and verification technique for electrical characterization of textiles. In *Middle East Conference on Antennas and Propagation* (pp. 1–4). doi:10.1109/MECAP.2012.6618190

Hassanalieragh, M., Page, A., Soyata, T., Sharma, G., Aktas, M., Mateos, G., . . . Andreescu, S. (2015). Health monitoring and management using internet-of-things (iot) sensing with cloud-based processing: Opportunities and challenges. *Services Computing (SCC), 2015 IEEE International Conference on*, 285–292. doi:10.1109/SCC.2015.47

Hauck, R. V., Thatcher, S. M., & Weisband, S. P. (2012). Temporal aspects of information technology use: Increasing shift work effectiveness. In J. Wang (Ed.), *Advancing the service sector with evolving technologies: Techniques and principles* (pp. 87–104). Hershey, PA: IGI Global. doi:10.4018/978-1-4666-0044-7.ch006

Havenith, G. (1999). Heat balance when wearing protective clothing. *The Annals of Occupational Hygiene*, *43*(5), 289–296. doi:10.1016/S0003-4878(99)00051-4 PMID:10481628

Havenith, G. (2002). Interaction of clothing and thermoregulation. *Exogenous Dermatology*, *1*(5), 221–230. doi:10.1159/000068802

Hawk, S., & Witt, T. (2006). Telecommunications courses in information systems programs. *International Journal of Information and Communication Technology Education*, *2*(1), 79–92. doi:10.4018/jicte.2006010107

Haykin, S. (1999). *Neural Networks - A Comprehensive Foundation* (2nd ed.). Prentice Hall.

He, W., Guo, Y., Gao, C., & Li, X. (2012). Recognition of human activities with wearable sensors. EURASIP Journal on Advances in Signal Processing. doi:10.1186/1687-6180-2012-108

He, Z.-Y., & Jin, L.-W. (2008). Activity recognition from acceleration data using AR model representation and SVM. In *Machine Learning and Cybernetics, 2008 International Conference on* (Vol. 4, pp. 2245–2250). IEEE. Retrieved from http://ieeexplore.ieee.org/xpls/abs_all.jsp?arnumber=4620779

Head, H., & Holmes, G. (1911). Sensory Disturbances from Cerebral Lesions. *Brain*, *34*(2-3), 102–254. doi:10.1093/brain/34.2-3.102

Helms, M. M., Moore, R., & Ahmadi, M. (2009). Information technology (IT) and the healthcare industry: A SWOT analysis. In J. Tan (Ed.), *Medical informatics: Concepts, methodologies, tools, and applications* (pp. 134–152). Hershey, PA: IGI Global. doi:10.4018/978-1-60566-050-9.ch012

Hendrickson, S. M., & Young, M. E. (2014). Electronic records management at a federally funded research and development center. In J. Krueger (Ed.), *Cases on electronic records and resource management implementation in diverse environments* (pp. 334–350). Hershey, PA: IGI Global. doi:10.4018/978-1-4666-4466-3.ch020

Henman, P. (2010). Social policy and information communication technologies. In J. Martin & L. Hawkins (Eds.), *Information communication technologies for human services education and delivery: Concepts and cases* (pp. 215–229). Hershey, PA: IGI Global. doi:10.4018/978-1-60566-735-5.ch014

Hertel, T. W., & Smith, G. S. (2003). On the dispersive properties of the conical spiral antenna and its use for pulsed radiation. *IEEE Transactions on Antennas and Propagation*, *51*(7), 1426–1433. doi:10.1109/TAP.2003.813602

Hertleer, C., Rogier, H., Member, S., Vallozzi, L., & Van Langenhove, L. (2009). A Textile Antenna for Off-Body Communication Integrated Into Protective Clothing for Firefighters. *IEEE Transactions on Antennas and Propagation*, *57*(4), 919–925. doi:10.1109/TAP.2009.2014574

Hiremath, S., Yang, G., & Mankodiya, K. (2014). Wearable Internet of Things. *Proceedings of the Fourth International Conference on Wireless Mobile Communication and Healthcare*, 304 – 307. Doi:10.4108/icst.mobihealth.2014.257440

Hismanoglu, M. (2011). Important issues in online education: E-pedagogy and marketing. In U. Demiray & S. Sever (Eds.), *Marketing online education programs: Frameworks for promotion and communication* (pp. 184–209). Hershey, PA: IGI Global. doi:10.4018/978-1-60960-074-7.ch012

Ho, K. K. (2008). The e-government development, IT strategies, and portals of the Hong Kong SAR government. In A. Anttiroiko (Ed.), *Electronic government: Concepts, methodologies, tools, and applications* (pp. 715–733). Hershey, PA: IGI Global. doi:10.4018/978-1-59904-947-2.ch060

Holden, S. H. (2003). The evolution of information technology management at the federal level: Implications for public administration. In G. Garson (Ed.), *Public information technology: Policy and management issues* (pp. 53–73). Hershey, PA: IGI Global. doi:10.4018/978-1-59140-060-8.ch003

Holden, S. H. (2007). The evolution of federal information technology management literature: Does IT finally matter? In G. Garson (Ed.), *Modern public information technology systems: Issues and challenges* (pp. 17–34). Hershey, PA: IGI Global. doi:10.4018/978-1-59904-051-6.ch002

Holland, J. W. (2009). Automation of American criminal justice. In M. Khosrow-Pour (Ed.), *Encyclopedia of information science and technology* (2nd ed.; pp. 300–302). Hershey, PA: IGI Global. doi:10.4018/978-1-60566-026-4.ch051

Holloway, K. (2013). Fair use, copyright, and academic integrity in an online academic environment. In *Digital rights management: Concepts, methodologies, tools, and applications* (pp. 917–928). Hershey, PA: IGI Global. doi:10.4018/978-1-4666-2136-7.ch044

Holmér, I. (2005). Protection against the cold. In R. Shishoo (Ed.), *Textiles in sport* (pp. 262–286). Cambridge, UK: Woodhead Publishing. doi:10.1533/9781845690885.4.262

Holmér, I. (2009). Evaluation of cold workplaces: An overview of standards for assessment of cold stress. *Industrial Health*, *47*(3), 228–234. doi:10.2486/indhealth.47.228 PMID:19531908

Horiuchi, C. (2005). E-government databases. In L. Rivero, J. Doorn, & V. Ferraggine (Eds.), *Encyclopedia of database technologies and applications* (pp. 206–210). Hershey, PA: IGI Global. doi:10.4018/978-1-59140-560-3.ch035

Horiuchi, C. (2006). Creating IS quality in government settings. In E. Duggan & J. Reichgelt (Eds.), *Measuring information systems delivery quality* (pp. 311–327). Hershey, PA: IGI Global. doi:10.4018/978-1-59140-857-4.ch014

Hsiao, N., Chu, P., & Lee, C. (2012). Impact of e-governance on businesses: Model development and case study. In *Digital democracy: Concepts, methodologies, tools, and applications* (pp. 1407–1425). Hershey, PA: IGI Global. doi:10.4018/978-1-4666-1740-7.ch070

Huang, J., Badam, A., Chandra, R., & Nightingale, E. B. (2015). *WearDrive: Fast and Energy efficient storage for Wearables.* Available online: http://research.microsoft.com/pubs/24461/weardrive.pdf

Huang, F. C., Luebke, D., & Wetzstein, G. (2015). The light field stereoscope. In *Proceedings of ACM SIGGRAPH 2015 Emerging Technologies* (pp. 24-24). New York, NY: ACM.

Huang, T., & Lee, C. (2010). Evaluating the impact of e-government on citizens: Cost-benefit analysis. In C. Reddick (Ed.), *Citizens and e-government: Evaluating policy and management* (pp. 37–52). Hershey, PA: IGI Global. doi:10.4018/978-1-61520-931-6.ch003

Hunter, M. G., Diochon, M., Pugsley, D., & Wright, B. (2002). Unique challenges for small business adoption of information technology: The case of the Nova Scotia ten. In S. Burgess (Ed.), *Managing information technology in small business: Challenges and solutions* (pp. 98–117). Hershey, PA: IGI Global. doi:10.4018/978-1-930708-35-8.ch006

Hurskainen, J. (2003). Integration of business systems and applications in merger and alliance: Case metso automation. In T. Reponen (Ed.), *Information technology enabled global customer service* (pp. 207–225). Hershey, PA: IGI Global. doi:10.4018/978-1-59140-048-6.ch012

Iazzolino, G., & Pietrantonio, R. (2011). The soveria.it project: A best practice of e-government in southern Italy. In D. Piaggesi, K. Sund, & W. Castelnovo (Eds.), *Global strategy and practice of e-governance: Examples from around the world* (pp. 34–56). Hershey, PA: IGI Global. doi:10.4018/978-1-60960-489-9.ch003

i. Martinez, A. C. (2008). Accessing administration's information via internet in Spain. In F. Tan (Ed.), *Global information technologies: Concepts, methodologies, tools, and applications* (pp. 2558–2573). Hershey, PA: IGI Global. doi:10.4018/978-1-59904-939-7.ch186

Imran, A., & Gregor, S. (2012). A process model for successful e-government adoption in the least developed countries: A case of Bangladesh. In F. Tan (Ed.), *International comparisons of information communication technologies: Advancing applications* (pp. 321–350). Hershey, PA: IGI Global. doi:10.4018/978-1-61350-480-2.ch014

Inoue, Y., & Bell, S. T. (2005). Electronic/digital government innovation, and publishing trends with IT. In M. Khosrow-Pour (Ed.), *Encyclopedia of information science and technology* (pp. 1018–1023). Hershey, PA: IGI Global. doi:10.4018/978-1-59140-553-5.ch180

Islam, M. M., & Ehsan, M. (2013). Understanding e-governance: A theoretical approach. In M. Islam & M. Ehsan (Eds.), *From government to e-governance: Public administration in the digital age* (pp. 38–49). Hershey, PA: IGI Global. doi:10.4018/978-1-4666-1909-8.ch003

ISO 7730. (1995). *Moderate Thermal Environments – Determination of the PMV and PPD Indices and Specification of the Conditions for Thermal Comfort.* International Organization for Standardization.

Itoh, Y., & Klinker, G. (2014). Interaction-free calibration for optical see-through head-mounted displays based on 3d eye localization. In *IEEE Symposium on 3D User Interfaces* (pp. 75-82). Minneapolis, MN: IEEE.

Itoh, Y., Orlosky, J., Huber, M., Kiyokawa, K., & Klinker, G. (2016). OST Rift: Temporally consistent augmented reality with a consumer optical see-through head-mounted display. In *IEEE Virtual Reality Conference* (pp. 189-190). Greenville, SC: IEEE. doi:10.1109/VR.2016.7504717

Iwaya, L. H., Gomes, M. A. L., Simplício, M. A., Carvalho, T. C. M. B., Dominicini, C. K., Sakuragui, R. R. M., & Håkansson, P. et al. (2013). Mobile health in emerging countries: A survey of research initiatives in Brazil. *International Journal of Medical Informatics*, *82*(5), 283–298. doi:10.1016/j.ijmedinf.2013.01.003 PMID:23410658

J., R.-T., Roa, L. M., & Prado, M. (2006). Design of antennas for a wearable sensor for homecare movement monitoring. *IEEE Engineering in Medicine and Biology Society*, 5972–5976.

Jabbar, H., Song, Y. S., & Jeong, T. T. (2010). RF energy harvesting system and circuits for charging of mobile devices. *IEEE Transactions on Consumer Electronics*, *56*(1), 247–253. doi:10.1109/TCE.2010.5439152

Jacobs, A. (2009). The pathologies of big data. *Communications of the ACM*, *52*(8), 36–44. doi:10.1145/1536616.1536632

Jaeger, B. (2009). E-government and e-democracy in the making. In M. Khosrow-Pour (Ed.), *Encyclopedia of information science and technology* (2nd ed.; pp. 1318–1322). Hershey, PA: IGI Global. doi:10.4018/978-1-60566-026-4.ch208

Jain, R. B. (2007). Revamping the administrative structure and processes in India for online diplomacy. In A. Anttiroiko & M. Malkia (Eds.), *Encyclopedia of digital government* (pp. 1418–1423). Hershey, PA: IGI Global. doi:10.4018/978-1-59140-789-8.ch217

Jain, R. B. (2008). Revamping the administrative structure and processes in India for online diplomacy. In A. Anttiroiko (Ed.), *Electronic government: Concepts, methodologies, tools, and applications* (pp. 3142–3149). Hershey, PA: IGI Global. doi:10.4018/978-1-59904-947-2.ch233

Jauhiainen, J. S., & Inkinen, T. (2009). E-governance and the information society in periphery. In C. Reddick (Ed.), *Handbook of research on strategies for local e-government adoption and implementation: Comparative studies* (pp. 497–514). Hershey, PA: IGI Global. doi:10.4018/978-1-60566-282-4.ch026

Jensen, M. J. (2009). Electronic democracy and citizen influence in government. In C. Reddick (Ed.), *Handbook of research on strategies for local e-government adoption and implementation: Comparative studies* (pp. 288–305). Hershey, PA: IGI Global. doi:10.4018/978-1-60566-282-4.ch015

Jiang, P., Winkley, J., Zhao, C., Munnoch, R., Min, G., & Yang, L. T. (2014). *An intelligent information forwarder for healthcare big data systems with distributed wearable sensors*. Retrieved from http://ieeexplore.ieee.org/xpls/abs_all.jsp?arnumber=6775278

Jiao, Y., Hurson, A. R., Potok, T. E., & Beckerman, B. G. (2009). Integrating mobile-based systems with healthcare databases. In J. Erickson (Ed.), *Database technologies: Concepts, methodologies, tools, and applications* (pp. 484–504). Hershey, PA: IGI Global. doi:10.4018/978-1-60566-058-5.ch031

Joia, L. A. (2002). A systematic model to integrate information technology into metabusinesses: A case study in the engineering realms. In F. Tan (Ed.), *Advanced topics in global information management* (Vol. 1, pp. 250–267). Hershey, PA: IGI Global. doi:10.4018/978-1-930708-43-3.ch016

Jones, T. H., & Song, I. (2000). Binary equivalents of ternary relationships in entity-relationship modeling: A logical decomposition approach. *Journal of Database Management*, *11*(2), 12–19. doi:10.4018/jdm.2000040102

Jr, H., & Williams, C. S. (1965). Transients in wide-angle conical antennas. *IEEE Transactions on Antennas and Propagation*, *13*(2), 236–246. doi:10.1109/TAP.1965.1138399

Juana-Espinosa, S. D. (2007). Empirical study of the municipalitites' motivations for adopting online presence. In L. Al-Hakim (Ed.), *Global e-government: Theory, applications and benchmarking* (pp. 261–279). Hershey, PA: IGI Global. doi:10.4018/978-1-59904-027-1.ch015

Jung, A., Schill, W. B., & Schuppe, H. C. (2005). Improvement of semen quality by nocturnal scrotal cooling in oligozoospermic men with a history of testicular maldescent. *International Journal of Andrology*, *28*(2), 93–98. doi:10.1111/j.1365-2605.2004.00517.x PMID:15811070

Jun, K., & Weare, C. (2012). Bridging from e-government practice to e-government research: Past trends and future directions. In K. Bwalya & S. Zulu (Eds.), *Handbook of research on e-government in emerging economies: Adoption, e-participation, and legal frameworks* (pp. 263–289). Hershey, PA: IGI Global. doi:10.4018/978-1-4666-0324-0.ch013

Junqueira, A., Diniz, E. H., & Fernandez, M. (2010). Electronic government implementation projects with multiple agencies: Analysis of the electronic invoice project under PMBOK framework. In J. Cordoba-Pachon & A. Ochoa-Arias (Eds.), *Systems thinking and e-participation: ICT in the governance of society* (pp. 135–153). Hershey, PA: IGI Global. doi:10.4018/978-1-60566-860-4.ch009

Juntunen, A. (2009). Joint service development with the local authorities. In C. Reddick (Ed.), *Handbook of research on strategies for local e-government adoption and implementation: Comparative studies* (pp. 902–920). Hershey, PA: IGI Global. doi:10.4018/978-1-60566-282-4.ch047

Kaija, T., Lilja, J., & Salonen, P. (2010). Exposing textile antennas for harsh environment. In *Military Communications Conference (MILCOM)* (pp. 737–742).

Kaivanto, E., Berg, M., Salonen, E., & Maagt, P. (2011). Wearable circularly polarized antenna for personal satellite communication and navigation. *IEEE Transactions on Antennas and Propagation*, *59*(12), 4490–4496. doi:10.1109/TAP.2011.2165513

Kaldeli, E., Warriach, E. U., Lazovik, A., & Aiello, M. (2013). Coordinating the web of services for a smart home. *ACM Transactions on the Web*, *7*(2), 10. doi:10.1145/2460383.2460389

Kamel, S. (2001). *Using DSS for crisis management*. Hershey, PA: IGI Global. doi:10.4018/978-1-87828-961-2.ch020

Kamel, S. (2006). DSS for strategic decision making. In M. Khosrow-Pour (Ed.), *Cases on information technology and organizational politics & culture* (pp. 230–246). Hershey, PA: IGI Global. doi:10.4018/978-1-59904-411-8.ch015

Kamel, S. (2009). The software industry in Egypt as a potential contributor to economic growth. In M. Khosrow-Pour (Ed.), *Encyclopedia of information science and technology* (2nd ed.; pp. 3531–3537). Hershey, PA: IGI Global. doi:10.4018/978-1-60566-026-4.ch562

Kamel, S., & Hussein, M. (2008). Xceed: Pioneering the contact center industry in Egypt. *Journal of Cases on Information Technology*, *10*(1), 67–91. doi:10.4018/jcit.2008010105

Kamel, S., & Wahba, K. (2003). The use of a hybrid model in web-based education: "The Global campus project. In A. Aggarwal (Ed.), *Web-based education: Learning from experience* (pp. 331–346). Hershey, PA: IGI Global. doi:10.4018/978-1-59140-102-5.ch020

Kardaras, D. K., & Papathanassiou, E. A. (2008). An exploratory study of the e-government services in Greece. In G. Garson & M. Khosrow-Pour (Eds.), *Handbook of research on public information technology* (pp. 162–174). Hershey, PA: IGI Global. doi:10.4018/978-1-59904-857-4.ch016

Kartsakli, E., Lalos, A S., Antonopoulos, A., Tennina, S., Di Renzo, M., Alonso, L., & Verikoukis, C. (2014). A Survey on M2M Systems for mHealth: A wireless communication perspective. *Sensors, 14*(10), 18009 – 18052. DOI: 10.3390/s141018009

Kassahun, A. E., Molla, A., & Sarkar, P. (2012). Government process reengineering: What we know and what we need to know. In *Digital democracy: Concepts, methodologies, tools, and applications* (pp. 1730–1752). Hershey, PA: IGI Global. doi:10.4018/978-1-4666-1740-7.ch086

Khaleel, H., Al-Rizzo, H. M., Rucker, D. G., & Mohan, S. (2012). A compact polyimide based UWB antenna for flexible electronics. *Antennas and Wireless Propagation Letters*, *11*, 564–567. doi:10.1109/LAWP.2012.2199956

Khalifa, S., Hassan, M., & Seneviratne, A. (2015c). Step detection from power generation pattern in energy-harvesting wearable devices. *Proceedings of the 8th IEEE International Conference on Internet of Things (iThings 2015)*. doi:10.1109/DSDIS.2015.102

Khalifa, S., Hassan, M., & Seneviratne, A. (2016b). *Feasibility and Accuracy of Hotword Detection using Vibration Energy Harvester*. In the 17th International Symposium on A World Of Wireless, Mobile And Multimedia Networks (WoWMoM), Coimbra, Portugal. doi:10.1109/WoWMoM.2016.7523555

Khalifa, S., Lan, G., Hassan, M., & Hu, W. (2016a). *A Bayesian framework for energy-neutral activity monitoring with self-powered wearable sensors*, In the 12th IEEE PerCom Workshop on Context and Activity Modeling and Recognition, Sydney, Australia. doi:10.1109/PERCOMW.2016.7457112

Khalifa, S., Hassan, M., & Seneviratne, A. (2013). *Adaptive pedestrian activity classification for indoor dead reckoning systems*. In *International Conference on Indoor Positioning and Indoor Navigation(IPIN13)*, Montbeliard- Belfort, France. doi:10.1109/IPIN.2013.6817868

Khalifa, S., Hassan, M., & Seneviratne, A. (2015a). Pervasive self-powered human activity recognition without the accelerometer. *Proceedings of the International Conference on Pervasive Computing and Communication (PerCom)*. doi:10.1109/PERCOM.2015.7146512

Khalifa, S., Hassan, M., Seneviratne, A., & Das, S. K. (2015b). Energy harvesting wearables for activity-aware services. *IEEE Internet Computing*, *19*(5), 8–16. doi:10.1109/MIC.2015.115

Khan, A. M., Lee, Y., & Kim, T.-S. (2008). Accelerometer signal-based human activity recognition using augmented autoregressive model coefficients and artificial neural nets. *Proceedings 30th annual International Conference of the IEEE Engineering in Medicine and Biology Society*. doi:10.1109/IEMBS.2008.4650379

Khan, B. (2005). Technological issues. In B. Khan (Ed.), *Managing e-learning strategies: Design, delivery, implementation and evaluation* (pp. 154–180). Hershey, PA: IGI Global. doi:10.4018/978-1-59140-634-1.ch004

Khasawneh, A., Bsoul, M., Obeidat, I., & Al Azzam, I. (2012). Technology fears: A study of e-commerce loyalty perception by Jordanian customers. In J. Wang (Ed.), *Advancing the service sector with evolving technologies: Techniques and principles* (pp. 158–165). Hershey, PA: IGI Global. doi:10.4018/978-1-4666-0044-7.ch010

Khatibi, V., & Montazer, G. A. (2012). E-research methodology. In A. Juan, T. Daradoumis, M. Roca, S. Grasman, & J. Faulin (Eds.), *Collaborative and distributed e-research: Innovations in technologies, strategies and applications* (pp. 62–81). Hershey, PA: IGI Global. doi:10.4018/978-1-4666-0125-3.ch003

Kidd, T. (2011). The dragon in the school's backyard: A review of literature on the uses of technology in urban schools. In L. Tomei (Ed.), *Online courses and ICT in education: Emerging practices and applications* (pp. 242–257). Hershey, PA: IGI Global. doi:10.4018/978-1-60960-150-8.ch019

Kidd, T. T. (2010). My experience tells the story: Exploring technology adoption from a qualitative perspective - A pilot study. In H. Song & T. Kidd (Eds.), *Handbook of research on human performance and instructional technology* (pp. 247–262). Hershey, PA: IGI Global. doi:10.4018/978-1-60566-782-9.ch015

Kieley, B., Lane, G., Paquet, G., & Roy, J. (2002). e-Government in Canada: Services online or public service renewal? In Å. Grönlund (Ed.), Electronic government: Design, applications and management (pp. 340-355). Hershey, PA: IGI Global. doi:10.4018/978-1-930708-19-8.ch016

Kim, H., Kim, J., & Kim, J. (2011). A review of piezoelectric energy harvesting based on vibration. International Journal of Precision Engineering and Manufacturing, 12(6), 1129–1141. doi:10.1007/s12541-011-0151-3

Kim, J., & Rahmat-Samii, Y. (2004). Implanted antennas inside a human body: Simulations, designs, and characterizations. *IEEE Transactions on Microwave Theory and Techniques*, *52*(8), 1934–1943. doi:10.1109/TMTT.2004.832018

Kim, P. (2012). "Stay out of the way! My kid is video blogging through a phone!": A lesson learned from math tutoring social media for children in underserved communities. In *Wireless technologies: Concepts, methodologies, tools and applications* (pp. 1415–1428). Hershey, PA: IGI Global. doi:10.4018/978-1-61350-101-6.ch517

Kim, Y., Kim, H., & Yoo, H. (2010). Electrical characterization of screen-printed circuits on the fabric. *IEEE Transactions on Advanced Packaging*, *33*(1), 196–205. doi:10.1109/TADVP.2009.2034536

Kiourti, A., Member, S., Costa, J. R., Member, S., Fernandes, C. A., Member, S., & Member, S. et al. (2012). Miniature Implantable Antennas for Biomedical Telemetry : From Simulation to Realization. *IEEE Transactions on Bio-Medical Engineering*, *59*(11), 3140–3147. doi:10.1109/TBME.2012.2202659 PMID:22692865

Kirlidog, M. (2010). Financial aspects of national ICT strategies. In S. Kamel (Ed.), *E-strategies for technological diffusion and adoption: National ICT approaches for socioeconomic development* (pp. 277–292). Hershey, PA: IGI Global. doi:10.4018/978-1-60566-388-3.ch016

Kisielnicki, J. (2006). Transfer of information and knowledge in the project management. In E. Coakes & S. Clarke (Eds.), *Encyclopedia of communities of practice in information and knowledge management* (pp. 544–551). Hershey, PA: IGI Global. doi:10.4018/978-1-59140-556-6.ch091

Kittner, M., & Van Slyke, C. (2006). Reorganizing information technology services in an academic environment. In M. Khosrow-Pour (Ed.), *Cases on the human side of information technology* (pp. 49–66). Hershey, PA: IGI Global. doi:10.4018/978-1-59904-405-7.ch004

Kiyokawa, K. (2015). Head-mounted display technologies for augmented reality. In W. Barfield (Ed.), *Fundamentals of Wearable Computers and Augmented Reality* (Vol. 1, pp. 59–84). Boca Raton, FL: CRC Press. doi:10.1201/b18703-7

Kleinberger, T., Becker, M., Ras, E., Holzinger, A., & Müller, P. (2007). Ambient Intelligence in Assisted Living: Enable Elderly People to Handle Future Interfaces. In Universal access in human-computer interaction. ambient interaction (pp. 103–112). Academic Press.

Klingeberg, T., & Schilling, M. (2012). Mobile wearable device for long term monitoring of vital signs. *Computer Methods and Programs in Biomedicine*, *106*(2), 89–96. doi:10.1016/j.cmpb.2011.12.009 PMID:22285459

Knoell, H. D. (2008). Semi virtual workplaces in German financial service enterprises. In P. Zemliansky & K. St.Amant (Eds.), *Handbook of research on virtual workplaces and the new nature of business practices* (pp. 570–581). Hershey, PA: IGI Global. doi:10.4018/978-1-59904-893-2.ch041

Koh, S. L., & Maguire, S. (2009). Competing in the age of information technology in a developing economy: Experiences of an Indian bank. In S. Koh & S. Maguire (Eds.), *Information and communication technologies management in turbulent business environments* (pp. 326–350). Hershey, PA: IGI Global. doi:10.4018/978-1-60566-424-8.ch018

Kollmann, T., & Häsel, M. (2009). Competence of information technology professionals in internet-based ventures. In A. Cater-Steel (Ed.), *Information technology governance and service management: Frameworks and adaptations* (pp. 239–253). Hershey, PA: IGI Global. doi:10.4018/978-1-60566-008-0.ch013

Kollmann, T., & Häsel, M. (2009). Competence of information technology professionals in internet-based ventures. In I. Lee (Ed.), *Electronic business: Concepts, methodologies, tools, and applications* (pp. 1905–1919). Hershey, PA: IGI Global. doi:10.4018/978-1-60566-056-1.ch118

Kollmann, T., & Häsel, M. (2010). Competence of information technology professionals in internet-based ventures. In *Electronic services: Concepts, methodologies, tools and applications* (pp. 1551–1565). Hershey, PA: IGI Global. doi:10.4018/978-1-61520-967-5.ch094

Kong, J. H., Ong, J. J., Ang, L. M., & Seng, K. P. (2012). Low Complexity processor design for Energy – Efficient security and error correction in wireless sensor networks. In *Wireless Sensor Networks and Energy Efficiency: Protocols, Routing and Management*. IGI Global. DOI: 10.4018/978-1-4666-0101-7.ch017

Koohestani, M., Zurcher, J.-F., Moreira, A. A., & Skrivervik, A. K. (2014). A Novel, Low-Profile, Vertically-Polarized UWB Antenna for WBAN. *IEEE Transactions on Antennas and Propagation*, *62*(4), 1888–1894. doi:10.1109/TAP.2014.2298886

Koo, T., Kim, D., Ryu, J., Seo, H., Yook, J., & Kim, J. (2011). Design of a label typed UHF RFID tag antenna for metallic objects. *Antennas and Wireless Propagation Letters*, *10*, 1010–1014. doi:10.1109/LAWP.2011.2166370

Kosir, S. (2015). *Wearables in Healthcare*. Available online at: https://www.wearable-technologies.com/2015/04/wearables-in-healthcare/

Kraemer, K., & King, J. L. (2006). Information technology and administrative reform: Will e-government be different? *International Journal of Electronic Government Research*, *2*(1), 1–20. doi:10.4018/jegr.2006010101

Kraemer, K., & King, J. L. (2008). Information technology and administrative reform: Will e-government be different? In D. Norris (Ed.), *E-government research: Policy and management* (pp. 1–20). Hershey, PA: IGI Global. doi:10.4018/978-1-59904-913-7.ch001

Kramida, G. (2016). Resolving the vergence-accommodation conflict in head-mounted displays. *IEEE Transactions on Visualization and Computer Graphics*, *22*(7), 1912-1931.

Kress, B. (2015). Optics for smart glasses, smart eyewear, augmented reality, and virtual reality headsets. In W. Barfield (Ed.), *Fundamentals of Wearable Computers and Augmented Reality* (Vol. 1, pp. 85–124). Boca Raton, FL: CRC Press. doi:10.1201/b18703-8

Kwapisz J. R., Weiss G. M., & Moore S. A. (2011). Activity recognition using cell phone accelerometers. *ACM SigKDD Explorations Newsletter, 12*(2).

L. Foundation Yocto. (2016, September). *Yocto project.* Retrieved September 21, 2016 from https://www.yoctoproject.org/

L. Foundation Zephyr. (2016, September). *Zephyr project.* Retrieved September 19, 2016, from https://www.zephyrproject.org/

Lampathaki, F., Tsiakaliaris, C., Stasis, A., & Charalabidis, Y. (2011). National interoperability frameworks: The way forward. In Y. Charalabidis (Ed.), *Interoperability in digital public services and administration: Bridging e-government and e-business* (pp. 1–24). Hershey, PA: IGI Global. doi:10.4018/978-1-61520-887-6.ch001

Lan, G., Khalifa, S., Hassan, M., & Hu, W. (2015). Estimating calorie expenditure from output voltage of piezoelectric energy harvester - an experimental feasibility study. *Proceedings of the 10th EAI International Conference on Body Area Networks (BodyNets).* doi:10.4108/eai.28-9-2015.2261453

Lan, G., Xu, W., Khalifa, S., Hassan, M., & Hu, W. (2016). *Transportation Mode Detection Using Kinetic Energy Harvesting Wearables.* In *WiP of the International Conference on Pervasive Computing and Communication (PerCom)*, Sydney, Australia. doi:10.1109/PERCOMW.2016.7457048

Lan, Z., & Scott, C. R. (1996). The relative importance of computer-mediated information versus conventional non-computer-mediated information in public managerial decision making. *Information Resources Management Journal, 9*(1), 27–0. doi:10.4018/irmj.1996010103

Lapinski, M., Feldmeier, M., & Paradiso, J. A. (2011). Wearable wireless sensing for sports and ubiquitous interactivity. *IEEE Sensors, 1425–1428.* doi:10.1109/ICSENS.2011.6126902

Law, W. (2004). *Public sector data management in a developing economy.* Hershey, PA: IGI Global. doi:10.4018/978-1-59140-259-6.ch034

Law, W. K. (2005). Information resources development challenges in a cross-cultural environment. In M. Khosrow-Pour (Ed.), *Encyclopedia of information science and technology* (pp. 1476–1481). Hershey, PA: IGI Global. doi:10.4018/978-1-59140-553-5.ch259

Law, W. K. (2009). Cross-cultural challenges for information resources management. In M. Khosrow-Pour (Ed.), *Encyclopedia of information science and technology* (2nd ed.; pp. 840–846). Hershey, PA: IGI Global. doi:10.4018/978-1-60566-026-4.ch136

Law, W. K. (2011). Cross-cultural challenges for information resources management. In *Global business: Concepts, methodologies, tools and applications* (pp. 1924–1932). Hershey, PA: IGI Global. doi:10.4018/978-1-60960-587-2.ch704

Lee, D., Dulai, G., & Karanassios, V. (2013). Survey of energy harvesting and energy scavenging approaches for on-site powering of wireless sensor and microinstrument-networks. In *Proceedings SPIE 8728, Energy Harvesting and Stor age.* Materials, Devices, and Applications.

Lee, G. Y., Psychoudakis, D., Chen, C. C., & Volakis, J. L. (2011). Omnidirectional vest-mounted body-worn antenna system for UHF operation. *IEEE Antennas and Wireless Propagation Letters, 10*, 581–583. doi:10.1109/LAWP.2011.2158381

Lefeuvre, E., Badel, A., Richard, C., Petit, L., & Guyomar, D. (2006). A comparison between several vibration-powered piezoelectric generators for standalone systems. *Sensors andActuators, 126*(2), 405–416. doi:10.1016/j.sna.2005.10.043

Leonardi-Bee, J., Bath, P. M. W., Phillips, S. J., & Sandercock, P. A. G. (2002). Blood pressure and clinical outcomes in the International Stroke Trial. *Stroke, 33*(5), 1315–1320. doi:10.1161/01.STR.0000014509.11540.66 PMID:11988609

Lewis, N. (2012). Remote Patient Monitoring Market to Double By 2016. *InformationWeek Healthcare.* Available Online: http://www.informationweek.com/mobile/remote-patient-monitoring-market-to-double-by-2016/d/d-id/1105484

Lilja, J., & Salonen, P. (2009). Textile material characterization for software antennas. In *IEEE Military Communications Conference* (pp. 1–7).

Lin, C. H., Li, Z., Ito, K., Takahashi, M., & Saito, K. (2012). A small tunable and wearable planar inverted-F antenna (PIFA). In *6th European Conference on Antennas and Propagation, EuCAP* (pp. 742–745). http://doi.org/ doi:10.1109/EuCAP.2012.6206554

Liu, R., & Lin, F. X. (2016). Understanding the characteristics of android wear os. *Proceedings of the 14th Annual International Conference on Mobile Systems, Applications, and Services, ser. MobiSys '16.* New York, NY: ACM. doi:10.1145/2906388.2906398

López-de-Ipiña, D., Lorido, T., & López, U. (2011). Blindshopping: enabling accessible shopping for visually impaired people through mobile technologies. In *Toward Useful Services for Elderly and People with Disabilities* (pp. 266–270). Springer. Retrieved from http://link-springer-com.ez67.periodicos.capes.gov.br/chapter/10.1007/978-3-642-21535-3_39

Lowe, D. G. (2004). Distinctive image features from scale invariant keypoints. *International Journal of Computer Vision, 60*(2), 91-110.

Lucas, I. R. B., & Aguiar, C. S. R. (2013). *Sistema de Reconhecimento de Ações - Uso para Monitoramento de Pacientes Hemiplégicos Vítimas de Acidente Vascular Cerebral (Monograph).* Universidade de Brasília.

Lymberis, A., & Dittmar, A. (2007, May). Advanced Wearable Health Systems and Applications. IEEE Engineering in Medicine and Biology Magazine, 29 – 33.

Lynch, W. C. (1972, July). Operating system performance. *Communications of the ACM, 15*(7), 579–585. doi:10.1145/361454.361476

Ma, L., Edwards, R., & Bashir, S. (2008). A wearable monopole antenna for ultra wideband with notching function. In *IET Seminar on Wideband and Ultrawideband Systems and Technologies: Evaluating current Research and Development* (pp. 1–5). doi:10.1049/ic.2008.0695

Machado, A., Padoin, E. L., Salvadori, F., Righi, L., Campos, M., Sausen, P. S., & Dill, S. (2008). Mobilidade no monitoramento de pacientes através de Serviços Web. In VII Simpósio de Informática da Região Centro do RS, 8.

Malkia, M., & Savolainen, R. (2004). eTransformation in government, politics and society: Conceptual framework and introduction. In M. Malkia, A. Anttiroiko, & R. Savolainen (Eds.), eTransformation in governance: New directions in government and politics (pp. 1-21). Hershey, PA: IGI Global. doi:10.4018/978-1-59140-130-8.ch001

Mandal, S., Turicchia, L., & Sarpeshkar, R. (2010). A low-power, battery-free tag for body sensor networks. *IEEE Pervasive Computing / IEEE Computer Society [and] IEEE Communications Society*, *9*(1), 71–77. doi:10.1109/MPRV.2010.1

Mandujano, S. (2011). Network manageability security. In D. Kar & M. Syed (Eds.), *Network security, administration and management: Advancing technology and practice* (pp. 158–181). Hershey, PA: IGI Global. doi:10.4018/978-1-60960-777-7.ch009

Mann, S. (2013). Vision 2.0. *IEEE Spectrum*, *50*(3), 64-102.

Mann, S. (1996, August). Smart clothing: The shift to wearable computing. *Communications of the ACM*, *39*(8), 23–24. doi:10.1145/232014.232021

Mann, S. (1997, February). Wearable computing: A first step toward personal imaging. *Computer*, *30*(2), 25–32. doi:10.1109/2.566147

Mann, S. (2002). The eyetap principle: Effectively locating the camera inside the eye as an alternative to wearable camera systems. In *Intelligent Image Processing* (pp. 64–102). New York, NY: John Wiley & Sons, Inc.

Mantash, M., Tarot, A., Collardey, S., & Mahdjoubi, K. (2011). Wearable monopole zip antenna. *Electronics Letters*, *47*(23), 1266–1267. doi:10.1049/el.2011.2784

Marich, M. J., Schooley, B. L., & Horan, T. A. (2012). A normative enterprise architecture for guiding end-to-end emergency response decision support. In M. Jennex (Ed.), *Managing crises and disasters with emerging technologies: Advancements* (pp. 71–87). Hershey, PA: IGI Global. doi:10.4018/978-1-4666-0167-3.ch006

Markets and Markets. (2015). *Reports*. Available online at: http://www.marketsandmarkets.com/Market-Reports/wearable-computing-market-125877882.html

Markov, R., & Okujava, S. (2008). Costs, benefits, and risks of e-government portals. In G. Putnik & M. Cruz-Cunha (Eds.), *Encyclopedia of networked and virtual organizations* (pp. 354–363). Hershey, PA: IGI Global. doi:10.4018/978-1-59904-885-7.ch047

Martin, W., Dancer, K., Rock, K., Zeleny, C., & Yelamarthi, K. (2009). The smart cane: an electrical engineering design project. In *ASEE North Central Section Conference*. Retrieved from http://people.cst.cmich.edu/yelam1k/CASE/Publications_files/Yelamarthi_ASEE_NCS_2009.pdf

Martin, N., & Rice, J. (2013). Evaluating and designing electronic government for the future: Observations and insights from Australia. In V. Weerakkody (Ed.), *E-government services design, adoption, and evaluation* (pp. 238–258). Hershey, PA: IGI Global. doi:10.4018/978-1-4666-2458-0.ch014

Matthews, J., & Pettitt, G. (2009). Development of flexible, wearable antennas. In *3rd European Conference on Antennas and Propagation (EuCAP)* (pp. 273–277).

Mazilu, S., Hardegger, M., Zhu, Z., Roggen, D., Troster, G., Plotnik, M., & Hausdorff, J. M. (2012). Online detection of freezing of gait with smartphones and machine learning techniques. In *Pervasive Computing Technologies for Healthcare (PervasiveHealth), 2012 6th International Conference on* (pp. 123–130). IEEE. Retrieved from http://ieeexplore.ieee.org/xpls/abs_all.jsp?arnumber=6240371

Mbarika, V. W., Meso, P. N., & Musa, P. F. (2006). A disconnect in stakeholders' perceptions from emerging realities of teledensity growth in Africa's least developed countries. In M. Hunter & F. Tan (Eds.), *Advanced topics in global information management* (Vol. 5, pp. 263–282). Hershey, PA: IGI Global. doi:10.4018/978-1-59140-923-6.ch012

Mbarika, V. W., Meso, P. N., & Musa, P. F. (2008). A disconnect in stakeholders' perceptions from emerging realities of teledensity growth in Africa's least developed countries. In F. Tan (Ed.), *Global information technologies: Concepts, methodologies, tools, and applications* (pp. 2948–2962). Hershey, PA: IGI Global. doi:10.4018/978-1-59904-939-7.ch209

McCann, J., Hurford, R., & Martin, A. (2005). A design process for the development of innovative smart clothing that addresses end-user needs from technical, functional, aesthetic and cultural view points. In *Ninth IEEE International Symposium on Wearable Computers (ISWC'05)* (pp. 70–77). http://doi.org/ doi:10.1109/ISWC.2005.3

Mcheick, H., Nasser, H., Dbouk, M., & Nasser, A. (2016). Stroke Prediction Context-Aware Health Care System. *2016 IEEE First International Conference on Connected Health: Applications, Systems and Engineering Technologies (CHASE)*, 30–35.

McLean, J. S. (1996). A re-examination of the fundamental limits on the radiation Q of\ nelectrically small antennas. *IEEE Transactions on Antennas and Propagation, 44*(5), 672–676. doi:10.1109/8.496253

Means, T., Olson, E., & Spooner, J. (2013). Discovering ways that don't work on the road to success: Strengths and weaknesses revealed by an active learning studio classroom project. In A. Benson, J. Moore, & S. Williams van Rooij (Eds.), *Cases on educational technology planning, design, and implementation: A project management perspective* (pp. 94–113). Hershey, PA: IGI Global. doi:10.4018/978-1-4666-4237-9.ch006

Medical Device Radiocommunications Service (MedRadio). (2009). Retrieved from www.fcc.gov

Melitski, J., Holzer, M., Kim, S., Kim, C., & Rho, S. (2008). Digital government worldwide: An e-government assessment of municipal web sites. In G. Garson & M. Khosrow-Pour (Eds.), *Handbook of research on public information technology* (pp. 790–804). Hershey, PA: IGI Global. doi:10.4018/978-1-59904-857-4.ch069

Memmola, M., Palumbo, G., & Rossini, M. (2009). Web & RFID technology: New frontiers in costing and process management for rehabilitation medicine. In L. Al-Hakim & M. Memmola (Eds.), *Business web strategy: Design, alignment, and application* (pp. 145–169). Hershey, PA: IGI Global. doi:10.4018/978-1-60566-024-0.ch008

Meng, Z., Fahong, Z., & Lei, L. (2008). Information technology and environment. In Y. Kurihara, S. Takaya, H. Harui, & H. Kamae (Eds.), *Information technology and economic development* (pp. 201–212). Hershey, PA: IGI Global. doi:10.4018/978-1-59904-579-5.ch014

Mentzingen de Moraes, A. J., Ferneda, E., Costa, I., & Spinola, M. D. (2011). Practical approach for implementation of governance process in IT: Information technology areas. In N. Shi & G. Silvius (Eds.), *Enterprise IT governance, business value and performance measurement* (pp. 19–40). Hershey, PA: IGI Global. doi:10.4018/978-1-60566-346-3.ch002

Merkouris, A., & Chorianopoulos, K. (2015). *Introducing Computer Programming to Children through Robotic and Wearable Devices.* Paper presented at WiPSCE 2015: The 10th Workshop in Primary and Secondary Computing Education. doi:10.1145/2818314.2818342

Merli, F. (2008). Implantable antennas for biomedical applications. In *Antennas and propagation society international Symposium* (Vol. 5110, pp. 1642–1649). doi:<ALIGNMENT.qj></ ALIGNMENT>10.5075/epfl-thesis-5110

Merwin, G. A. Jr, McDonald, J. S., & Odera, L. C. (2008). Economic development: Government's cutting edge in IT. In M. Raisinghani (Ed.), *Handbook of research on global information technology management in the digital economy* (pp. 1–37). Hershey, PA: IGI Global. doi:10.4018/978-1-59904-875-8.ch001

Meso, P. N., & Duncan, N. B. (2002). Can national information infrastructures enhance social development in the least developed countries? In F. Tan (Ed.), *Advanced topics in global information management* (Vol. 1, pp. 207–226). Hershey, PA: IGI Global. doi:10.4018/978-1-930708-43-3.ch014

Meso, P., & Duncan, N. (2002). Can national information infrastructures enhance social development in the least developed countries? An empirical investigation. In M. Dadashzadeh (Ed.), *Information technology management in developing countries* (pp. 23–51). Hershey, PA: IGI Global. doi:10.4018/978-1-931777-03-2.ch002

Mezghani, E., Exposito, E., Drira, K., Da Silveira, M., & Pruski, C. (2015). A Semantic Big Data Platform for Integrating Heterogeneous Wearable Data in Healthcare. *Journal of Medical Systems, 39*(12), 1–8. doi:10.1007/s10916-015-0344-x PMID:26490143

Michahelles, F., Matter, P., Schmidr, A., & Schiele, B. (2003). Applying Wearable Sensors to Avalanche Rescue: First experiences with a Novel Avalanche Beacon. *Computers and Graphics, 27*(6), 839 – 847. DOI: 10.1016/j.cag.2003.08.008

Micoach. (2014). Retrieved from http://micoach.adidas.com/

Microsoft. (2016, September). *Hololens*. Retrieved September 20, 2016 from https://www.microsoft.com/microsoft-hololens/en-us

Microsoft. (n.d.). *Microsoft HoloLens*. Retrieved February 21, 2017, from https://www.microsoft.com/microsoft-hololens/en-us

Middleton, M. (2008). Evaluation of e-government web sites. In G. Garson & M. Khosrow-Pour (Eds.), *Handbook of research on public information technology* (pp. 699–710). Hershey, PA: IGI Global. doi:10.4018/978-1-59904-857-4.ch063

Mingers, J. (2010). Pluralism, realism, and truth: The keys to knowledge in information systems research. In D. Paradice (Ed.), *Emerging systems approaches in information technologies: Concepts, theories, and applications* (pp. 86–98). Hershey, PA: IGI Global. doi:10.4018/978-1-60566-976-2.ch006

Miorandi, D., Sicari, S., De Pellegrini, F., & Chlamtac, I. (2012). Internet of Things: Vision, applications and research Challenges. *Ad Hoc Networks, 10*(7), 1497 – 1516. DOI:10.1016/j.adhoc.2012.02.016

Mirza, K., & Sarayeddine, K. (2015). *Key Challenges to Affordable See Through Wearable Displays: The Missing Link for Mobile AR Mass Deployment*. Retrieved February 21, 2017, from: http://www.optinvent.com/publications

Mital, K. M. (2012). ICT, unique identity and inclusive growth: An Indian perspective. In A. Manoharan & M. Holzer (Eds.), *E-governance and civic engagement: Factors and determinants of e-democracy* (pp. 584–612). Hershey, PA: IGI Global. doi:10.4018/978-1-61350-083-5.ch029

Mitcheson, P. D., Yeatman, E. M., Rao, G. K., Holmes, A. S., & Green, T. C. (2008, September). Energy harvesting from human and machine motion for wireless electronic devices. *Proceedings of the IEEE, 96*(9), 1457–1486. doi:10.1109/JPROC.2008.927494

Mizell, A. P. (2008). Helping close the digital divide for financially disadvantaged seniors. In F. Tan (Ed.), *Global information technologies: Concepts, methodologies, tools, and applications* (pp. 2396–2402). Hershey, PA: IGI Global. doi:10.4018/978-1-59904-939-7.ch173

Molinari, F., Wills, C., Koumpis, A., & Moumtzi, V. (2011). A citizen-centric platform to support networking in the area of e-democracy. In H. Rahman (Ed.), *Cases on adoption, diffusion and evaluation of global e-governance systems: Impact at the grass roots* (pp. 282–302). Hershey, PA: IGI Global. doi:10.4018/978-1-61692-814-8.ch014

Molinari, F., Wills, C., Koumpis, A., & Moumtzi, V. (2013). A citizen-centric platform to support networking in the area of e-democracy. In H. Rahman (Ed.), *Cases on progressions and challenges in ICT utilization for citizen-centric governance* (pp. 265–297). Hershey, PA: IGI Global. doi:10.4018/978-1-4666-2071-1.ch013

Monteverde, F. (2010). The process of e-government public policy inclusion in the governmental agenda: A framework for assessment and case study. In J. Cordoba-Pachon & A. Ochoa-Arias (Eds.), *Systems thinking and e-participation: ICT in the governance of society* (pp. 233–245). Hershey, PA: IGI Global. doi:10.4018/978-1-60566-860-4.ch015

Monti, G., Corchia, L., & Tarricone, L. (2013). UHF wearable rectenna on textile materials. *IEEE Transactions on Antennas and Propagation*, *61*(7), 3869–3873. doi:10.1109/TAP.2013.2254693

Moodley, S. (2008). Deconstructing the South African government's ICT for development discourse. In A. Anttiroiko (Ed.), *Electronic government: Concepts, methodologies, tools, and applications* (pp. 622–631). Hershey, PA: IGI Global. doi:10.4018/978-1-59904-947-2.ch053

Moodley, S. (2008). Deconstructing the South African government's ICT for development discourse. In C. Van Slyke (Ed.), *Information communication technologies: Concepts, methodologies, tools, and applications* (pp. 816–825). Hershey, PA: IGI Global. doi:10.4018/978-1-59904-949-6.ch052

Mora, M., Cervantes-Perez, F., Gelman-Muravchik, O., Forgionne, G. A., & Mejia-Olvera, M. (2003). DMSS implementation research: A conceptual analysis of the contributions and limitations of the factor-based and stage-based streams. In G. Forgionne, J. Gupta, & M. Mora (Eds.), *Decision-making support systems: Achievements and challenges for the new decade* (pp. 331–356). Hershey, PA: IGI Global. doi:10.4018/978-1-59140-045-5.ch020

Morrissey, M., & Rossi, R. M. (2013). Clothing systems for outdoor activities. *Textile Progress*, *45*(2-3), 145–181. doi:10.1080/00405167.2013.845540

Mörtberg, C., & Elovaara, P. (2010). Attaching people and technology: Between e and government. In S. Booth, S. Goodman, & G. Kirkup (Eds.), *Gender issues in learning and working with information technology: Social constructs and cultural contexts* (pp. 83–98). Hershey, PA: IGI Global. doi:10.4018/978-1-61520-813-5.ch005

Moustafa, H., Kenn, H., Sayrafian, K., Scanlon, W., & Zhang, Y. (2015). Mobile wearable communications [Guest Editorial]. *IEEE Wireless Communications*, *22*(1), 10–11. doi:10.1109/MWC.2015.7054713

Murphy, K. P. (2012). *Machine learning: a probabilistic perspective*. MIT Press. Retrieved from https://books.google.com.br/books?hl=pt-BR&lr=&id=NZP6AQAAQBAJ&oi=fnd&pg=PR7&dq=Machine+Learning+A+Probabilistic+Perspective&ots=KQTfz1Blgu&sig=IU4rkH5LF1lnicHaA1Ta0kvYccM

Murphy, J., Harper, E., Devine, E. C., Burke, L. J., & Hook, M. L. (2011). Case study: Lessons learned when embedding evidence-based knowledge in a nurse care planning and documentation system. In A. Cashin & R. Cook (Eds.), *Evidence-based practice in nursing informatics: Concepts and applications* (pp. 174–190). Hershey, PA: IGI Global. doi:10.4018/978-1-60960-034-1.ch014

Mutula, S. M. (2013). E-government's role in poverty alleviation: Case study of South Africa. In H. Rahman (Ed.), *Cases on progressions and challenges in ICT utilization for citizen-centric governance* (pp. 44–68). Hershey, PA: IGI Global. doi:10.4018/978-1-4666-2071-1.ch003

Nassani, A., Bai, H., Lee, G., & Billinghurst, M. (2015). Tag It! AR Annotation Using Wearable Sensors. *Proceedings of SIGGRAPH Asia 2015 Mobile Graphics and Interactive Applications*, 12:1—-12:4. doi:10.1145/2818427.2818438

Nath, R., & Angeles, R. (2005). Relationships between supply characteristics and buyer-supplier coupling in e-procurement: An empirical analysis. *International Journal of E-Business Research*, *1*(2), 40–55. doi:10.4018/jebr.2005040103

National Swiss Plan. (2014). *Swiss National Frequency Allocation Plan and Specific Assignments*. Author.

Nesbitt, P., Lam, Y., & Thompson, L. (1999). Human metabolism of mammalian lignan precursors in raw and processed flaxseed. *The American Journal of Clinical Nutrition*, *69*(3), 549–555. PMID:10075344

Nguyen, K. D., Chen, I. M., Luo, Z., Yeo, S. H., & Duh, H. B. L. (2011). A wearable sensing system for tracking and monitoring of functional arm movement. *IEEE/ASME Transactions on Mechatronics*, *16*(2), 213–220. doi:10.1109/TMECH.2009.2039222

Nielsen, B. (1978). Physiology of thermoregulation during swimming. In B. Eriksson & B. Furberg (Eds.), *Swimming Medicine IV* (pp. 297–303). Baltimore, MD: University Park Press.

Nikolaou, S., Ponchak, G., Papapolymerou, J., & Tentzeris, M. (2006). Conformal double exponentially tapered slot antenna (DETSA) on LCP for UWB applications. *IEEE Transactions on Antennas and Propagation*, *54*(6), 1663–1669. doi:10.1109/TAP.2006.875915

Nikolov, N. M., & Cunningham, A. J. (2003). Mild therapeutic hypothermia to improve the neurologic outcome after cardiac arrest. *Survey of Anesthesiology*, *47*(4), 219–220. doi:10.1097/01.sa.0000087691.31092.12

Nintanavongsa, P., Muncuk, U., Lewis, D. R., & Chowdhury, K. R. (2012). Design optimization and implementation for rf energy harvesting circuits. IEEE Journal on Emerging and Selected Topics in Circuits and Systems, 2(1), 24–33. doi:10.1109/JETCAS.2012.2187106

Nissen, M. E. (2006). Application cases in government. In M. Nissen (Ed.), *Harnessing knowledge dynamics: Principled organizational knowing & learning* (pp. 152–181). Hershey, PA: IGI Global. doi:10.4018/978-1-59140-773-7.ch008

Norris, D. F. (2003). Leading-edge information technologies and American local governments. In G. Garson (Ed.), *Public information technology: Policy and management issues* (pp. 139–169). Hershey, PA: IGI Global. doi:10.4018/978-1-59140-060-8.ch007

Norris, D. F. (2008). Information technology among U.S. local governments. In G. Garson & M. Khosrow-Pour (Eds.), *Handbook of research on public information technology* (pp. 132–144). Hershey, PA: IGI Global. doi:10.4018/978-1-59904-857-4.ch013

Northrop, A. (1999). The challenge of teaching information technology in public administration graduate programs. In G. Garson (Ed.), *Information technology and computer applications in public administration: Issues and trends* (pp. 1–22). Hershey, PA: IGI Global. doi:10.4018/978-1-87828-952-0.ch001

Northrop, A. (2003). Information technology and public administration: The view from the profession. In G. Garson (Ed.), *Public information technology: Policy and management issues* (pp. 1–19). Hershey, PA: IGI Global. doi:10.4018/978-1-59140-060-8.ch001

Northrop, A. (2007). Lip service? How PA journals and textbooks view information technology. In G. Garson (Ed.), *Modern public information technology systems: Issues and challenges* (pp. 1–16). Hershey, PA: IGI Global. doi:10.4018/978-1-59904-051-6.ch001

Null, E. (2013). Legal and political barriers to municipal networks in the United States. In A. Abdelaal (Ed.), *Social and economic effects of community wireless networks and infrastructures* (pp. 27–56). Hershey, PA: IGI Global. doi:10.4018/978-1-4666-2997-4.ch003

Numakura, D. (2007). Flexible Circuit Applications and Materials. In Printed Circuit Handbook (6th ed.). McGraw Hill.

Okamoto-Mizuno, K., Tsuzuki, K., Ohshiro, Y., & Mizuno, K. (2005). Effects of an electric blanket on sleep stages and body temperature in young men. *Ergonomics*, *48*(7), 749–757. doi:10.1080/00140130500120874 PMID:16076735

Okunoye, A., Frolick, M., & Crable, E. (2006). ERP implementation in higher education: An account of pre-implementation and implementation phases. *Journal of Cases on Information Technology*, *8*(2), 110–132. doi:10.4018/jcit.2006040106

Olasina, G. (2012). A review of egovernment services in Nigeria. In A. Tella & A. Issa (Eds.), *Library and information science in developing countries: Contemporary issues* (pp. 205–221). Hershey, PA: IGI Global. doi:10.4018/978-1-61350-335-5.ch015

Olivares, A., Grriz, J. M., & Olivares, G., (2010). A study of vibration-based energy harvesting in activities of daily living. *Proc. 4th International Conference on Pervasive Computing Technologies for Healthcare (PervasiveHealth).*

OPC Privacy Research Papers. (2014). Office of Privacy Commissioner of Canada. Available online: https://www.priv.gc.ca/information/research-recherche/2014/wc_201401_e.asp

Orgeron, C. P. (2008). A model for reengineering IT job classes in state government. In G. Garson & M. Khosrow-Pour (Eds.), *Handbook of research on public information technology* (pp. 735–746). Hershey, PA: IGI Global. doi:10.4018/978-1-59904-857-4.ch066

Orlando, F. (2015). *Gartner identifies top 10 strategic technology trends for 2016.* Available online www.gartner.com/newsroom/id/3143521

Orsi, G., & Tanca, L. (2010). Context modelling and context-aware querying: can datalog be of help*? Proceedings of the Datalog 2.0 Workshop.*

Osmani, V., Balasubramaniam, S., & Botvich, D. (2008). Human activity recognition in pervasive health-care: Supporting efficient remote collaboration. *Journal of Network and Computer Applications*, *31*(4), 628–655. doi:10.1016/j.jnca.2007.11.002

Ouyang, Y., & Chappell, W. (2008). High-frequency properties of electro-textiles for wearable antenna applications. *IEEE Transactions on Antennas and Propagation*, *56*(2), 381–389. doi:10.1109/TAP.2007.915435

Owsinski, J. W., & Pielak, A. M. (2011). Local authority websites in rural areas: Measuring quality and functionality, and assessing the role. In Z. Andreopoulou, B. Manos, N. Polman, & D. Viaggi (Eds.), *Agricultural and environmental informatics, governance and management: Emerging research applications* (pp. 39–60). Hershey, PA: IGI Global. doi:10.4018/978-1-60960-621-3.ch003

Owsiński, J. W., Pielak, A. M., Sęp, K., & Stańczak, J. (2014). Local web-based networks in rural municipalities: Extension, density, and meaning. In Z. Andreopoulou, V. Samathrakis, S. Louca, & M. Vlachopoulou (Eds.), *E-innovation for sustainable development of rural resources during global economic crisis* (pp. 126–151). Hershey, PA: IGI Global. doi:10.4018/978-1-4666-4550-9.ch011

Özdemir, A. T., & Barshan, B. (2014). Detecting falls with wearable sensors using machine learning techniques. *Sensors (Basel, Switzerland)*, *14*(6), 10691–10708. doi:10.3390/s140610691 PMID:24945676

Pagani, M., & Pasinetti, C. (2008). Technical and functional quality in the development of t-government services. In A. Anttiroiko (Ed.), *Electronic government: Concepts, methodologies, tools, and applications* (pp. 2943–2965). Hershey, PA: IGI Global. doi:10.4018/978-1-59904-947-2.ch220

Palma, R., Hartmann, J., & Haase, P. (2008). *Ontology Metadata Vocabulary for the Semantic Web*. OMV Consortium.

Pani, A. K., & Agrahari, A. (2005). On e-markets in emerging economy: An Indian experience. In M. Khosrow-Pour (Ed.), *Advanced topics in electronic commerce* (Vol. 1, pp. 287–299). Hershey, PA: IGI Global. doi:10.4018/978-1-59140-819-2.ch015

Papadopoulos, T., Angelopoulos, S., & Kitsios, F. (2011). A strategic approach to e-health interoperability using e-government frameworks. In A. Lazakidou, K. Siassiakos, & K. Ioannou (Eds.), *Wireless technologies for ambient assisted living and healthcare: Systems and applications* (pp. 213–229). Hershey, PA: IGI Global. doi:10.4018/978-1-61520-805-0.ch012

Papadopoulos, T., Angelopoulos, S., & Kitsios, F. (2013). A strategic approach to e-health interoperability using e-government frameworks. In *User-driven healthcare: Concepts, methodologies, tools, and applications* (pp. 791–807). Hershey, PA: IGI Global. doi:10.4018/978-1-4666-2770-3.ch039

Papakostas, G. I., Shelton, R. C., Kinrys, G., Henry, M. E., Bakow, B. R., Lipkin, S. H., & Bilello, J. A. et al. (2013). Assessment of a multi-assay, serum-based biological diagnostic test for major depressive disorder: A pilot and replication study. *Molecular Psychiatry*, *18*(3), 332–339. doi:10.1038/mp.2011.166 PMID:22158016

Papaleo, G., Chiarella, D., Aiello, M., & Caviglione, L. (2012). Analysis, development and deployment of statistical anomaly detection techniques for real e-mail traffic. In T. Chou (Ed.), *Information assurance and security technologies for risk assessment and threat management: Advances* (pp. 47–71). Hershey, PA: IGI Global. doi:10.4018/978-1-61350-507-6.ch003

Papp, R. (2003). Information technology & FDA compliance in the pharmaceutical industry. In M. Khosrow-Pour (Ed.), *Annals of cases on information technology* (Vol. 5, pp. 262–273). Hershey, PA: IGI Global. doi:10.4018/978-1-59140-061-5.ch017

Paris, D. G., & Miller, K. R. (2016). Wearables and People with Disabilities: Socio-Cultural and Vocational Implications. *Wearable Technology and Mobile Innovations for Next-Generation Education*, 167.

Park, S., & Jayaraman, S. (2003). Enhancing the Quality of Life Through Wearable Technology. *IEEE Engineering in Medicine and Biology Magazine*, *22*(June), 41–48. doi:10.1109/MEMB.2003.1213625 PMID:12845818

Park, S., & Jayaraman, S. (2003). Smart textiles: Wearable electronic systems. *MRS Bulletin*, *28*(8), 585–591. doi:10.1557/mrs2003.170

Parsons, K. (2003). *Human Thermal Environments. The Effects of Hot, Moderate and Cold Environments on Human Health, Comfort and Performance*. London, UK: Taylor and Francis.

Parsons, T. W. (2007). Developing a knowledge management portal. In A. Tatnall (Ed.), *Encyclopedia of portal technologies and applications* (pp. 223–227). Hershey, PA: IGI Global. doi:10.4018/978-1-59140-989-2.ch039

Passaris, C. E. (2007). Immigration and digital government. In A. Anttiroiko & M. Malkia (Eds.), *Encyclopedia of digital government* (pp. 988–994). Hershey, PA: IGI Global. doi:10.4018/978-1-59140-789-8.ch148

Pavlichev, A. (2004). The e-government challenge for public administration. In A. Pavlichev & G. Garson (Eds.), *Digital government: Principles and best practices* (pp. 276–290). Hershey, PA: IGI Global. doi:10.4018/978-1-59140-122-3.ch018

Pebble. (2016, September). *Pebble developer*. Retrieved September 19, 2016, from https://developer.pebble.com/

Pellegrini, A., Brizzi, A., Zhang, L., Ali, K., Hao, Y., Wu, X., & Sauleau, R. et al. (2013). Antennas and propagation for body-centric wireless communications at millimeter-wave frequencies: A review. *IEEE Antennas and Propagation Magazine*, *55*(4), 262–287. doi:10.1109/MAP.2013.6645205

Penrod, J. I., & Harbor, A. F. (2000). Designing and implementing a learning organization-oriented information technology planning and management process. In L. Petrides (Ed.), *Case studies on information technology in higher education: Implications for policy and practice* (pp. 7–19). Hershey, PA: IGI Global. doi:10.4018/978-1-878289-74-2.ch001

Pérez-Martínez, P. A., Martínez-Ballesté, A., & Solanas, A. (2013). *Privacy in Smart Cities-A Case Study of Smart Public Parking* (pp. 55–59). PECCS.

Perlini, N. M. O. G., & Mancussi e Faro, A. C. (2005). Taking care of persons handicapped by cerebral vascular accident at home: The familial caregiver activity. *Revista da Escola de Enfermagem da U S P.*, *39*, 154–163. doi:10.1590/S0080-62342005000200005 PMID:16060302

Petersen, M. K., Stahlhut, C., Stopczynski, A., Larsen, J. E., & Hansen, L. K. (2011). Smartphones Get Emotional: Mind Reading Images and Reconstructing the Neural Sources. *Proceedings of Affective Computing and Intelligent Interaction Conference*, 578 - 587. Available online: http://www2.imm.dtu.dk/pubdb/views/edoc_download.php/6124/pdf/ imm6124.pdf

Planas-Silva, M. D., & Joseph, R. C. (2011). Perspectives on the adoption of electronic resources for use in clinical trials. In M. Guah (Ed.), *Healthcare delivery reform and new technologies: Organizational initiatives* (pp. 19–28). Hershey, PA: IGI Global. doi:10.4018/978-1-60960-183-6.ch002

Plopski, A., Itoh, Y., Nitschke, C., Kiyokawa, K., Klinker, G., & Takemura, H. (2015). Corneal-imaging calibration for optical see-through head-mounted displays. *IEEE Transactions on Visualization and Computer Graphics, 21*(4), 481-490.

Pomazalová, N., & Rejman, S. (2013). The rationale behind implementation of new electronic tools for electronic public procurement. In N. Pomazalová (Ed.), *Public sector transformation processes and internet public procurement: Decision support systems* (pp. 85–117). Hershey, PA: Engineering Science Reference. doi:10.4018/978-1-4666-2665-2.ch006

Popovic, Z., Momenroodaki, P., & Scheeler, R. (2014). Toward wearable wireless thermometers for internal body temperature measurements. *IEEE Communications Magazine*, *52*(10), 118–125. doi:10.1109/MCOM.2014.6917412

Poppe, R. (2010). A survey on vision-based human action recognition. Image and Vision Computing Journal, 28(6). doi:10.1016/j.imavis.2009.11.014

Postorino, M. N. (2012). City competitiveness and airport: Information science perspective. In M. Bulu (Ed.), *City competitiveness and improving urban subsystems: Technologies and applications* (pp. 61–83). Hershey, PA: IGI Global. doi:10.4018/978-1-61350-174-0.ch004

Poupa, C. (2002). Electronic government in Switzerland: Priorities for 2001-2005 - Electronic voting and federal portal. In Å. Grönlund (Ed.), *Electronic government: Design, applications and management* (pp. 356–369). Hershey, PA: IGI Global. doi:10.4018/978-1-930708-19-8.ch017

Powell, S. R. (2010). Interdisciplinarity in telecommunications and networking. In *Networking and telecommunications: Concepts, methodologies, tools and applications* (pp. 33–40). Hershey, PA: IGI Global. doi:10.4018/978-1-60566-986-1.ch004

Priya, P. S., & Mathiyalagan, N. (2011). A study of the implementation status of two e-governance projects in land revenue administration in India. In M. Shareef, V. Kumar, U. Kumar, & Y. Dwivedi (Eds.), *Stakeholder adoption of e-government services: Driving and resisting factors* (pp. 214–230). Hershey, PA: IGI Global. doi:10.4018/978-1-60960-601-5.ch011

Prysby, C. L., & Prysby, N. D. (2003). Electronic mail in the public workplace: Issues of privacy and public disclosure. In G. Garson (Ed.), *Public information technology: Policy and management issues* (pp. 271–298). Hershey, PA: IGI Global. doi:10.4018/978-1-59140-060-8.ch012

Prysby, C. L., & Prysby, N. D. (2007). You have mail, but who is reading it? Issues of e-mail in the public workplace. In G. Garson (Ed.), *Modern public information technology systems: Issues and challenges* (pp. 312–336). Hershey, PA: IGI Global. doi:10.4018/978-1-59904-051-6.ch016

Prysby, C., & Prysby, N. (2000). Electronic mail, employee privacy and the workplace. In L. Janczewski (Ed.), *Internet and intranet security management: Risks and solutions* (pp. 251–270). Hershey, PA: IGI Global. doi:10.4018/978-1-878289-71-1.ch009

Psychoudakis, D., Lee, G., Chen, C., & Volakis, J. (2010). Military UHF body-worn antennas for armored vests. In *6th European Conference on Antennas and Propagation (EuCAP) IEEE* (pp. 1–4).

Psychoudakis, D., & Volakis, J. L. (2009). Conformal Asymmetric Meandered Flare (AMF) Antenna for Body-Worn Applications. *IEEE Antennas and Wireless Propagation Letters, 8*, 931–934. doi:10.1109/LAWP.2009.2028662

Qi, X., Keally, M., Zhou, G., Li, Y., & Ren, Z. (2013). Adasense: Adapting sampling rates for activity recognition in body sensor networks. *Proceedings of IEEE 19th Real-Time and Embedded Technology and Applications Symposium (RTAS).*

Radl, A., & Chen, Y. (2005). Computer security in electronic government: A state-local education information system. *International Journal of Electronic Government Research, 1*(1), 79–99. doi:10.4018/jegr.2005010105

Rahman, H. (2008). Information dynamics in developing countries. In C. Van Slyke (Ed.), *Information communication technologies: Concepts, methodologies, tools, and applications* (pp. 104–114). Hershey, PA: IGI Global. doi:10.4018/978-1-59904-949-6.ch008

Rahmat-Samii, Y. (2007). Wearable and implantable antennas in body-centric communications. In *2nd European Conference on Antennas and Propagation (EuCAP)* (pp. 1–5).

Rais, N., Soh, P., Malek, F., Ahmad, S., Hashim, N., & Hall, P. (2009). A review of wearable antenna. In *Loughborough Antennas and Propagation Conference(LAPC)* (pp. 225–228).

Raju, M. (2008). *Energy Harvesting ULP Meets Esnergy Harvesting: A Game-Changing Combination for Design Engineers*. Texas Instrument. Retrieved June 11, 2016, from http://www.ti.com/corp/docs/landing/cc430/graphics/slyy018_20081031.pdf

Ramanathan, J. (2009). Adaptive IT architecture as a catalyst for network capability in government. In P. Saha (Ed.), *Advances in government enterprise architecture* (pp. 149–172). Hershey, PA: IGI Global. doi:10.4018/978-1-60566-068-4.ch007

Ramos, I., & Berry, D. M. (2006). Social construction of information technology supporting work. In M. Khosrow-Pour (Ed.), *Cases on information technology: Lessons learned* (Vol. 7, pp. 36–52). Hershey, PA: IGI Global. doi:10.4018/978-1-59140-673-0.ch003

Ranck, J. (n.d.). *The wearable computing market: A global analysis*. Retrieved from http://go.gigaom.com/rs/gigaom/images/ wearable-computing-the-next-big-thing-in-tech.pdf

Randell, C., & Muller, H. (2000). Context awareness by analysing accelerometer data. *Digest of Papers. Fourth International Symposium on Wearable Computers*, 175–176. doi:10.1109/ISWC.2000.888488

Rao, Y., Cheng, S., & Arnold, D. P. (2013). An energy harvesting system for passively generating power from human activities. *Journal of Micromechanics and Microengineering*, *23*(11), 114012. doi:10.1088/0960-1317/23/11/114012

Ravi, N., Dandekar, N., Mysore, P., & Littman, M. L. (2005). Activity recognition from accelerometer data. *IAAI'05 Proceedings of the 17th conference on Innovative applications of artificial intelligence*.

Rawassizadeh, R., Price, B. A., & Petre, M. (2014). Wearables: Has the Age of Smartwatches Finally Arrived? *Communications of the ACM*, *58*(1), 45–47. doi:10.1145/2629633

Ray, D., Gulla, U., Gupta, M. P., & Dash, S. S. (2009). Interoperability and constituents of interoperable systems in public sector. In V. Weerakkody, M. Janssen, & Y. Dwivedi (Eds.), *Handbook of research on ICT-enabled transformational government: A global perspective* (pp. 175–195). Hershey, PA: IGI Global. doi:10.4018/978-1-60566-390-6.ch010

Reades, J., Calabrese, F., Sevtsuk, A., & Ratti, C. (2007). Cellular census: Explorations in urban data collection. *IEEE Pervasive Computing / IEEE Computer Society [and] IEEE Communications Society*, *6*(3), 30–38. doi:10.1109/MPRV.2007.53

Reddick, C. G. (2007). E-government and creating a citizen-centric government: A study of federal government CIOs. In G. Garson (Ed.), *Modern public information technology systems: Issues and challenges* (pp. 143–165). Hershey, PA: IGI Global. doi:10.4018/978-1-59904-051-6.ch008

Reddick, C. G. (2010). Citizen-centric e-government. In C. Reddick (Ed.), *Homeland security preparedness and information systems: Strategies for managing public policy* (pp. 45–75). Hershey, PA: IGI Global. doi:10.4018/978-1-60566-834-5.ch002

Reddick, C. G. (2010). E-government and creating a citizen-centric government: A study of federal government CIOs. In C. Reddick (Ed.), *Homeland security preparedness and information systems: Strategies for managing public policy* (pp. 230–250). Hershey, PA: IGI Global. doi:10.4018/978-1-60566-834-5.ch012

Reddick, C. G. (2010). Perceived effectiveness of e-government and its usage in city governments: Survey evidence from information technology directors. In C. Reddick (Ed.), *Homeland security preparedness and information systems: Strategies for managing public policy* (pp. 213–229). Hershey, PA: IGI Global. doi:10.4018/978-1-60566-834-5.ch011

Reddick, C. G. (2012). Customer relationship management adoption in local governments in the United States. In S. Chhabra & M. Kumar (Eds.), *Strategic enterprise resource planning models for e-government: Applications and methodologies* (pp. 111–124). Hershey, PA: IGI Global. doi:10.4018/978-1-60960-863-7.ch008

Reed, F. (2007). 60 GHz WPAN Standardization within IEEE. In *International Symposium on Signals, Systems and Electronics, (ISSSE '07)* (pp. 103–105).

Reeder, F. S., & Pandy, S. M. (2008). Identifying effective funding models for e-government. In A. Anttiroiko (Ed.), *Electronic government: Concepts, methodologies, tools, and applications* (pp. 1108–1138). Hershey, PA: IGI Global. doi:10.4018/978-1-59904-947-2.ch083

Resch, B. (2013). People as sensors and collective sensing-contextual observations complementing geo-sensor network measurements. In *Progress in Location-Based Services* (pp. 391–406). Springer. doi:10.1007/978-3-642-34203-5_22

Riesco, D., Acosta, E., & Montejano, G. (2003). An extension to a UML activity graph from workflow. In L. Favre (Ed.), *UML and the unified process* (pp. 294–314). Hershey, PA: IGI Global. doi:10.4018/978-1-93177-744-5.ch015

Riot, O. S. (2016, September). *Operating system for the internet of things*. Retrieved September 21, 2016 from https://riot-os.org/

Rishani, N. R., AI-Husseini, A., E.-H., & Kabalan, K. Y. (2012). Design and relative permittivity determination of an EBG-based wearable antenna. In Progress in Electromagnetics and Radio Frequency (pp. pp. 96–99). Moscow, Russia: PIERS.

Ritzhaupt, A. D., & Gill, T. G. (2008). A hybrid and novel approach to teaching computer programming in MIS curriculum. In S. Negash, M. Whitman, A. Woszczynski, K. Hoganson, & H. Mattord (Eds.), *Handbook of distance learning for real-time and asynchronous information technology education* (pp. 259–281). Hershey, PA: IGI Global. doi:10.4018/978-1-59904-964-9.ch014

Roche, E. M. (1993). International computing and the international regime. *Journal of Global Information Management*, *1*(2), 33–44. doi:10.4018/jgim.1993040103

Rocheleau, B. (2007). Politics, accountability, and information management. In G. Garson (Ed.), *Modern public information technology systems: Issues and challenges* (pp. 35–71). Hershey, PA: IGI Global. doi:10.4018/978-1-59904-051-6.ch003

Rodrigues Filho, J. (2010). E-government in Brazil: Reinforcing dominant institutions or reducing citizenship? In C. Reddick (Ed.), *Politics, democracy and e-government: Participation and service delivery* (pp. 347–362). Hershey, PA: IGI Global. doi:10.4018/978-1-61520-933-0.ch021

Rodriguez, S. R., & Thorp, D. A. (2013). eLearning for industry: A case study of the project management process. In A. Benson, J. Moore, & S. Williams van Rooij (Eds.), Cases on educational technology planning, design, and implementation: A project management perspective (pp. 319-342). Hershey, PA: IGI Global. doi:10.4018/978-1-4666-4237-9.ch017

Rofouei, M., Sinclair, M., Bittner, R., Blank, T., Saw, N., DeJean, G., & Heffron, J. (2011). A Non-invasive Wearable Neck-Cuff System for Real-Time Sleep Monitoring. *2011 International Conference on Body Sensor Networks*, 156–161. doi:10.1109/BSN.2011.38

Rolim, C. O., Koch, F. L., Westphall, C. B., Werner, J., Fracalossi, A., & Salvador, G. S. (2010). A Cloud Computing solution for patient's data collection in healthcare institutions. *Proceedings of the International Conference on eHealth, Telemedicine and Social Medicine*, 95 – 99. Doi:10.1109/eTELEMED.2010.19

Rolland, J. P., & Fuchs, H. (2000). Optical versus Video See-Through Head-Mounted Displays in Medical Visualization. *Presence*, *9*(3), 287-309.

Roman, A. V. (2013). Delineating three dimensions of e-government success: Security, functionality, and transformation. In J. Gil-Garcia (Ed.), *E-government success factors and measures: Theories, concepts, and methodologies* (pp. 171–192). Hershey, PA: IGI Global. doi:10.4018/978-1-4666-4058-0.ch010

Rosamond, W., Flegal, K., Furie, K., Go, A., & Greenlund, K. (2008). Heart Disease and Stroke Statistics—2008 Update. *Circulation*, *2008*(117), e25–e146. doi:10.1161/CIRCULATIONAHA.107.187998 PMID:18086926

Ross, S. C., Tyran, C. K., & Auer, D. J. (2008). Up in smoke: Rebuilding after an IT disaster. In H. Nemati (Ed.), *Information security and ethics: Concepts, methodologies, tools, and applications* (pp. 3659–3675). Hershey, PA: IGI Global. doi:10.4018/978-1-59904-937-3.ch248

Ross, S. C., Tyran, C. K., Auer, D. J., Junell, J. M., & Williams, T. G. (2005). Up in smoke: Rebuilding after an IT disaster. *Journal of Cases on Information Technology*, *7*(2), 31–49. doi:10.4018/jcit.2005040103

Roy, J. (2008). Security, sovereignty, and continental interoperability: Canada's elusive balance. In T. Loendorf & G. Garson (Eds.), *Patriotic information systems* (pp. 153–176). Hershey, PA: IGI Global. doi:10.4018/978-1-59904-594-8.ch007

Rubeck, R. F., & Miller, G. A. (2009). vGOV: Remote video access to government services. In A. Scupola (Ed.), Cases on managing e-services (pp. 253-268). Hershey, PA: IGI Global. doi:10.4018/978-1-60566-064-6.ch017

Running G. P. S. Watches. (2014). Retrieved from http://sports.tomtom.com/en_us/

Saekow, A., & Boonmee, C. (2011). The challenges of implementing e-government interoperability in Thailand: Case of official electronic correspondence letters exchange across government departments. In Y. Charalabidis (Ed.), *Interoperability in digital public services and administration: Bridging e-government and e-business* (pp. 40–61). Hershey, PA: IGI Global. doi:10.4018/978-1-61520-887-6.ch003

Saekow, A., & Boonmee, C. (2012). The challenges of implementing e-government interoperability in Thailand: Case of official electronic correspondence letters exchange across government departments. In *Digital democracy: Concepts, methodologies, tools, and applications* (pp. 1883–1905). Hershey, PA: IGI Global. doi:10.4018/978-1-4666-1740-7.ch094

Sagl, G., Resch, B., Hawelka, B., & Beinat, E. (2012). From social sensor data to collective human behaviour patterns: Analysing and visualising spatio-temporal dynamics in urban environments. In *Proceedings of the GI-Forum* (pp. 54–63). Retrieved from http://gispoint.de/fileadmin/user_upload/paper_gis_open/537521043.pdf

Sagsan, M., & Medeni, T. (2012). Understanding "knowledge management (KM) paradigms" from social media perspective: An empirical study on discussion group for KM at professional networking site. In M. Cruz-Cunha, P. Gonçalves, N. Lopes, E. Miranda, & G. Putnik (Eds.), *Handbook of research on business social networking: Organizational, managerial, and technological dimensions* (pp. 738–755). Hershey, PA: IGI Global. doi:10.4018/978-1-61350-168-9.ch039

Sahi, G., & Madan, S. (2013). Information security threats in ERP enabled e-governance: Challenges and solutions. In *Enterprise resource planning: Concepts, methodologies, tools, and applications* (pp. 825–837). Hershey, PA: IGI Global. doi:10.4018/978-1-4666-4153-2.ch048

Salman, L. K. H., & Talbi, L. (2010). G-shaped wearable cuff button antenna for 2.45 GHZ ISM band applications. In *2010 14th International Symposium on Antenna Technology and Applied Electromagnetics and the American Electromagnetics Conference, ANTEM/AMEREM 2010* (pp. 14–17). doi:<ALIGNMENT.qj></ALIGNMENT>10.1109/ANTEM.2010.5552573

Salonen, P., Rahmat-Samii, Y., Hurme, H., & Kivikoski, M. (2004). Effect of conductive material on wearable antenna performance: A case study of WLAN antennas. *Antennas and Propagation Society, 1*, 455–458. doi:10.1109/APS.2004.1329672

Salonen, P., Sydanheimo, L., Keskilammi, M., & Kivikoski, M. (1999). A Small Planar Inverted-F Antenna for Wearable Applications. In *IEEE Conference Publications* (pp. 95–100). doi:10.1109/ISWC.1999.806679

Samsung. (2016, September). *Tizen project*. Retrieved September 19, 2016, from https://www.tizen.org/

Sanford, C., & Bhattacherjee, A. (2008). IT implementation in a developing country municipality: A sociocognitive analysis. *International Journal of Technology and Human Interaction, 4*(3), 68–93. doi:10.4018/jthi.2008070104

Sankaralingam, S., & Gupta, B. (2009). A circular disk microstrip WLAN antenna for wearable applications. In *IEEE India Council Conference* (pp. 3–6). doi:10.1109/INDCON.2009.5409355

Santas, J., Alomainy, A., & Hao, Y. (2007). Textile Antennas for On-Body Communications: Techniques and Properties. In *2nd European Conference on Antennas and Propagation (EuCAP)* (pp. 1–4.). doi:10.1049/ic.2007.1064

Sanz-Izquierdo, B., Huang, F., & Batchelor, J. C. (2006). Small size wearable button antenna. In *First European Conference on Antennas and Propagation* (pp. 1–4).

Sanz-Izquierdo, B., Miller, J. A., Batchelor, J. C., & Sobhy, M. I. (2010). Dual-Band Wearable Metallic Button Antennas and Transmission in Body Area Networks. *IET Microw. Antennas Propag., 4*(2), 182–190. doi:10.1049/iet-map.2009.0010

Saponara, S., Donati, M., Bacchillone, T., Sanchez-Tato, I., Carmona, C., Fanucci, L., & Barba, P. (2012). Remote monitoring of vital signs in patients with Chronic Heart Failure: Sensor devices and data analysis perspective. *Sensors Applications Symposium (SAS)*, 1–6. doi:10.1109/SAS.2012.6166310

Schelin, S. H. (2003). E-government: An overview. In G. Garson (Ed.), *Public information technology: Policy and management issues* (pp. 120–138). Hershey, PA: IGI Global. doi:10.4018/978-1-59140-060-8.ch006

Schelin, S. H. (2004). Training for digital government. In A. Pavlichev & G. Garson (Eds.), *Digital government: Principles and best practices* (pp. 263–275). Hershey, PA: IGI Global. doi:10.4018/978-1-59140-122-3.ch017

Schelin, S. H. (2007). E-government: An overview. In G. Garson (Ed.), *Modern public information technology systems: Issues and challenges* (pp. 110–126). Hershey, PA: IGI Global. doi:10.4018/978-1-59904-051-6.ch006

Schelin, S. H., & Garson, G. (2004). Theoretical justification of critical success factors. In G. Garson & S. Schelin (Eds.), *IT solutions series: Humanizing information technology: Advice from experts* (pp. 4–15). Hershey, PA: IGI Global. doi:10.4018/978-1-59140-245-9.ch002

Scime, A. (2002). Information systems and computer science model curricula: A comparative look. In M. Dadashzadeh, A. Saber, & S. Saber (Eds.), *Information technology education in the new millennium* (pp. 146–158). Hershey, PA: IGI Global. doi:10.4018/978-1-931777-05-6.ch018

Scime, A. (2009). Computing curriculum analysis and development. In M. Khosrow-Pour (Ed.), *Encyclopedia of information science and technology* (2nd ed.; pp. 667–671). Hershey, PA: IGI Global. doi:10.4018/978-1-60566-026-4.ch108

Scime, A., & Wania, C. (2008). Computing curricula: A comparison of models. In C. Van Slyke (Ed.), *Information communication technologies: Concepts, methodologies, tools, and applications* (pp. 1270–1283). Hershey, PA: IGI Global. doi:10.4018/978-1-59904-949-6.ch088

Seidman, S. B. (2009). An international perspective on professional software engineering credentials. In H. Ellis, S. Demurjian, & J. Naveda (Eds.), *Software engineering: Effective teaching and learning approaches and practices* (pp. 351–361). Hershey, PA: IGI Global. doi:10.4018/978-1-60566-102-5.ch018

Seifert, J. W. (2007). E-government act of 2002 in the United States. In A. Anttiroiko & M. Malkia (Eds.), *Encyclopedia of digital government* (pp. 476–481). Hershey, PA: IGI Global. doi:10.4018/978-1-59140-789-8.ch072

Seifert, J. W., & Relyea, H. C. (2008). E-government act of 2002 in the United States. In A. Anttiroiko (Ed.), *Electronic government: Concepts, methodologies, tools, and applications* (pp. 154–161). Hershey, PA: IGI Global. doi:10.4018/978-1-59904-947-2.ch013

Sengupta, A., Rajan, V., Bhattacharya, S., & Sarma, G. R. K. (2016). A statistical model for stroke outcome prediction and treatment planning. *38th Annual International Conference of the IEEE Engineering in Medicine and Biology Society (EMBC)*, 2516–2519. doi:10.1109/EMBC.2016.7591242

Serra, A. A., Nepa, P., & Manara, G. (2011). Cospas Sarsat rescue applications. *IEEE International Symposium on Antennas and Propagation (APSURSI)*, *3*, 1319–1322. doi:10.1109/APS.2011.5996532

Seufert, S. (2002). E-learning business models: Framework and best practice examples. In M. Raisinghani (Ed.), *Cases on worldwide e-commerce: Theory in action* (pp. 70–94). Hershey, PA: IGI Global. doi:10.4018/978-1-930708-27-3.ch004

Shareef, M. A., & Archer, N. (2012). E-government initiatives: Review studies on different countries. In M. Shareef, N. Archer, & S. Dutta (Eds.), *E-government service maturity and development: Cultural, organizational and technological perspectives* (pp. 40–76). Hershey, PA: IGI Global. doi:10.4018/978-1-60960-848-4.ch003

Shareef, M. A., & Archer, N. (2012). E-government service development. In M. Shareef, N. Archer, & S. Dutta (Eds.), *E-government service maturity and development: Cultural, organizational and technological perspectives* (pp. 1–14). Hershey, PA: IGI Global. doi:10.4018/978-1-60960-848-4.ch001

Shareef, M. A., Kumar, U., & Kumar, V. (2011). E-government development: Performance evaluation parameters. In M. Shareef, V. Kumar, U. Kumar, & Y. Dwivedi (Eds.), *Stakeholder adoption of e-government services: Driving and resisting factors* (pp. 197–213). Hershey, PA: IGI Global. doi:10.4018/978-1-60960-601-5.ch010

Shareef, M. A., Kumar, U., Kumar, V., & Niktash, M. (2012). Electronic-government vision: Case studies for objectives, strategies, and initiatives. In M. Shareef, N. Archer, & S. Dutta (Eds.), *E-government service maturity and development: Cultural, organizational and technological perspectives* (pp. 15–39). Hershey, PA: IGI Global. doi:10.4018/978-1-60960-848-4.ch002

Shaw, A. (2014). *Energy harvested from body, environment could power wearables, IoT devices.* Available online: http://www.pcworld.com/article/2463500/energy-harvested-from-body-environment-could-power-wearables-iot-devices.html

Shi, B., Yang, J., Huang, Z., & Hui, P. (2015). Offloading Guidelines for Augmented Reality Applications on Wearable Devices. *Proceedings of the 23rd ACM International Conference on Multimedia*, 1271–1274. doi:10.1145/2733373.2806402

Shukla, P., Kumar, A., & Anu Kumar, P. B. (2013). Impact of national culture on business continuity management system implementation. *International Journal of Risk and Contingency Management*, *2*(3), 23–36. doi:10.4018/ijrcm.2013070102

Shulman, S. W. (2007). The federal docket management system and the prospect for digital democracy in U S rulemaking. In G. Garson (Ed.), *Modern public information technology systems: Issues and challenges* (pp. 166–184). Hershey, PA: IGI Global. doi:10.4018/978-1-59904-051-6.ch009

Simonovic, S. (2007). Problems of offline government in e-Serbia. In A. Anttiroiko & M. Malkia (Eds.), *Encyclopedia of digital government* (pp. 1342–1351). Hershey, PA: IGI Global. doi:10.4018/978-1-59140-789-8.ch205

Simonovic, S. (2008). Problems of offline government in e-Serbia. In A. Anttiroiko (Ed.), *Electronic government: Concepts, methodologies, tools, and applications* (pp. 2929–2942). Hershey, PA: IGI Global. doi:10.4018/978-1-59904-947-2.ch219

Singh, A. M. (2005). Information systems and technology in South Africa. In M. Khosrow-Pour (Ed.), *Encyclopedia of information science and technology* (pp. 1497–1502). Hershey, PA: IGI Global. doi:10.4018/978-1-59140-553-5.ch263

Singh, S., & Naidoo, G. (2005). Towards an e-government solution: A South African perspective. In W. Huang, K. Siau, & K. Wei (Eds.), *Electronic government strategies and implementation* (pp. 325–353). Hershey, PA: IGI Global. doi:10.4018/978-1-59140-348-7.ch014

Singla, G., Cook, D. J., & Schmitter-Edgecombe, M. (2010). Recognizing independent and joint activities among multiple residents in smart environments. Journal of Ambient Intelligence and Humanized Computing, 1(1). doi:10.1007/s12652-009-0007-1

Skrivervik, A. K., & Merli, F. (2011). Design strategies for implantable antennas. In *Loughborough Antennas and Propagation Conference* (pp. 1–5). doi:10.1109/LAPC.2011.6114011

Skrivervik, A. K., & Zürcher, J. F. (2007). Miniature antenna design at LEMA. In *19th International Conference on Applied Electromagnetics and Communications* (pp. 4–7). doi:<ALIGNMENT.qj></ALIGNMENT>10.1109/ICECOM.2007.4544410

Snoke, R., & Underwood, A. (2002). Generic attributes of IS graduates: An analysis of Australian views. In F. Tan (Ed.), *Advanced topics in global information management* (Vol. 1, pp. 370–384). Hershey, PA: IGI Global. doi:10.4018/978-1-930708-43-3.ch023

Snow flaik tracks skiers. (2011). Retrieved from http://news.discovery.com/tech/ snow-flaik-tracks-skiers.htm

Sodano, H. A., Inman, D. J., & Park, G. (2005). Comparison of piezoelectric energy harvesting devices for recharging batteries. *Journal of Intelligent Material Systems and Structures*, *16*(10), 799–807. doi:10.1177/1045389X05056681

Soh, P. J., Vandenbosch, G. A. E., Volski, V., & Nurul, H. M. R. (2010). Characterization of a Simple Broadband Textile Planar Inverted-F Antenna (PIFA) for on Body Communications. In *22nd International Conference on Applied Electromagnetics and Communications (ICECom)* (pp. 1–4).

Soh, P., Boyes, S., Vandenbosch, G., Huang, Y., & Ma, Z. (2012). Dual-band Sierpinski textile PIFA efficiency measurements. In *6th European Conference on Antennas and Propagation (EuCAP)*, (pp. 3322–3326).

Solanas, A., Patsakis, C., Conti, M., Vlachos, I., Ramos, V., Falcone, F., ... Perrea, D. (2014). Smart health: A context-aware health paradigm within smart cities. *Communications Magazine, IEEE*, *52*(8), 74–81.

Sommer, L. (2006). Revealing unseen organizations in higher education: A study framework and application example. In A. Metcalfe (Ed.), *Knowledge management and higher education: A critical analysis* (pp. 115–146). Hershey, PA: IGI Global. doi:10.4018/978-1-59140-509-2.ch007

Song, H., Kidd, T., & Owens, E. (2011). Examining technological disparities and instructional practices in English language arts classroom: Implications for school leadership and teacher training. In L. Tomei (Ed.), *Online courses and ICT in education: Emerging practices and applications* (pp. 258–274). Hershey, PA: IGI Global. doi:10.4018/978-1-60960-150-8.ch020

Soua, R., & Minet, P. (2011). A Survey of Energy Efficient techniques in Wireless Sensor Networks. *Proceedings of Wireless and Mobile Networking Conference*, 1 – 9. doi:10.1109/ WMNC.2011.6097244

Speaker, P. J., & Kleist, V. F. (2003). Using information technology to meet electronic commerce and MIS education demands. In A. Aggarwal (Ed.), *Web-based education: Learning from experience* (pp. 280–291). Hershey, PA: IGI Global. doi:10.4018/978-1-59140-102-5.ch017

Spitler, V. K. (2007). Learning to use IT in the workplace: Mechanisms and masters. In M. Mahmood (Ed.), *Contemporary issues in end user computing* (pp. 292–323). Hershey, PA: IGI Global. doi:10.4018/978-1-59140-926-7.ch013

State, A., Keller, K. P., & Fuchs, H. (2005). Simulation-based design and rapid prototyping of a parallax-free, orthoscopic video see-through head-mounted display. In *Fourth IEEE and ACM International Symposium on Mixed and Augmented Reality Proceedings* (pp. 28-31). Vienna, Austria: IEEE. doi:10.1109/ISMAR.2005.52

Staub, O., Zürcher, J.-F., & Skrivervik, A. (1998). Some considerations on the correct measurement of the gain and bandwidth of electrically small antennas. *Microwave and Optical Technology Letters*, *17*(3), 156–160. doi:10.1002/(SICI)1098-2760(19980220)17:3<156::AID-MOP2>3.0.CO;2-I

Staub, O., Zurcher, J.-F., Skrivervik, A. K., & Mosig, J. R. (1999). PCS antenna design: the challenge of miniaturisation. *IEEE Antennas and Propagation Society International Symposium*, 1. doi:10.1109/APS.1999.789198

Stellefson, M. (2011). Considerations for marketing distance education courses in health education: Five important questions to examine before development. In U. Demiray & S. Sever (Eds.), *Marketing online education programs: Frameworks for promotion and communication* (pp. 222–234). Hershey, PA: IGI Global. doi:10.4018/978-1-60960-074-7.ch014

Straub, D. W., & Loch, K. D. (2006). Creating and developing a program of global research. *Journal of Global Information Management*, *14*(2), 1–28. doi:10.4018/jgim.2006040101

Straub, D. W., Loch, K. D., & Hill, C. E. (2002). Transfer of information technology to the Arab world: A test of cultural influence modeling. In M. Dadashzadeh (Ed.), *Information technology management in developing countries* (pp. 92–134). Hershey, PA: IGI Global. doi:10.4018/978-1-931777-03-2.ch005

Straub, D. W., Loch, K. D., & Hill, C. E. (2003). Transfer of information technology to the Arab world: A test of cultural influence modeling. In F. Tan (Ed.), *Advanced topics in global information management* (Vol. 2, pp. 141–172). Hershey, PA: IGI Global. doi:10.4018/978-1-59140-064-6.ch009

Streitz, N. A. (2011). Smart cities, ambient intelligence and universal access. In *Universal Access in Human-Computer Interaction. Context Diversity* (pp. 425–432). Springer. Retrieved from http://link.springer.com/10.1007%2F978-3-642-21666-4_47

Suki, N. M., Ramayah, T., Ming, M. K., & Suki, N. M. (2013). Factors enhancing employed job seekers intentions to use social networking sites as a job search tool. In A. Mesquita (Ed.), *User perception and influencing factors of technology in everyday life* (pp. 265–281). Hershey, PA: IGI Global. doi:10.4018/978-1-4666-1954-8.ch018

Suomi, R. (2006). Introducing electronic patient records to hospitals: Innovation adoption paths. In T. Spil & R. Schuring (Eds.), *E-health systems diffusion and use: The innovation, the user and the use IT model* (pp. 128–146). Hershey, PA: IGI Global. doi:10.4018/978-1-59140-423-1.ch008

Sutherland, I. E. (1968). A head-mounted three dimensional display. In *Proceedings of the Fall Joint Computer Conference*. New York, NY: ACM.

Swan, M. (2012). Sensor mania! The Internet of Things, wearable computing, objective metrics and quantified self 2.0. *Journal of Sensor and Actuator Networks*, *1*(3), 217–253. doi:10.3390/jsan1030217

Swim, J., & Barker, L. (2012). Pathways into a gendered occupation: Brazilian women in IT. *International Journal of Social and Organizational Dynamics in IT*, *2*(4), 34–51. doi:10.4018/ijsodit.2012100103

Takagi, A., Yamazaki, S., Saito, Y., & Taniguchi, N. (2000). Development of a stereo video see-through HMD for AR systems. In *IEEE and ACM International Symposium on Augmented Reality Proceedings* (pp. 68-77). Munich, Germany: IEEE. doi:10.1109/ISAR.2000.880925

Tamura, T., Sekine, M., Tang, Z., Yoshida, M., Takeuchi, Y., & Imai, M. (2015). Preliminary study of a new home healthcare monitoring to prevent the recurrence of stroke. *37th Annual International Conference of the IEEE Engineering in Medicine and Biology Society (EMBC)*, 5489–5492. doi:10.1109/EMBC.2015.7319634

Tan, S. L., & Nguyen, B. A. T. (2009). Survey and performance evaluation of real-time operating systems (rtos) for small microcontrollers. IEEE Micro.

Tarafdar, M., & Vaidya, S. D. (2006). Adoption and implementation of IT in developing nations: Experiences from two public sector enterprises in India. In M. Khosrow-Pour (Ed.), *Cases on information technology planning, design and implementation* (pp. 208–233). Hershey, PA: IGI Global. doi:10.4018/978-1-59904-408-8.ch013

Tarafdar, M., & Vaidya, S. D. (2008). Adoption and implementation of IT in developing nations: Experiences from two public sector enterprises in India. In G. Garson & M. Khosrow-Pour (Eds.), *Handbook of research on public information technology* (pp. 905–924). Hershey, PA: IGI Global. doi:10.4018/978-1-59904-857-4.ch076

Tarlochan, F., & Ramesh, S. (2005). Heat transfer model for predicting survival time in cold water immersion. *Biomedical Engineering: Applications. Basis and Communications*, *17*(4), 159–166. doi:10.4015/S1016237205000251

Tesla, N. (1892). Experiments with alternate currents of high potential and high frequency. *Journal of the Institution of Electrical Engineers*, *21*(97), 51–162. doi:10.1049/jiee-1.1892.0002

Tetrapol. (2015). *Tetrapol Factsheet Trunked radio system for emergency services*. Author.

Thesing, Z. (2007). Zarina thesing, pumpkin patch. In M. Hunter (Ed.), *Contemporary chief information officers: Management experiences* (pp. 83–94). Hershey, PA: IGI Global. doi:10.4018/978-1-59904-078-3.ch007

Thierer, A. (2015). The Internet of Things and Wearable Technology: Addressing Privacy and Security concerns without derailing Innovation. *Richmond Journal of Law and Technology*, *21*(2), 1 – 118. Available online: http://jolt.richmond.edu/v21i2/article6.pdf

Thomas, J. C. (2004). Public involvement in public administration in the information age: Speculations on the effects of technology. In M. Malkia, A. Anttiroiko, & R. Savolainen (Eds.), *eTransformation in governance: New directions in government and politics* (pp. 67–84). Hershey, PA: IGI Global. doi:10.4018/978-1-59140-130-8.ch004

Thrift, A. G., Cadilhac, D. A., Thayabaranathan, T., Howard, G., Howard, V. J., Rothwell, P. M., & Donnan, G. A. (2014). Global stroke statistics. *International Journal of Stroke*, *9*(1), 6–18. doi:10.1111/ijs.12245 PMID:24350870

Trajkovikj, J., Fuchs, B., & Skrivervik, A. (2011). *LEMA internal report*. Academic Press.

Treiblmaier, H., & Chong, S. (2013). Trust and perceived risk of personal information as antecedents of online information disclosure: Results from three countries. In F. Tan (Ed.), *Global diffusion and adoption of technologies for knowledge and information sharing* (pp. 341–361). Hershey, PA: IGI Global. doi:10.4018/978-1-4666-2142-8.ch015

Ullah, S., & Kwak, K. S. (2011). *Body Area Network for Ubiquitous Healthcare Applications: Theory and Implementation*. 10.1007/s10916-011-9787-x

United Nations. (2006). *Convention on the rights of persons with disabilities*. Retrieved February 25, 2016, from http://www.un.org/disabilities/convention/conventionfull.shtml

Valenti, R., & Gevers, T. (2012). Accurate eye center location through invariant isocentric patterns. *IEEE Transactions on Pattern Analysis and Machine Intelligence*, *34*(9), 1785-1798.

Valenti, R., & Gevers, T. (2008). Accurate eye center location and tracking using isophote curvature. In *IEEE Conference on Computer Vision and Pattern Recognition* (pp. 1-8). Anchorage, AK: IEEE. doi:10.1109/CVPR.2008.4587529

Valenti, R., Staiano, J., Sebe, N., & Gevers, T. (2009). Webcam-based visual gaze estimation. In *International Conference on Image Analysis and Processing Proceedings* (pp. 662-671). Vietri sul Mare, Italy: Springer.

Vallozzi, L., Van Torre, P., Hertleer, C., Rogier, H., Moeneclaey, M., & Verhaevert, J. (2010). Wireless communication for firefighters using dual-polarized textile antennas integrated in their garment. *IEEE Transactions on Antennas and Propagation*, *58*(4), 1357–1368. doi:10.1109/TAP.2010.2041168

Vallozzi, L., Vandendriessche, W., Rogier, H., Hertleer, C., & Scarpello, M. L. (2010). Wearable textile GPS antenna for integration in protective garments. In *4th European Conference on Antennas and Propagation (EuCAP)* (pp. 1–4).

van Grembergen, W., & de Haes, S. (2008). IT governance in practice: Six case studies. In W. van Grembergen & S. De Haes (Eds.), *Implementing information technology governance: Models, practices and cases* (pp. 125–237). Hershey, PA: IGI Global. doi:10.4018/978-1-59904-924-3.ch004

Van Langenhove, L., & Hertleer, C. (2004). Smart textiles in vehicles: A foresight. *Journal of Textiles and Apparel. Technology and Management*, *3*(4), 1–6.

van Os, G., Homburg, V., & Bekkers, V. (2013). Contingencies and convergence in European social security: ICT coordination in the back office of the welfare state. In M. Cruz-Cunha, I. Miranda, & P. Gonçalves (Eds.), *Handbook of research on ICTs and management systems for improving efficiency in healthcare and social care* (pp. 268–287). Hershey, PA: IGI Global. doi:10.4018/978-1-4666-3990-4.ch013

Velloso, A. B., Gassenferth, W., & Machado, M. A. (2012). Evaluating IBMEC-RJ's intranet usability using fuzzy logic. In M. Cruz-Cunha, P. Gonçalves, N. Lopes, E. Miranda, & G. Putnik (Eds.), *Handbook of research on business social networking: Organizational, managerial, and technological dimensions* (pp. 185–205). Hershey, PA: IGI Global. doi:10.4018/978-1-61350-168-9.ch010

Verb for shoe - very intelligent shoes. (2004). Retrieved from http://www.gizmag. com/go/3565/picture/7504/

Villablanca, A. C., Baxi, H., & Anderson, K. (2009). Novel data interface for evaluating cardiovascular outcomes in women. In A. Dwivedi (Ed.), *Handbook of research on information technology management and clinical data administration in healthcare* (pp. 34–53). Hershey, PA: IGI Global. doi:10.4018/978-1-60566-356-2.ch003

Villablanca, A. C., Baxi, H., & Anderson, K. (2011). Novel data interface for evaluating cardiovascular outcomes in women. In *Clinical technologies: Concepts, methodologies, tools and applications* (pp. 2094–2113). Hershey, PA: IGI Global. doi:10.4018/978-1-60960-561-2.ch806

Virkar, S. (2011). Information and communication technologies in administrative reform for development: Exploring the case of property tax systems in Karnataka, India. In J. Steyn, J. Van Belle, & E. Mansilla (Eds.), *ICTs for global development and sustainability: Practice and applications* (pp. 127–149). Hershey, PA: IGI Global. doi:10.4018/978-1-61520-997-2.ch006

Virkar, S. (2013). Designing and implementing e-government projects: Actors, influences, and fields of play. In S. Saeed & C. Reddick (Eds.), *Human-centered system design for electronic governance* (pp. 88–110). Hershey, PA: IGI Global. doi:10.4018/978-1-4666-3640-8.ch007

Virtanen, J., Björninen, T., Ukkonen, L., & Sydänheimo, L. (2010). Passive UHF inkjet-printed narrow-line RFID tags. *Antennas and Wireless Propagation Letters*, *9*, 440–443. doi:10.1109/LAWP.2010.2050050

Vocca, H., & Cottone, F. (2014). Kinetic energy harvesting. In *ICT* (pp. 25–48). Energy - Concepts Towards Zero - Power Information and Communication Technology.

Wallace, A. (2009). E-justice: An Australian perspective. In A. Martínez & P. Abat (Eds.), *E-justice: Using information communication technologies in the court system* (pp. 204–228). Hershey, PA: IGI Global. doi:10.4018/978-1-59904-998-4.ch014

Wang, S., Yang, J., Chen, N., Chen, X., & Zhang, Q. (2005). *Human activity recognition with user-free accelerometers in the sensor networks.* In Neural Networks and Brain, In International Conference on Neural Networks and Brain.

Wang, G. (2012). E-democratic administration and bureaucratic responsiveness: A primary study of bureaucrats' perceptions of the civil service e-mail box in Taiwan. In K. Kloby & M. D'Agostino (Eds.), *Citizen 2.0: Public and governmental interaction through web 2.0 technologies* (pp. 146–173). Hershey, PA: IGI Global. doi:10.4018/978-1-4666-0318-9.ch009

Wangpipatwong, S., Chutimaskul, W., & Papasratorn, B. (2011). Quality enhancing the continued use of e-government web sites: Evidence from e-citizens of Thailand. In V. Weerakkody (Ed.), *Applied technology integration in governmental organizations: New e-government research* (pp. 20–36). Hershey, PA: IGI Global. doi:10.4018/978-1-60960-162-1.ch002

Wang, Y., Li, L., Wang, B., & Wang, L. (2009). A body sensor network platform for in-home health monitoring application. In *4th International Conference on Ubiquitous Information Technologies Applications* (pp. 1–5). doi:10.1109/ICUT.2009.5405731

Wedemeijer, L. (2006). Long-term evolution of a conceptual schema at a life insurance company. In M. Khosrow-Pour (Ed.), *Cases on database technologies and applications* (pp. 202–226). Hershey, PA: IGI Global. doi:10.4018/978-1-59904-399-9.ch012

Weinberg, H. (2002). Minimizing power consumption of imems accelerometers. In *Applications AN-601*. Analog Devices.

Weiner, M. M. (2003). *Monopole Antennas* (1st ed.). Taylor and Francis. doi:10.1201/9780203912676

Weiser, M. (1999). The Computer for the 21st Century. *SIGMOBILE Mob. Comput. Commun. Rev.*, *3*(3), 3–11. doi:10.1145/329124.329126

Wheeler, H. A. (1975). Small antennas. *IEEE Transactions on Antennas and Propagation*, *23*(4), 462–469. doi:10.1109/TAP.1975.1141115

Whybrow, E. (2008). Digital access, ICT fluency, and the economically disadvantages: Approaches to minimize the digital divide. In F. Tan (Ed.), *Global information technologies: Concepts, methodologies, tools, and applications* (pp. 1409–1422). Hershey, PA: IGI Global. doi:10.4018/978-1-59904-939-7.ch102

Whybrow, E. (2008). Digital access, ICT fluency, and the economically disadvantages: Approaches to minimize the digital divide. In C. Van Slyke (Ed.), *Information communication technologies: Concepts, methodologies, tools, and applications* (pp. 764–777). Hershey, PA: IGI Global. doi:10.4018/978-1-59904-949-6.ch049

Wickramasinghe, N., & Geisler, E. (2010). Key considerations for the adoption and implementation of knowledge management in healthcare operations. In M. Saito, N. Wickramasinghe, M. Fuji, & E. Geisler (Eds.), *Redesigning innovative healthcare operation and the role of knowledge management* (pp. 125–142). Hershey, PA: IGI Global. doi:10.4018/978-1-60566-284-8.ch009

Wickramasinghe, N., & Geisler, E. (2012). Key considerations for the adoption and implementation of knowledge management in healthcare operations. In *Organizational learning and knowledge: Concepts, methodologies, tools and applications* (pp. 1316–1328). Hershey, PA: IGI Global. doi:10.4018/978-1-60960-783-8.ch405

Wickramasinghe, N., & Goldberg, S. (2007). A framework for delivering m-health excellence. In L. Al-Hakim (Ed.), *Web mobile-based applications for healthcare management* (pp. 36–61). Hershey, PA: IGI Global. doi:10.4018/978-1-59140-658-7.ch002

Wickramasinghe, N., & Goldberg, S. (2008). Critical success factors for delivering m-health excellence. In N. Wickramasinghe & E. Geisler (Eds.), *Encyclopedia of healthcare information systems* (pp. 339–351). Hershey, PA: IGI Global. doi:10.4018/978-1-59904-889-5.ch045

Wong, K. (2002). *Compact and Broadband Microstrip Antennas* (1st ed.). Wiley-VCH. doi:10.1002/0471221112

Wongpatikaseree, K., Ikeda, M., Buranarach, M., Supnithi, T., Lim, A. O., & Tan, Y. (2012). *Activity recognition using context-aware infrastructure ontology in smart home domain.* In *Seventh International Conference on Knowledge, Information and Creativity Support Systems (KICSS).* doi:10.1109/KICSS.2012.26

Woznowski, P., Fafoutis, X., Song, T., Hannuna, S., Camplani, M., Tao, L., & Craddock, I. et al. (2015). A multi-modal sensor infrastructure for healthcare in a residential environment. *2015 IEEE International Conference on Communication Workshop (ICCW)*, 271–277. doi:10.1109/ICCW.2015.7247190

Wu, X. Y., Akhoondzadeh-Asl, L., Wang, Z. P., & Hall, P. S. (2010). Novel Yagi-Uda antennas for on-body communication at 60GHz. In *Loughborough Antennas* (pp. 153–156). Propagation Conference. doi:10.1109/LAPC.2010.5666188

Wyld, D. (2009). Radio frequency identification (RFID) technology. In J. Symonds, J. Ayoade, & D. Parry (Eds.), *Auto-identification and ubiquitous computing applications* (pp. 279–293). Hershey, PA: IGI Global. doi:10.4018/978-1-60566-298-5.ch017

Xia, W., Saito, K., Takahashi, M., & Ito, K. (2009). Performances of an implanted cavity slot antenna embedded in the human arm. In IEEE Transactions on Antennas and Propagation (Vol. 57, pp. 894–899). doi:10.1109/TAP.2009.2014579

Xu, G., Yang, Y., Zhou, Y., & Liu, J. (2013). Wearable thermal energy harvester powered by human foot. *Frontiers in Energy*, *7*(1), 26–38. doi:10.1007/s11708-012-0215-9

Yaghmaei, F. (2010). Understanding computerised information systems usage in community health. In J. Rodrigues (Ed.), *Health information systems: Concepts, methodologies, tools, and applications* (pp. 1388–1399). Hershey, PA: IGI Global. doi:10.4018/978-1-60566-988-5.ch088

Yamamoto, T., Terada, T., & Tsukamoto, M. (2011). Designing Gestures for Hands and Feet in Daily Life. In *Proceedings of the 9th International Conference on Advances in Mobile Computing and Multimedia* (pp. 285–288). New York, NY: ACM. doi:10.1145/2095697.2095757

Yan, Z., Subbaraju, V., Chakraborty, D., Misra, A., & Aberer, K. (2012). Energy-efficient continuous activity recognition on mobile phones: An activity-adaptive approach. *Proceedings of the 16th Annual International Symposium on Wearable Computers (ISWC).* doi:10.1109/ISWC.2012.23

Ye, H., Malu, M., Oh, U., & Findlater, L. (2014). Current and future mobile and wearable device use by people with visual impairments. *In Proceedings of the SIGCHI Conference on Human Factors in Computing Systems*. New York, NY: ACM. doi:10.1145/2556288.2557085

Yee, G., El-Khatib, K., Korba, L., Patrick, A. S., Song, R., & Xu, Y. (2005). Privacy and trust in e-government. In W. Huang, K. Siau, & K. Wei (Eds.), *Electronic government strategies and implementation* (pp. 145–190). Hershey, PA: IGI Global. doi:10.4018/978-1-59140-348-7.ch007

Yeh, S., & Chu, P. (2010). Evaluation of e-government services: A citizen-centric approach to citizen e-complaint services. In C. Reddick (Ed.), *Citizens and e-government: Evaluating policy and management* (pp. 400–417). Hershey, PA: IGI Global. doi:10.4018/978-1-61520-931-6.ch022

Yin, J., Yang, Q., & Pan, J. J. (2008). Sensor-based abnormal human-activity detection. *Knowledge and Data Engineering. IEEE Transactions on, 20*(8), 1082–1090.

Young-Jin, S., & Seang-tae, K. (2008). E-government concepts, measures, and best practices. In A. Anttiroiko (Ed.), *Electronic government: Concepts, methodologies, tools, and applications* (pp. 32–57). Hershey, PA: IGI Global. doi:10.4018/978-1-59904-947-2.ch004

Yun, H. J., & Opheim, C. (2012). New technology communication in American state governments: The impact on citizen participation. In K. Bwalya & S. Zulu (Eds.), *Handbook of research on e-government in emerging economies: Adoption, e-participation, and legal frameworks* (pp. 573–590). Hershey, PA: IGI Global. doi:10.4018/978-1-4666-0324-0.ch029

Yun, J., Patel, S., Reynolds, M., & Abowd, G. (2008). A quantitative investigation of inertial power harvesting for human-powered devices. *Proceedings of the 10th International Conference on Ubiquitous Computing, UbiComp '08*. doi:10.1145/1409635.1409646

Zappi, P., Lombriser, C., Stiefmeier, T., Farella, E., Roggen, D., Benini, L., & Trster, G. (2008). *Activity recognition from on-body sensors: Accuracy-power trade-off by dynamic sensor selection.* In *European Conference on Wireless Sensor Networks (EWSN)*, Bologna, Italy. doi:10.1007/978-3-540-77690-1_2

Zegna Sport Bluetooth iJacket incorporates smart fabric. (2007). Retrieved from http://www.gizmag.com/go/7856/

Zhang, Q., Ren, L., & Shi, W. (2013). Honey: A multimodality fall detection and telecare system. *Telemedicine and E-Health, 19*(5), 415 – 429. DOI: 10.1089/tmj.2012.0109

Zhang, N., Guo, X., Chen, G., & Chau, P. Y. (2011). User evaluation of e-government systems: A Chinese cultural perspective. In F. Tan (Ed.), *International enterprises and global information technologies: Advancing management practices* (pp. 63–84). Hershey, PA: IGI Global. doi:10.4018/978-1-60960-605-3.ch004

Ziviani, A., Correa, B., Gonçalves, B., Teixeira, I., & Gomes, A. (2011). AToMS: A ubiquitous teleconsultation system for supporting ami patients with prehospital thrombolysis. *International Journal of Telemedicine and Applications*, 12.

Zungeru, A. M., Ang, L. M., Prabaharan, S., & Seng, K. P. (2012). Radio frequency energy harvesting and management for wireless sensor networks. In *Green Mobile Devices and Networks* (pp. 341–368). CRC Press. doi:10.1201/b10081-16

Zuo, Y., & Hu, W. (2011). Trust-based information risk management in a supply chain network. In J. Wang (Ed.), *Supply chain optimization, management and integration: Emerging applications* (pp. 181–196). Hershey, PA: IGI Global. doi:10.4018/978-1-60960-135-5.ch013

Zürcher, J. F., Staub, O., & Skrivervik, A. K. (2000). SMILA: A compact and efficient antenna for mobile communications. *Microwave and Optical Technology Letters*, *27*(3), 155–157. doi:10.1002/1098-2760(20001105)27:3<155::AID-MOP1>3.0.CO;2-P

Zurcher, J.-F., Skrivewik, A. K., & Staub, O. (2000). SMILA: a miniaturized antenna for PCS applications. *IEEE Antennas and Propagation Society International Symposium*, *3*, 1646–1649. doi:10.1109/APS.2000.874556

Index

IGI Global
DISSEMINATOR OF KNOWLEDGE

Printed in the United States
By Bookmasters